Handbook of Soil Science and Agricultural Chemistry

THE AUTHORS

Dr. Ashok K. Patra, Vice Chancellor of Bidhan Chandra Krishi Viswavidyalaya, Mohanpur, West Bengal, obtained his B.Sc. in Agriculture from Banaras Hindu University, and his M.Sc. and Ph.D. in Soil Science and Agricultural Chemistry from the Indian Agricultural Research Institute (IARI), New Delhi. He served as Director of the ICAR–Indian Institute of Soil Science, Bhopal (2014–2022), Emeritus Scientist at ICAR–IISS, Bhopal (2023–2024), and Principal Scientist at ICAR-IARI, New Delhi (1999–2014). His research focuses on soil fertility, nutrient cycling, carbon sequestration, climate change, and sustainable agriculture. A Fellow of NAAS, NASI, WAST, and other scientific bodies, Dr. Patra has held key leadership roles, including President of the Indian Society of Soil Science and the Agriculture & Forestry Sciences Section of the Indian Science Congress. His notable honours include the UN FAO Glinka World Soil Prize (2022), ICAR Rafi Ahmed Kidwai Award (2021), ISSS Platinum Jubilee Commemoration Award (2022), IPNI–FAI Award (2018), and ICAR–IARI Hooker Award (2013), among others.

Dr. Nishant K. Sinha graduated from Banaras Hindu University and obtained his M.Sc. and Ph.D. degrees from the Indian Agricultural Research Institute (IARI), New Delhi. He is currently serving as a Senior Scientist in the Division of Soil Physics at ICAR–Indian Institute of Soil Science, Bhopal, Madhya Pradesh. His research focuses on assessing the impacts of climate change on soil properties, greenhouse gas emissions, and key soil processes such as carbon and nitrogen cycling, with implications for crop productivity. He is also engaged in developing crop adaptation strategies to strengthen food security under changing climatic conditions. Notably, he has developed a software tool, **SQICAL**, for rapid assessment of soil quality. Dr. Sinha has published over 95 research papers in reputed national and international journals.

Handbook of Soil Science and Agricultural Chemistry

– Authors –

Ashok K. Patra

Nishant K. Sinha

2026

Daya Publishing House®

A Division of

Astral International Pvt. Ltd.

New Delhi – 110 002

ISBN 978-93-59195-25-4

Published by : **Daya Publishing House®**
A Division of
Astral International Pvt. Ltd.
– ISO 9001:2015 Certified Company –
4736/23, Ansari Road, Darya Ganj
New Delhi-110 002
Ph. 011-43549197, 8130496929
e-mail: info@astralint.com
Website: www.astralint.com

Printed at : **Replika Press Pvt Ltd.**

Bidhan Chandra Krishi Viswavidyalaya
P.O. Krishi Viswavidyalaya, Mohanpur 741252
District: Nadia, West Bengal, India
Email: vc@bckv.edu.in/bckvvc@gmail.com
Cell: + 91-9830071278
Website: www.bckv.edu.in

Dr. Ashok K. Patra, Ph.D (IARI)
FNASc, FNAAS, FAScT, FISSS, FRMSI
Vice Chancellor

Dated: 09/05/2025

FOREWORD

It is my pleasure to write about this book entitled ***'Handbook of Soil Science & Agricultural Chemistry'*** which is authored by Dr. Ashok K. Patra and Dr. Nishant K. Sinha. This book has been meticulously developed with the primary objective of imparting robust foundational knowledge in *Soil Science and Agricultural Chemistry*—a core discipline underpinning agricultural education, research, and practice. Tailored specifically for undergraduate students of agricultural universities and institutes, it presents a well-organized and scientifically rigorous exploration of essential topics including soil genesis, classification, physical and chemical properties, soil fertility, and nutrient management.

The content is closely aligned with contemporary academic curricula, while also emphasizing the practical application of theoretical concepts to diverse agro-ecological conditions in India. In addition to core principles, the book addresses emerging issues such as soil health assessment, integrated nutrient management, environmental sustainability, and climate-resilient agricultural practices.

By providing a sound scientific foundation, this book equips students with the analytical tools and applied knowledge necessary to address the complex challenges facing modern agriculture. It is our earnest hope that this volume will serve not only as a reliable academic resource but also as an effective guide for nurturing the next generation of soil scientists, educators, and agricultural professionals committed to sustainable soil management.

Vice Chancellor
Bidhan Chandra Krishi Viswavidyalaya

Acknowledgement

The preparation of this book has been made possible through the generous support and institutional backing provided by the Indian Council of Agricultural Research (ICAR), New Delhi. Ashok K. Patra sincerely expresses his gratitude to ICAR for granting him the position of *ICAR–Emeritus Scientist* during the period 2023–2024, a tenure during which the manuscript of this book was conceptualized and developed.

Authors ae deeply thankful to the Director, ICAR–Indian Institute of Soil Science (ICAR–IISS), Bhopal, for extending all necessary facilities, encouragement, and continuous cooperation throughout the course of this work. The conducive academic environment and access to institutional resources at ICAR–IISS greatly facilitated the writing and refinement of this manuscript.

We extend our sincere gratitude to the scientists and staff of the ICAR–Indian Institute of Soil Science (ICAR–IISS), Bhopal, for their direct and indirect contributions to the preparation of this manuscript. Every effort has been made to ensure the accuracy of the content and the proper attribution of all sources throughout this book. Nevertheless, should any inadvertent errors or omissions be identified, we respectfully request readers to bring them to our attention so they may be addressed in future editions.

It is our hope that this book will contribute meaningfully to the academic growth of students and professionals in the field of Soil Science and Agricultural Chemistry and further strengthen the scientific foundations of sustainable agriculture in India.

Preface

Soil science and agricultural chemistry is foundational discipline in the pursuit of sustainable agriculture. It offers essential insights into the physical, chemical, and biological characteristics of soil, enabling informed decision-making for crop production, nutrient management, and environmental stewardship. Soil science helps us understand soil formation, classification, and fertility, as well as the dynamic interactions among soil components. Agricultural chemistry complements this by focusing on the chemical aspects of crop production, such as fertilizers, pesticides, and plant growth regulators, thereby supporting efficient and environmentally responsible farming. This integrated field is critical for developing effective strategies that enhance crop productivity while preserving the health of soil and ecosystems. This synergy supports innovations like precision agriculture, which enables site-specific management of inputs and reduces the overuse of chemicals. By improving our understanding of soil processes and chemical interactions, we can reduce degradation, optimize input use, and ensure the resilience of agricultural systems in the face of climate change and resource constraints.

This volume, *Handbook of Soil Science and Agricultural Chemistry*, is designed to serve as a comprehensive and accessible resource for undergraduate and postgraduate students, researchers, teachers, extension personnel, and practitioners in agriculture and allied fields. It presents a structured exploration of key topics such as soil health, taxonomy, nutrient cycles, and the chemistry of soil amendments, while also addressing critical issues like wasteland reclamation, environmental soil chemistry, and the application of modern geospatial technologies such as GIS, GPS, and remote sensing. Each chapter is crafted to build a strong conceptual foundation, supported by practical examples, relevant terminology, and references for further reading.

We sincerely hope that this book will serve not only as a valuable academic resource but also as a practical guide for field applications, ultimately contributing to improved agricultural productivity and the well-being of farming communities. It is our firm belief that students, educators, researchers, and practitioners alike will find this work to be a useful and insightful companion in their academic and professional endeavours.

Ashok K. Patra
Nishant K. Sinha

Contents

Chapter 1

Soil Science and Agricultural Chemistry: An Introduction

- ***Soil science*** is a discipline within agricultural science that examines soil as a natural entity and as a crucial medium for nutrient provision in plant growth and food production systems.
- ***Agricultural chemistry*** is the examination of chemistry, particularly organic chemistry and biochemistry, in relation to agriculture, encompassing agricultural production, the transformation of raw materials into food and drinks, and environmental assessment and restoration.
- As a form of applied sciences, some of the main aims of agricultural chemistry are:
  - Increase yield of crops and livestock
  - Improving the quality of products
  - Minimizing product expenses

Uses

Agricultural chemistry employs numerous approaches to attain the aforementioned objectives. Examples include:

- Agricultural chemists focus on both protecting and enhancing soil quality.
- Creating efficacious substances include fungicides, insecticides, herbicides, and other pesticides. Certain chemical substances, including animal feed additives, fertilizers, and plant growth regulators.

- Agricultural chemists aim to sustain our resources to prevent depletion and assure replenishment for future generations.

Definitions of Soil

- ***Dokuchaiev,*** (Father of Soil Science) viewed the soil as a natural body composed of mineral and organic constituents, having a definite genesis and distinct nature of its own.
- Soil is defined as a dynamic natural body on the surface of the earth in which plants grow, composed of mineral and organic materials and living forms. ***-Buckman and Bardy***
- Soil may be defined as the three dimensional natural body on the surface of earth, composed of mineral and organic materials and living forms, derived from the parent material due to influence of climate and organisms as conditioned by relief, over a period of time which is a medium for plant growth. ***- USDA***
- *As soil provides nutrients, water, air and anchorage and supports life on Earth, it can be called as* Soul of Infinite Life (SOIL).
- The terms Earth, Land, and Soil are frequently conflated by pupils as synonyms. However, there exists a significant disparity among them. The Universe is described as the entirety of all that exists, encompassing all physical space, time, matter, energy, planets, stars, galaxies, and more. The Earth is a little component of the Universe.
- **Soil:** Soil is a superficial layer of the Earth's crust that functions as a natural medium for plant growth. Rocks serve as crucial sources for the parent materials from which soils are formed through weathering and various soil-forming processes.
- All components of Earth's system can be categorized into four primary subsystems: lithosphere, hydrosphere, biosphere, and atmosphere. The four subsystems are referred to as spheres. Specifically, they comprise the Lithosphere (the solid layer of the Earth), Hydrosphere (the water enveloping the Earth), Biosphere (the realm of living organisms), and Atmosphere (the gaseous layer above the Earth's surface). All four spheres can be, and frequently are, co-located in a single area. A sample of soil will comprise minerals from the lithosphere.
  - Biosphere
  - Atmosphere
  - Hydrosphere
  - Lithosphere
- ***Parent material:*** It is the regolith, or at least its superficial layer. It may be defined as the unconsolidated and more or less chemically weathered mineral materials from which soil are developed OR it is the mass (consolidated material) from which the soil has originated.

- ***Soil as a natural body:*** Soil is a natural body at the surface of earth because it is a composed of naturally occurring entities *i.e.* rocks and minerals, organic material, air, water and organisms. Besides, soil formation is a result of natural biogeochemical processes which are influenced by natural soil forming factors by taking several decades and millennia. Therefore, soil is a natural body; it cannot be synthesized artificially in the laboratory.
- ***Soil is a three dimensional natural body:*** Soil (a field, at a particular location/area) is a thin layer to cover the land as the peel covers an orange. Nonetheless, although the peel exhibits remarkable uniformity across the orange, the soil demonstrates significant variability across different locations on Earth. Therefore, along with length and breath, it has a certain depth, below which bed rocks/parent material are deposited. The depth of soils are varies with location from a few centimeter to few meter as affected by soil forming factors.

Characteristics of Soil

- Soil is a natural body and cannot be synthesized artificially
- It is varies with location
- Differentiated in to distinct/characteristic horizons
- Different soil properties *i.e.* physical, chemical and biological
- It is a living entity

Pedology: The science that deals with the formation (genesis), morphology, survey and classification of soils for land use planning.

Edaphology: It is the discipline of soil science that examines the relationship between soil and plant growth, including crop nutrition and yield, specifically crop output.

Soil geology: Soil geology is the study of the geological materials from which the soil is derived and its processes of formation.

Soil chemistry: An examination of the soil's chemical constituents, their interactions, and the influence of the soil's chemical environment on agricultural requirements.

Soil physics: Soil physics is a division of soil science that involves the study of the effects of physical laws on the evolution of chemical properties and physical properties of soil.

Soil biology: Soil biology deals with the study of the effect of plants, animals and soil micro organisms on the evolution, chemical composition and physical condition of the soil.

Soil conservation: Soil conservation pertains to safeguarding soil from physical erosion and chemical degradation.

Soil Constituents (by volumetric proportion)

- ☆ Soils comprise four primary components: mineral matter, organic matter, soil air, and soil water. The relative composition of these components significantly affects the characteristics and productivity of soils.
- ☆ The four components of soil are intricately interwoven; yet, the volume proportion of each component can be illustrated in a straightforward pie chart.

Major Components of Soil

- ☆ Mineral matter (45 per cent)
- ☆ Organic matter (5 per cent)
- ☆ Soil air (20-30 per cent)
- ☆ Soil water (20-30 per cent)

Soil comprises approximately 50 percent solid matter and 50 percent pore space. The solid volume of the soil comprises approximately 45 percent mineral stuff and 5 percent organic content.

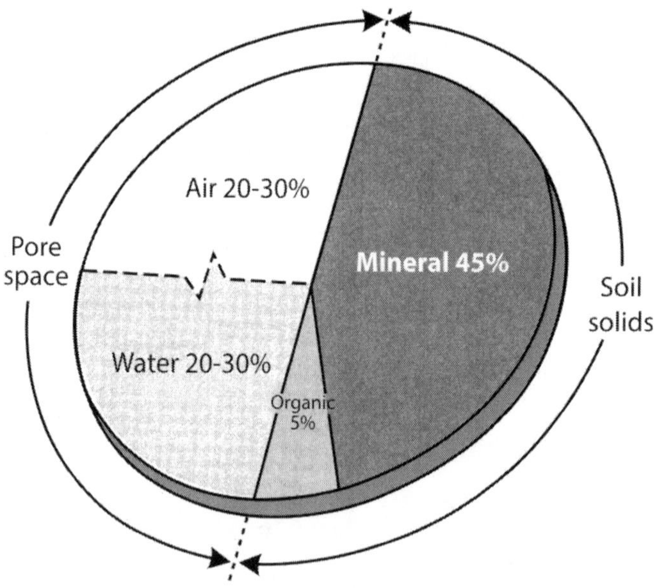

Composition of the Earth

The Earth's crust, the outermost thin layer, is primarily formed of solid substances. The mineral substance consists of diverse components amalgamated to create compounds. Nearly all elements are found in the Earth's crust.

Element	*O*	*Si*	*Al*	*Fe*	*Ca*	*Mg*	*Na*	*K*	*Others*
Per cent	46.6	27.7	8.1	5	3.6	2.1	2.8	2.6	1.5

Minerals: Minerals are naturally occurring inorganic entities with specific physical qualities and chemical compositions. It often comprises two or more parts in specific proportions. Examples include Feldspar, Mica, and Quartz.

Rocks: The Earth's rocks consist of minerals. A consolidated mass of one or more minerals or rocks can be characterized as a mixture of two or more minerals.

Soil Composition

- ***Solid Phase:*** It includes minerals (sand, silt and clay) and organic matter both dead and living micro and macro-organisms.
- ***Liquid Phase:*** Soil water residing in the pore spaces between mineral particles transports plant nutrients to the roots.
- ***Gaseous Phase:*** Also called as soil air, occupies pore spaces similar to soil water. It fills these voids when soil water is absent. It carries respiratory products of roots and soil-organisms. It has a higher concentration of carbon dioxide (CO_2) (0.3 per cent) than atmospheric air (0.03 per cent).

Soil Constituents

- Soil is a dynamic substance composed of minerals, organic matter, water, air, and organisms, including bacteria and earthworms.
- It was established and continually evolves due to five principal physical elements, collectively referred to as 'clorpt' (cl, r, o, p, t): parent material (p), time (t), climate (cl), organisms (o), and topography/relief (r). The management of soil is a significant component affecting its characteristics.

Characteristics, Attributes, and Significance of Soil

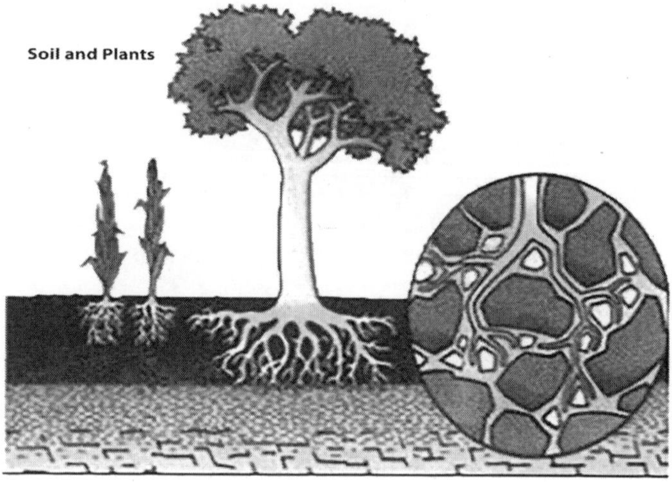

Characteristics and Properties of Soil

Soil Colour

- ✰ The dark hue typically signifies medium to high fertility, attributed to a substantial presence of organic matter. Soil color is often characterized by hue, value, and chroma. These soils often possess a high nutrient content, excellent water retention capacity, optimal structure, and adequate aeration.
- ✰ The light hue signifies moderate to poor fertility. These soils may experience leaching issues, wherein water accelerates the downward movement of organic matter and other nutrients.

Soil Depth

- ✰ Soil depth is crucial for plant growth, as it allows roots to effectively penetrate the soil to access water and nutrients.
- ✰ A depth of 3-5 feet is preferable, as deeper soils possess a greater capacity for retaining nutrients and water.

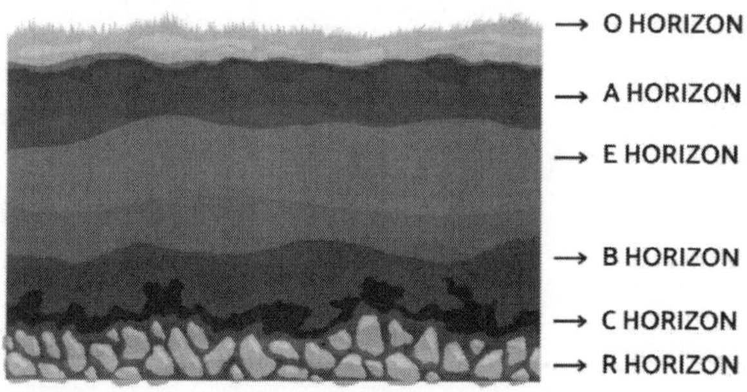

Soil Texture

- Texture denotes the relative proportions of mineral particles (sand, silt, and clay) inside soil. The features of soils, such as drainage, water retention, aeration, and nutrient availability, are predominantly influenced by soil texture.
  - ***Sandy:*** Reduced fertility and water retention capacity, although excellent aeration.
  - ***Loamy:*** Moderate fertility and adequate aeration.
  - ***Clayey:*** Elevated fertility and inadequate aeration, challenging to cultivate.

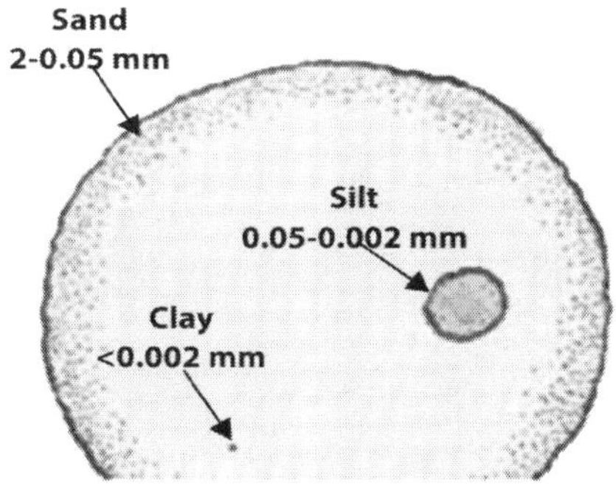

Soil Constituents

- Farmers may describe their soil as heavy or light, indicating the ease of cultivation. Dense soils are typically difficult to till and necessitate significantly greater effort than looser soils. Organic matter can be incorporated to enhance soil texture.

Land Slope

- The soil gradient refers to the angle of inclination of the soil surface relative to the underlying soil.
- It is given as a percentage, representing the vertical change in feet per 100 feet of horizontal distance.

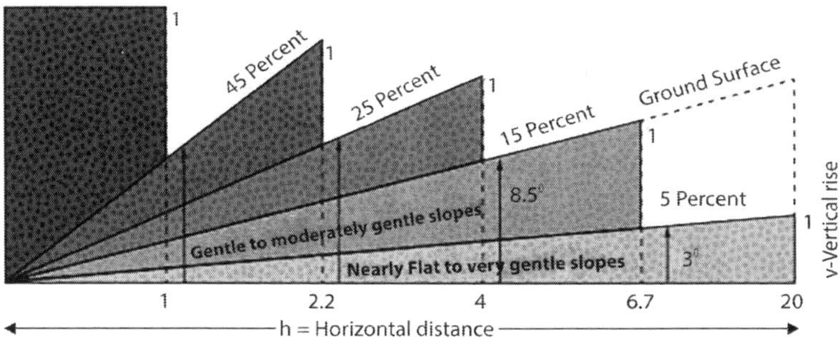

Land Slope

- A gentle gradient of up to 1 percent is preferable.
- Steeper gradients are undesirable as they result in soil and water erosion.
- Perfect leveling is necessary solely for paddy cultivation.

Soil pH

- Soil pH is critically significant for plant growth as it affects nutrient availability, toxicity levels, and the activity of soil organisms.

pH Range	*Soil Reaction Rating*
< 4.6	Extremely acid
4.6 - 5.5	Strongly acid
5.6-6.5	Moderately acid
6.6-6.9	Slightly acid
7.0	Neutral
7.1-8.5	Moderately alkaline
>8.5	Strongly alkaline

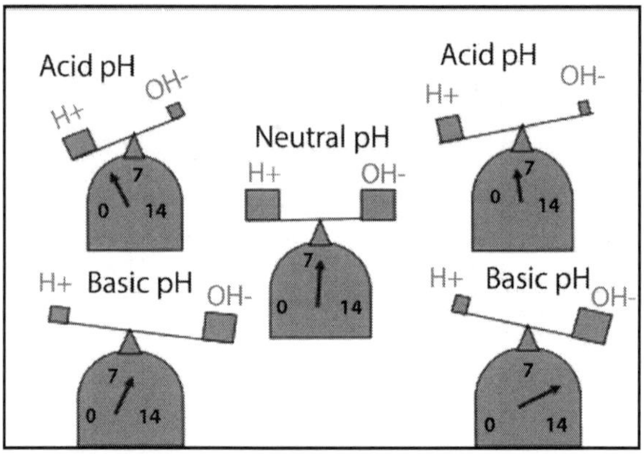

Guidelines for Soil pH Regulation

- Acidic soils should be amended with lime, with the quantity of lime and organic fertilizers determined by the soil test report.
- Sodic soils require correction through the use of gypsum or sulfur, with the quantity of organic manure determined by the soil test report.
- Saline-alkali soils require treatment with gypsum and enhanced drainage.

Soil Organic Matter

- Soil organic matter comprises decomposed plant and animal material at various stages of decomposition.
- Soil organic matter is essential for the biological, physical, and chemical functions of soil.

Soil Organic Matter Contributes by:

- Supplying nutrients for soil microorganisms.
- Functioning as a primary store of plant nutrients
- The nutrient interaction between soil and plant roots is facilitated.
- Enhancing soil structure Enhancing water retention capacity
- Modulating soil temperature
- Mitigating the risk of soil erosion and land deterioration

Enhancement of Soil Organic Matter can be achieved through

- Reintroducing crop leftovers to the field without incineration or waste.
- Utilizing compost

Microbial Activity and Soil Organic Matter

Soil Organic Matter

- Soil organic matter (SOM) comprises the many carbon compounds in soil, including plant and animal leftovers at different decomposition stages, cellular structures of soil organisms, and substances produced by the soil biota.
- These chemicals serve multiple functions in soil nutrient, water, and biological cycles.

Soil Organic Matter Improves Soil Properties

- Enhancing soil structure increases friability, reduces compaction, facilitates tillage, and promotes aeration..
- An increased water infiltration rate diminishes runoff and enhances the water availability for crops. Minimizing runoff also diminishes the flow of silt and soluble nutrients from soils to surface waters.
- Enhanced water infiltration rates and diminished runoff volumes may result in decreased nutrient loss from fields treated with manure compared to equal applications of nitrogen and phosphorus as commercial fertilizers, owing to the incorporation of organic matter present in the manure.
- Water holding capacity – Enhanced water retention is particularly significant in sandy soils and regions with limited precipitation. The organic component in highly sandy soil significantly contributes to its overall water retention and available water capacity.
- Cation exchange capacity (CEC) – Soil organic matter possesses a cation exchange capacity ranging from approximately 200 to 250 cmol/kg, which is roughly two to three times greater than the maximum cation exchange capacity of clay minerals. Coarse sandy soils, deficient in organic matter,

exhibit cation exchange capacities ranging from 2 to 5 cmolec/kg. The proportion of organic matter in soil significantly affects cation exchange capacity, especially in sandy soils.

- Buffering pH – SOM functions as a pH buffer or stabilizer, releasing H^+ upon the introduction of alkaline substances and adsorbing H^+ when acidic substances are added.
- Chelation of micronutrients and trace elements – Certain micronutrients and trace elements associate with soluble organic molecules to create chelates and other soluble organic complexes, facilitating the dissolution and mobilization of micronutrients and trace metals.
- Chelation enhances nutrient solubility, inhibits the development of insoluble precipitates, and reduces the toxicity of some micronutrients. Chelated nutrients, while not immediately accessible, exhibit mobility and can rapidly transform into forms useful to plants at the root zone.
- The enhancement of soil organic matter (SOM) typically leads to improved soil characteristics, hence augmenting soil productivity. However, in numerous soils, appropriate soil characteristics manifest at comparatively low organic matter levels (2-4 percent).
- The primary source of organic matter incorporated into soil is the residue left after crop cultivation. Fertilization indirectly enhances soil organic matter levels by augmenting crop yields and, subsequently, the volume of residues returned to the soil.
- The enhancement of soil organic matter is a gradual process, as only a minor proportion of the agricultural wastes incorporated into the soil persists as soil organic matter after decomposition.

Impact of Manure on Soil Organic Matter (SOM)

- Manure is a good source of nutrients and organic matter. Manure indirectly enhances soil organic matter by augmenting crop yields and the volume of leftovers returned to the soil. Organic matter is immediately incorporated into the soil through the application of manure, especially solid manure.
- The organic matter composition of manure exhibits significant variability. Solid and semi-solid manures possess greater organic matter content than liquid manure due to the inclusion of bedding materials. Prolonged application of solid manure can lead to significant increases in soil organic matter (SOM).
- Liquid manure contributes less organic matter to soil compared to solid manure; yet, these contributions exceed those from synthetic fertilizers. Three annual applications of liquid pig manure on pasture enhanced the light fraction of organic matter in soil, despite no substantial impact on total organic carbon content within this relatively brief timeframe.

Soil Microbial Activity

- Soil microbial activity is crucial as microbial processes facilitate the breakdown of crop wastes and the conversion of organic nutrients into inorganic forms for plant absorption.
- The size and diversity of the soil microbial population are directly correlated with the potential for nutrient mineralization and cycling.
- The application of manure to soil influences microbial populations. The incorporation of manure into soil enhances nutrient cycling, microbial biomass, microbial activity, and enzymatic activities, if the manure is devoid of elevated levels of undesired components (e.g., potentially poisonous trace metals).

Soil and Plant Nutrition

- Utilizing diverse organic fertilizers
- Composting organic waste
- Utilizing green manures and cover crops
- Appropriate crop rotation
- Minimizing soil tillage
- Mitigating soil erosion

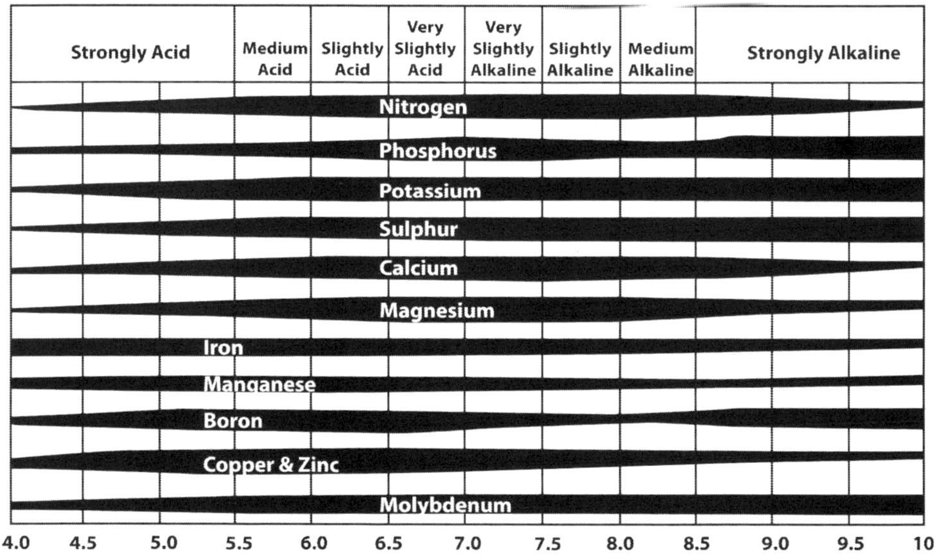

Impact of Soil pH on Nutrient Accessibility

Carbon - Nitrogen Ratio (C:N Ratio)

- Chemical elements in organic matter are critically significant, particularly with their relationships or proportions to one another. They are carbon and nitrogen. The relationship is referred to as the Carbon-Nitrogen Ratio (C:N Ratio).
- For instance, decomposed manure has a carbon to nitrogen ratio of 20:1, while sawdust has a ratio of 400:1. Legumes often possess elevated nitrogen content and a low carbon-to-nitrogen ratio, which is particularly advantageous. Agriculturists may utilize blood meal, bone meal, poultry manure, cottonseed meal/cake, soybean meal/cake, and other nitrogen-rich substances as organic matter to promote the decomposition and recycling of plant nutrients.

Electrical Conductivity (EC)

Electrical conductivity (EC) is typically regarded as an indicator of the concentration of dissolved salts in a solution.

Comprehensive Analysis of EC Values

- The application of gypsum can modify saline/sodic soils. The amount of Gypsum to be administered is determined by the EC value.
- Farmers experiencing issues with salty or sodic soils should do soil testing and consult extension staff for additional assistance.

Soil	*EC (mS/cm)*	*Crop reaction*
Salt free	0-2	The impact of salinity is minimal, especially for crops that are more sensitive.
Slightly saline	4 – 8	Production of numerous crops constrained
Moderately saline	8 - 15	Only tolerant crops produce satisfactory yields.
Highly saline	>15	Only highly tolerant crops produce satisfactory yields.

Soil Fertility

- Soil fertility is typically defined as the capacity of soil to provide essential nutrients for plants. Soil structure, texture, temperature, moisture, light, and air significantly contribute to the preservation of soil fertility.
- Nitrogen (N), potassium (K), and phosphorus (P) are essential plant nutrients that are frequently deficient in soil due to their substantial consumption by plants for development and survival.

Chapter 2

Soil Health and its Management

Introduction

- ☆ Soil health, or soil quality, is described as the ability of soil to sustain plant and animal productivity, preserve or improve water and air quality, and support human health and habitation over a human time scale.
- ☆ In a specific terms, a healthy soil must have good tilth and drainage, sufficient depth for crop growth, sufficient exchangeable nutrient supply (not excessive or prone to leaching), small population of weeds, insect pests or plant pathogens, large population of beneficial organisms, no toxins, and resilience to adverse conditions.
- ☆ A number of individual soil tests may be used to assess soil health; however, a comprehensive evaluation should include a suite of complementary tests to measure chemical, physical, and biological properties of soil.

Soil health, also referred to as soil quality, is defined as the continued capacity of soil to function as a vital living ecosystem that sustains plants, animals, and humans. This definition speaks to the importance of managing soils so they are sustainable for future generations. To do this, we need to remember that soil contains living organisms that when provided the basic necessities of life - food, shelter, and water - perform functions required to produce food and fiber. Only "living" things can have health, so viewing soil as a living ecosystem reflects a fundamental shift in the way we care for our nation's soils. Soil isn't an inert growing medium, but rather is teaming with billions of bacteria, fungi, and other microbes that are the foundation of an elegant symbiotic ecosystem.

Characteristics Feature of Healthy Soils

- Sufficient soil depth is crucial for better plant root development; a soil depth of 8 inches or greater is preferred in the case of berry crops.
- Optimal soil should possess excellent tilth, effective water retention, and adequate drainage. It must have adequate, yet not excessive, nutrients and be devoid of plant-toxic substances, like heavy metals, pesticide residues, or other pollutants.
- Healthy soils should have low populations of plant disease and parasitic organisms such as fungi, bacteria, nematodes, springtails, and so on.
- Conversely, a healthy soil should contain high populations of beneficial organisms such as mycorrhizae and earthworms.
- Healthy soils should exhibit resistance to being degraded and along with that – resiliency or the ability to recover quickly from adverse conditions such as drought, hurricanes, flooding, *etc*.

Soil constitutes an ecosystem that can be managed to supply nutrients for plant growth, retain and store rainwater for utilization during arid periods, filter and mitigate potential pollutants from exiting agricultural fields, serve as a stable foundation for farming activities, and offer habitat for soil microbes to thrive and diversify, thereby ensuring the ecosystem's functionality.

Major Soil Health Processes

- Think of soil health then in terms of the three major realms that impact it:
  - the physical,
  - the chemical, and
  - the biological
- These three realms intercept and interact; thus it is important to view each of these more as processes than characteristics. If one process is compromised, the others are similarly impacted. Healthy soil is balanced, so facilitating improved growth conditions, enhanced crop resilience, and diminished inputs.

The Physical Processes

- In this processes soil may be limited by inherent or dynamic qualities; some of these may be remediated; others may not.
  - ***Internal drainage*** – poor internal drainage reduces root growth and function and may support disease development.

- ❖ ***Water availability*** – is a function of soil texture, soil organic matter content and effects rooting depth.
- ❖ ***Soil aggregate (crumbs) stability*** – The stability of soil aggregates (crumbs) is contingent upon sufficient soil organic matter, which produces humates and other compounds that bind soil aggregates and enhance soil tilth.
- ❖ ***Soil structure*** – dense soils with small pore spaces do not allow for adequate aeration or water penetration and reduces root growth and function.
- ❖ ***Compaction*** – compaction layers, either near the surface or deeper, can inhibit root penetration and also water drainage contributing to excess runoff or erosion.

The Chemical Processes

- ☆ In this processes soil provide essential nutrients for plants. Soil pH is a crucial factor in the chemical process, influencing nutrient availability. Any alterations in soil pH must be rectified prior to planting; neglecting to correct soil pH to ideal values for the crop may significantly affect plant establishment and subsequent crop yield.
- ☆ Soil pH adjustment is more difficult after a perennial crop is established and may reduce the success of the planting.
- ☆ The chemical process encompasses macronutrients (required in substantial amounts, such as nitrogen, phosphorus, and potassium), secondary nutrients (including calcium, magnesium, and sulfur), and micronutrients (needed in trace amounts, such as boron and zinc); tailored recommendations have been established to rectify deficiencies of these nutrients vital for berry crop production.

The Biological Processes

- ☆ Soil constitutes a multifaceted habitat characterized by very changeable conditions. The majority of biological activity transpires on the soil surface, where the bulk of organic matter is concentrated.
- ☆ Soil has three primary categories of organic matter: living, deceased, and profoundly decomposed. All three are crucial in facilitating the production of high yields of healthy crops. The incorporation of organic matter into soil yields numerous advantages.
  - ❖ ***Living Organic Matter*** - includes plant roots, bacteria, fungi, nematodes, earthworms, mites, springtails, collembolans, moles, and many other types of organisms. They utilize soil resources in multiple manners, digesting organic materials, cycling nutrients, affecting plants and other organisms, and reacting to their chemical and physical surroundings in intricate ways.

- ***Dead Organic Matter*** - is composed of recently dead soil organisms and crop residues that provide food (energy and nutrients) for soil organisms to live and function. Dead organic matter is referred to as "active" or "particulate" organic matter. This is the additional crucial partner in the mineralization of nutrients for plants, soil aggregation, and humus formation.
- ***Very Dead Organic Matter*** - is not a biologically active fraction; rather it consists of well-decomposed organic materials, also called humus. Humus facilitates the chemical processes within soil; it possesses substantial negative charges that retain nutrients and cations in the soil. Humus possesses a significant water retention capacity and sequesters carbon.

Soil Biology Plays a Major Role in Soil Health

- Nutritious soil provides us with pure air and water, abundant crops and forests, fertile grazing pastures, varied fauna, and picturesque sceneries.
- Soil does all this by performing five essential functions:
  - ***Regulating water*** - Soil helps control where rain, snowmelt, and irrigation water goes. Water and dissolved solutes traverse the land or permeate the soil.
  - ***Sustaining plant and animal life*** - The diversity and productivity of living things depends on soil.
  - ***Filtering and buffering potential pollutants*** - The minerals and microbes in soil are responsible for filtering, buffering, degrading, immobilizing, and detoxifying organic and inorganic materials, including industrial and municipal by-products and atmospheric deposits.
  - ***Cycling nutrients*** - Carbon, nitrogen, phosphorus, and many other nutrients are stored, transformed, and cycled in the soil.
  - ***Physical stability and support*** - Soil structure provides a medium for plant roots. Soils also provide support for human structures and protection for archeological treasures.

Soil: Inherent and Dynamic Qualities

- Soil has both inherent and dynamic qualities. Inherent soil quality refers to the intrinsic capacity of soil to perform its functions. Sandy soil has superior drainage compared to clayey soil. Deep soil provides greater space for root development compared to soils with bedrock situated near the surface. These traits are not easily altered.
- Dynamic soil quality is how soil changes depending on how it is managed. Management decisions influence soil organic matter content, soil structure, soil depth, and the ability for water and nutrient retention.

A primary objective of soil health research is to ascertain methods for managing soil to enhance its functionality. Soils have varied responses to management based on their intrinsic qualities and the adjacent terrain.

Management of Soil Health

Introduction

- Soil is a critical resource–the way in which it is managed can improve or degrade the quality of that resource. Soil is a complex ecosystem wherein living microbes and plant roots aggregate mineral particles and organic materials into a dynamic structure that manages water, air, and nutrients.
- In an agricultural context, soil health most often refers to the ability of the soil to sustain agricultural productivity and protect environmental resources.
- A healthy soil provides many functions that support plant growth, including nutrient cycling, biological control of plant pests, and regulation of water and air supply.
- These functions are influenced by the interrelated physical, chemical, and biological properties of soil, many of which are sensitive to soil management practices.

Healthy soil is the foundation for profitable, productive, and environmentally sounds agricultural systems. By understanding how the soil processes that support plant growth and regulate environmental quality are affected by management practices, it is possible to design a crop and soil management system that improves and maintains soil health over time.

Nutrient Cycling

- Nutrient cycling refers to the many pathways through which nutrients are added to, removed from, and changed within the soil. Nutrients are found in two basic forms in the soil: organic and inorganic. Organic forms of nutrients contain carbon in the structure of the molecule, while inorganic forms do not.
- Nutrients are retained in many reservoirs within the soil: as inorganic compounds in soil particles, as organic substances in soil organic matter, as inorganic entities on cation exchange sites, and as both organic and inorganic forms dissolved in the water surrounding soil particles, referred to as the soil solution.

Nutrients Location in the Soil

- **Soil solution:** Inorganic and a few types of organic nutrients dissolved in the soil pore water are immediately available to plants.

- **Cation exchange sites:** Nutrients with a positive charge are known as cations. Macronutrient cations, essential for plants in substantial amounts, comprise calcium, magnesium, and potassium. Negatively charged spots on clay and organic materials retain positively charged plant nutrients. Nutrients on cation exchange sites are readily accessible to plants in the short term. The number of cation exchange sites in a specific soil is referred to as cation exchange capacity (CEC).
- **Organic matter:** Organic matter is composed of living and once-living material (*e.g.*, plant residues, manure) in various stages of decomposition. The accessibility of nutrients to plants in the short to long term is contingent upon the type of organic matter and the activity of soil organisms.
- **Decomposition** is the breakdown of organic matter into simpler organic and inorganic compounds through processes carried out by soil organisms.
- **Mineralization** is the release of plant-available forms of nutrients that occurs when soil organisms decompose organic matter.
- **Soil minerals:** Nutrients in the mineral component of soils become available to plants in the very long term.
- The management goal for a healthy agricultural soil is to supply the nutrients needed for optimal plant growth in the right quantity and at the right time while minimizing nutrient losses to the surrounding environment.
- Soil organic matter serves as a reservoir for several plant nutrients, including nitrogen, phosphorus, and sulfur.
- Every 1 percentage point of organic matter in the top 6 inches of soil contains about 1,000 pounds of nitrogen, 230 pounds of phosphorus, and 165 pounds of sulfur per acre. Nevertheless, the majority of nutrients within organic matter are not readily accessible to plants.
- To be used by plants, nutrients in organic matter must be converted to inorganic forms through decomposition and mineralization by soil organisms.
- Soil organisms form a food web that decomposes organic matter and releases nutrients in the process. Bacteria and fungi, positioned at the foundation of the food web, derive energy by immediately digesting soil organic materials.
- Protozoa and some nematodes are organisms that graze on bacteria and fungi, releasing nitrogen that can then be utilized by plants.

Soil Food Web

- Soil is home to a complex assemblage of organisms that interact to significantly impact both aboveground and belowground processes.
- The soil food web comprises the community of creatures that inhabit the soil, either partially or entirely during their life cycles. Soil-dwelling organisms are essential for soil function, underpinning vital activities such as soil structure creation, decomposition and nutrient cycling, bioremediation, and the enhancement of plant health and diversity.
- Soil organic matter is the base resource that supplies energy and nutrients used by plants and other organisms. Soil organic matter encompasses all organic substances present in or on the soil, comprising both plant- and animal-derived materials at various stages of decomposition.
- Larger organisms–for example, small arthropods–some barely visible to the unaided eye, help mediate the decomposition of plant and other organic residues.
- Common insects and related organisms that actively contribute to decomposition in agricultural systems include millipedes, springtails, mites, fly larvae, and burying beetles. Besides facilitating the decomposition of organic waste, decomposers are frequently consumed by other arthropods (e.g., spiders) and can aid in sustaining populations of advantageous predatory arthropods.
- Nitrogen (N) is a nutrient that can undergo many transformations in the soil through microbial processes. The process of converting organic nitrogen to plant-available ammonium (NH_4^+) is called mineralization. A specific group of bacteria convert ammonium to nitrate (NO_3^-) in a process called nitrification.
- Nitrogen fixation is carried out by both free-living and root-symbiotic organisms. Nitrogen fixation refers to the transformation of atmospheric N_2 into ammonia (NH_3) and constitutes a crucial method for augmenting nitrogen in the soil ecosystem.
- In symbiotic nitrogen fixation, nitrogen is fixed by bacteria within nodules of the roots of plants in the legume family and ammonia is then taken up by the plant to be turned into an organic form of nitrogen.
- The rhizobium-legume symbiosis is the most well known of the nitrogen-fixing associations. Legume cover crops will fix more N when grown in a soil with low NH_4^+ and NO_3^-.
- In addition to serving as a source of stored nutrients, soil organic matter provides a significant portion of the cation exchange capacity (CEC) in soil. Cation exchange facilitates the retention of positively charged nutrients in the soil, safeguarding them from loss by leaching. Enhancing the organic matter content of soil is one of the limited methods to augment soil cation exchange capacity (CEC).

- Soil pH, a measurement of the activity of hydrogen ions (H^+) in soil solution, is a variable that drives many aspects of nutrient cycling and soil biology. A pH of 7 is deemed neutral; values below 7 are classified as acidic, while those over 7 are categorized as alkaline.
- Most crops do best in the soil pH range of 6–7, though there are some exceptions. When soil pH falls below 6, aluminum in the soil alters its shape and becomes detrimental to plant roots. Manganese may reach dangerous concentrations at a low soil pH. Soils in humid climates tend to be inherently acidic, necessitating the application of liming compounds to maintain an optimal pH level.
- Soil pH governs the accessibility of many micronutrients, with iron, manganese, and zinc becoming increasingly available as pH decreases. The availability of molybdenum rises with increasing alkalinity of pH.
- Crops that favor a pH outside the typical range of 6–7 generally do so due to particular micronutrient requirements. Legumes, necessitating molybdenum for the nitrogen-fixing enzyme, like a soil pH close to 7. Blueberries, which have elevated iron levels, prefer a pH range of 4.5 to 5.
- Soil pH also influences soil organisms. Earthworms and microorganisms favor a nearly neutral soil pH. Fungi thrive at various soil pH levels, resulting in their predominance in acidic soils within the microbial community. Soil pH affects the cation exchange capacity provided by organic matter. As soil pH rises, the cation exchange capacity of soil organic matter will correspondingly increase.

Biological Control of Pest Organisms

- One ecosystem service provided by soil organisms that is of particular interest in agricultural systems is biological control of arthropod pests.
- Biological control is the term for reduction of pest organisms by natural enemies, which include predators, parasites, and pathogens.
- Healthy agricultural soil communities typically include a wide range of predators, parasites, and pathogens that contribute to the suppression of agricultural pests. Spiders, harvest men, and ground beetles are important ground-dwelling natural enemies of insect pests.
- Ground beetles play a major role in agro-ecosystems by contributing to the mortality of insects, weed seeds, and slugs.

Arbuscular Mycorrhizal Fungi

- Arbuscular mycorrhizal fungi are beneficial soil organisms that contribute to many aspects of soil health. Mycorrhizal fungi establish a symbiotic relationship with plant roots. Symbiosis refers to a tight relationship between distinct organisms. This symbiotic relationship grants the fungus consistent and direct access to carbohydrates provided by the plant. The plant gains advantages from the fungus's capacity to extend into the soil, forming a filamentous network of fungal biomass referred to as hyphae or mycelium, hence enhancing root volume.

- Mycorrhizal fungi are dependent on the host plant for an energy source and cannot survive for long periods of time without a plant host. Approximately 80 percent of terrestrial plants establish a symbiotic connection with mycorrhizal fungus. Several significant nonmycorrhizal crops and weeds encompass brassicas (such as broccoli, cabbage, radish, and canola) and chenopods (including spinach, chard, and lambsquarters).
- Mycorrhizal fungi are particularly proficient in assisting plants in the acquisition of phosphorus, a nutrient that exhibits significant immobility in the soil. Due to restricted mobility, a phosphorus depletion zone forms around the root as plant roots take phosphorus from the soil. Mycorrhizal fungi function as an extension of the plant root system, extracting phosphorus from non-depleted areas and delivering it to the roots.
- The exterior hyphae of mycorrhizal fungus enhance soil aggregation by secreting a viscous substance known as glomalin. Glomalin facilitates the cohesion of soil particles into aggregates that withstand erosion and preserve soil porosity.
- Mycorrhizal symbioses enhance a plant's resilience to stress. The network of fungal hyphae around the roots can impede the invasion of plant roots by diseases. Mycorrhizal fungi can reduce plant infections by improving plant nutrition, augmenting root resilience, altering the chemical composition of plant tissues, mitigating abiotic stress, and modifying the microbial population associated with roots.
- Numerous factors influence the populations of mycorrhizal fungus in the soil. Tillage destroys the intricate network of fungal hyphae, diminishing their populations. Elevated phosphorus concentrations in the soil inhibit mycorrhizal populations, as plants are less inclined to sustain the symbiotic relationship. Ultimately, mycorrhizal fungi rely on a host plant for energy; prolonged absence of a host, as seen in barren fallow fields or during the cultivation of non-host crops in rotation, will result in a gradual collapse of their populations.
- Typically, indigenous soils possess sufficient populations of active mycorrhizal fungi or dormant spores that will activate upon the cultivation of a host crop. Consequently, the inoculation of field soil with mycorrhizal fungus is typically superfluous.

Air and Water Regulation in Soil

- Plants necessitate both oxygen and water in the root zone for optimal growth. Water and air are retained in the pore spaces between soil particles and aggregates within the soil. The dimensions of the pores present between and within soil aggregates dictate the movement and retention of water and gasses in the soil.

- Larger pores called as macropores, which are important to promote good aeration and rapid infiltration of rainfall. Smaller pores called as micropores, which are important for absorbing and holding water.
- Macropores are typically discernible to the unaided eye, however micropores located between and inside microaggregates are not. To ensure sufficient aeration and water availability for optimal plant growth, the presence of both macro- and micropores in the soil is essential.
- Pores in the soil are created when soil particles collect into a structured hierarchy. Soil organisms are crucial for the formation of soil aggregates and the enhancement of aggregate stability.
- Clay, organic matter, root hairs, organic compounds from bacteria and fungi, and fungal hyphae contribute to the cohesion of soil aggregates. Aggregate stability denotes the capacity of soil aggregates to remain intact against the erosive impacts of water.
- Optimal aggregate stability will preserve macropores in the soil, mitigate surface crusting, enhance aeration, decrease rainwater runoff, and diminish soil erosion. Aggregates aid in the preservation of soil organic matter, as organic matter particles contained within aggregates are physically safeguarded against microbial degradation.
- Numerous substantial soil organisms possess the ability to displace dirt and generate macropores inside it. These encompass species such as ants, dung beetles, and earthworms.
- Earthworms are probably the best-known soil organism that contributes to the development and maintenance of soil structure. The burrowing activity of earthworms benefits soil health through increased nutrient availability, better drainage, and a stable soil structure.

Soil Organic Matter (SOM)

- It plays a crucial role in integrating several facets of soil health. Soil organic matter (SOM) can be categorized into labile and stable pools, each exhibiting distinct characteristics and functions within the soil. In agricultural soils, organic matter may vary from 1 to 8 percent, influenced by climate, soil type, and management approaches.
- The labile pool of organic matter, comprising 5–20 percent of the total soil organic matter (SOM), encompasses the living biomass of soil organisms and plant roots, fine particles of organic detritus, and relatively simple organic compounds, including polysaccharides, organic acids, and other substances produced by microbial activity or resulting from decomposition processes. Labile organic matter is easily digested by microorganisms and serves as the primary energy source for the soil

food web. It serves as the primary source of organic nitrogen that can be easily mineralized and rendered accessible for plant utilization.

- Polysaccharides in labile organic materials improve aggregate stability. When microbial degradation of labile organic matter exceeds the influx of fresh organic matter into the soil, levels of labile organic matter will diminish. Excessive soil tillage can hasten the depletion of labile organic matter by aerating the soil, hence enhancing microbial activity, and by exposing organic matter previously safeguarded within soil aggregates.
- The stable pool of organic matter, which accounts for 60–95 percent of the total pool of soil organic matter, consists of organic compounds that are relatively resistant to decomposition because of either their chemical structure, their adsorption to clay particles, or their protection within micro aggregates.
- Stable organic matter contributes cation exchange capacity and water-holding capacity to soil. The pool of stable organic matter is increased or depleted slowly as only a small portion of the labile organic matter that cycles through the food web is stabilized into forms that are resistant to decomposition.
- The amount of organic matter in soil results from the equilibrium between organic matter inputs, including crop residues, manure, and compost, and the rate of organic matter breakdown. Inputs of organic matter can be affected by crop management practices, including the implementation of cover crops, crop rotations, and residue management, alongside soil management techniques, such as the application of organic fertilizer sources like compost and manure. Labile organic matter levels react more swiftly to alterations in management techniques than stable soil organic matter levels, therefore serving as a precursor for long-term trends in total organic matter levels.

Improvement Soil Health through Management Practices

Reduce Inversion Tillage and Soil Traffic

- Excessive tillage adversely affects soil health in multiple ways. Tillage enhances soil aeration, promoting microbial activity and leading to the breakdown of organic materials.
- Tillage also disturbs soil aggregates, exposing previously protected organic matter particles to microbial digestion. If the additions of organic matter are inadequate to offset the losses from decomposition, organic matter levels will diminish over time, so impairing soil health.
- Inversion tillage also reduces the soil coverage provided by crop residues, leaving soil more exposed to erosion.

Commonly Used Primary Tillage Implements

- **Moldboard Plow:** Inverts the soil to bury residues, terminate cover crops and perennial sod, and kill weeds.
- **Disk Plow:** Concave disks affixed in a gang configuration sever residue and laterally invert soil, therefore aerating the soil and integrating residue within it. Soil disturbance and residue incorporation are contingent upon the dimensions, configuration, and inclination angle of the disks.
- **Chisel Plow:** Curved shanks with chisel points are dragged through the soil without inversion. Loosens surface soil, mixes some residue into the soil. Soil disturbance and residue incorporation are contingent upon the breadth and angle of chisel tips. Tillage can disturb the hyphal network of mycorrhizal fungi, perhaps resulting in their reduction over time. If not meticulously controlled, both inversion and non-inversion tillage techniques can compact the subsoil, forming a plow pan that impedes root development and limits access to water and nutrients in the subsoil. Excessive vehicular and pedestrian activity can compact the surface soil, diminishing macro porosity and hindering root development.

Soil Compaction

- Soil compaction transpires when soil is subjected to high pedestrian and machinery traffic while in a moist and plastic state. This traffic compacts the soil, diminishing pore space and elevating bulk density. Macropores are subjected to greater compression than micropores, resulting in inadequate water infiltration and drainage, as well as heightened runoff. Soil compaction elevates soil density, hindering the growth of plant roots inside the substrate. The diminishment of pore space also impacts the habitat of numerous minuscule soil creatures that are immobile, unable to displace soil particles, and confined to the pre-existing pore spaces and channels within the soil.
- Physical disruptions like inversion tillage can significantly impact the biological characteristics of soil. The compaction and removal of surface debris may lead to a decrease in soil moisture and habitat for soil-dwelling species. The diversity and number of soil surface arthropod predators may be superior under conservation tillage management compared to traditional inversion tillage, thereby enhancing the natural control of pest insects in conservation tillage systems. Beneficial soil insects are more likely to endure in fields employing non-inversion tillage methods, such as chisel plowing. Compared to inversion tillage methods (e.g., moldboard plow), non-inversion tillage results in reduced soil disturbance and hence lower direct death of beneficial soil organisms.
- Some tillage is still a necessary practice in certain production systems, especially organic systems that do not use herbicides for weed control.

When employing tillage, it is crucial to counterbalance the heightened rate of organic matter decomposition by augmenting organic matter inputs via crop wastes, manure, and compost. Incorporating multiple years of a perennial fodder crop into a rotation with annual crops necessitating tillage is a method to diminish tillage intensity over time.

Increase Organic Matter Inputs

- To maintain soil organic matter levels, inputs of organic matter must meet or exceed the losses of organic matter due to decomposition.
- Healthy crops can be a valuable source of organic matter, and crop residues should be returned to the soil to the extent possible.
- The integration of cover crops or perennial crops, along with strategic applications of animal manure, green manure, and compost, can enhance or sustain soil organic matter levels.
- The organic matter content of soil can be tracked over time by requesting an organic matter analysis when submitting soil fertility samples to the soil testing laboratory.
- Ensure that your comparisons of organic matter across time are derived from data obtained from the same laboratory or from laboratories employing same methodologies for organic matter analysis, as findings may vary considerably between different analytical techniques.

Use Cover Crops

- Cover crops have multiple advantages to soil health. They provide soil coverage throughout winter and other non-growing months, hence mitigating erosion risk. The biomass generated by cover crops is often reintegrated into the soil, augmenting organic matter content. Cover crops possessing taproots can generate macropores and mitigate compaction.
- Fibrous-rooted cover crops can enhance aggregation and stabilize soil structure. Cover crop species that support mycorrhizal fungus can maintain and enhance the population of these advantageous fungi. Leguminous cover crops can enhance soil nitrogen levels via nitrogen fixation. Cover crops can sequester nitrate and other nutrients prone to leaching losses.

Reduce Pesticide Application and Provide Habitat for Beneficial Organisms

- Beneficial insects that contribute to biological control can be harmed by the application of broad-spectrum insecticides. Farmscaping is a whole-farm, ecological approach to increase and manage biodiversity with the goal of increasing the presence of beneficial organisms.

- Farmscaping techniques encompass the implementation of insectary plants, hedgerows, cover crops, and water reservoirs to attract and sustain populations of beneficial creatures, including insects, spiders, amphibians, reptiles, bats, and birds, which parasitize or feast on insect pests. Farmscapes situated in contours between fields, steep ditches, or areas prone to erosion provide soil stability.
- Farmscaping can also be used as a filter strip to prevent water runoff and soil erosion. Plants used in farmscapes contribute to healthy soil by adding organic matter, the base of the soil food web.

Crop Rotations

- Diverse crop rotations will disrupt the life cycles of soil-borne pests and diseases, enhancing crop health and nutrient cycling. Rotations can also aid with weed management.
- By cultivating a variety of crops in both temporal and spatial dimensions, pests that flourish in certain crops are prevented from establishing their populations throughout time. Crop rotation can also mitigate nutrient surpluses.

Nutrients Management

- Meticulously strategizing the timing, application technique, and volume of manure, compost, and other fertilizers can enable you to satisfy crop nutrient requirements while reducing nutrient surpluses.
- Robust, vigorous plants that exhibit rapid growth are more resilient to insect harm. Nevertheless, over fertilization of crops might exacerbate pest issues. Elevated soluble nitrogen levels in plants may diminish their pest resistance, leading to increased insect density and crop damage.
- Maintaining an optimal soil pH for the cultivated crop enhances nutrient availability and mitigates toxicity. Ensuring sufficient calcium levels will promote earthworm vitality and enhance soil aggregation.
- Utilizing varied nutrient sources can aid in preserving soil health. Manure and compost contribute organic matter and other nutrients; however, relying solely on compost or manure to fulfill the crop's nitrogen requirements annually may lead to elevated phosphorus concentrations in the soil. Integrating moderate applications of manure or compost to fulfill phosphorus requirements with supplementary nitrogen from leguminous cover or forage crops in a crop rotation can effectively equilibrate nitrogen and phosphorus inputs.

Chapter 3

Wasteland and its Reclamation

Wasteland and their Management in India

Definition (NRSA)

Wasteland is that land which is presently lying unused or which is not being used to its optimum potential due to some constraints. The land may be declared as wasteland due to some limiting factors which may be natural or anthropogenic.

Classification

The National Wastelands Development Board categorizes wastelands into two classifications:

I. **Cultivable wastelands:** Cultivable wasteland are the lands which has the potential for its use for various purposes like afforestation, pastoral development and agricultural production. At present due to several constraints it is not being used for production purpose, but such wastelands can be brought under vegetation with reasonable effort.

II. **Uncultivable wastelands:** Uncultivable wasteland are barren land and unproductive which can't be used for any purpose like agriculture, afforestation and pastoral development.

The cultivable wastelands have been classified into:

- Gullied and/or ravenous lands
- Undulating land without shrubs
- Surface waterlogging land and marsh

- Salt affected land
- Shifting cultivation area
- Degraded forestland
- Degraded pasture/grazing land
- Degraded forest plantations
- Strip lands
- Sand dunes
- Mining/industrial wastelands

Uncultivable wastelands that are unsuitable for cultivation are categorized as:

- Brown rocky/stony/shut of rocks
- Steep sloppy areas
- Snow covered and/or glacier lands

Extent of Wastelands

NRSA estimates put wastelands at 16.21 per cent of the total land area of the country. Of this, 16.74 per cent is culturable and rest 4.47 per cent is unculturable. Total wastelands in India: 129.57mha.

Major Causes of Wasteland

Wasteland is caused due to many factors:

- Factors contributing to wasteland include waterlogging, ravine and gully erosion, riverine lands, salinity and alkalinity, shifting sand dunes, wind erosion, severe moisture deficiency, overexploitation of natural resources, accumulation of industrial and sewage waste, soil erosion, and deforestation.
- Water, wind, man and other source are causing agents for Wasteland. There are two factors of wasteland – Utilisable Waste Land, Un- Utilisable Waste Land. Utilisable Wastelands include marshy lands, saline lands, and undulating lands. Rocky land, Glacier Land and Deserts are part of Un- utilizable wasteland.

Distribution of Wasteland

Wastelands are lands which are economically unproductive, ecologically unsuitable and subject to environmental deterioration. The official estimate of wasteland in India in 2004 was nearly 63.85 million ha – more than 20 per cent of the country's geographical area. Among the categories of wastelands, 30 per cent goes to snow covered lands, 22 per cent comes under degraded notified Forest land, 10 per cent Barren Rockyland and the rest are occupied by the other categories.

The category wise wasteland types of India is shown in Table 3.1.

Table 3.1: Category-wise Wastelands of India (2004)

Area in sq. km

Sl. No.	*Category*	*Total Wastelands*	*Per cent of Total Geographical Area Covered*
1	Gullied and/or Ravinous land	20553.35	0.65
2	Land with or without scrub	194014.29	6.13
3	Waterlogged and Marshy land	16568.45	0.52
4	Land affected by salinity/alkalinity- coastal/inland	20477.38	0.65
5	Shifting Cultivation Area	35142.20	1.11
6	Under utilized/degraded notified forest land	140652.31	4.44
7	Degraded pastures/grazing land	25978.91	0.82
8	Degraded land under plantation crop	5828.09	0.18
9	Sands Inland/Coastal	50021.65	1.58
10	Mining Industrial wastelands	1252.13	0.04
11	Barren rocky/stony waste/sheet rock area	64584.77	2.04
12	Steep sloping area	7656.29	0.24
13	Snow covered and/or glacial area	55788.49	1.76
	Total Wastcland Area	**638518.31**	**20.17**

Note: 1.20.849.00 sq. km. in Jammu & Kashmir is not mapped and hence not considered for calculating the percentage.

Source: 1:50,000 scale wasteland maps prepared from Landsat Thematic Mapper/IRS LISS II/III Data.

Rajasthan has the highest percentage of classified area among states, at 45 percent, followed by the hill states of Jammu & Kashmir, Tripura, Himachal Pradesh, Manipur, Sikkim, Mizoram, Meghalaya, and Nagaland. The preeminence of the hill states in the rankings is attributable to the vast expanse of snow-covered and glacial terrain in these regions (Table 3.2).

Formation of Wasteland

- Wasteland is created by either natural processes or anthropogenic activity. The undulating upland, sandy terrain, snow-covered regions, and coastal saline zones are natural formations, whereas gullied or ravinous land, jhum or forest-clearance areas, and barren hill ridges are the result of anthropogenic actions.
- Four primary anthropogenic actions contribute to the creation of wasteland areas: deforestation, overgrazing, overcultivation, and unskilled irrigation. The initial three actions remove the land's protecting

vegetation, accelerating soil erosion and land degradation, while improper irrigation leads to issues of waterlogging and salinity.

Table 3.2: Aerial Assessment of Wasteland, (By State) in 2000

Area in ha

1	Rajasthan	Salt affected Land	497.640
		Gullied or Ravinous Land	8766.120
		Undulating Upland with or without scrub	7656
		Sandy Area	130193.938
		Barren Hill Ridge or Rock outcrop	7770.840
		Total	154884.538
		Percentage of State area	45.258
2	Jammu & Kashmir	Salt affected Land	38.280
		Undulating Upland with or without scrub	76.560
		Jhum or Forest Blank	2373.380
		Barren Hill Ridge or Rock outcrop	153.120
		Snow covered or Glacial Area	70054.141
		Total	72695.461
		Percentage of State area	32.710
3	Tripura	Undulating Upland with or without scrub	382.800
		Jhum or Forest Blank	2641.320
		Snow covered or Glacial Area	229.680
		Total	3253.800
		Percentage of State area	31.056
4	Himachal Pradesh	Undulating Upland with or without scrub	459.360
		Jhum or Forest Blank	995.280
		Snow covered or Glacial Area	13589.800
		Total	15044.440
		Percentage of State area	27.022
5	Manipur	Undulating Upland with or without scrub	306.240
		Jhum or Forest Blank	5014.680
		Barren Hill Ridge or Rock outcrop	344.520
		Total	5665.440
		Percentage of State area	25.340
6	Sikkim	Undulating Upland with or without scrub	765.600
		Snow covered or Glacial Area	1077.840
		Total	1843.440
		Percentage	25.256

7	Mizoram	Jhum or Forest Blank	3981 120
		Snow covered or Glacial Area	497.640
		Total	4478.780
		Percentage of State area	21.238
8	Meghalaya	Jhum or Forest Blank	4555.320
		Barren Hill Ridge or Rock outcrop	191.400
		Total	4746.720
		Percentage of State area	21.106
9	Nagaland	Jhum cr Forest Blank	3253.800
		Total	3253.800
		Percentage of State area	19.686
10	Gujarat	Salt affected Land	1110.120
		Gullied or Ravinous Land	191. 400
		Jhum or Forest Blank	76.560
		Sandy Area	1492.920
		Barren Hill ridge or Rock outcrop	842.160
		Total	3713.160
		Percentage of State area	8.396
11	Arunachal Pradesh	Undulating Upland with or without scrub	191.400
		Jhum or Forest Blank	5512.320
		Snow covered or Glacial Area	7732.560
		Total	13436.280
		Percentage of State area	16.076
12	Uttar Pradesh	Salt affected Land	19407.961
		Gullied or Ravinous Land	8842.680
		Undulating Upland with or without scrub	680.440
		Jhum or Forest Blank	2794.440
		Sandy Area	918.720
		Barren Hill Ridge or Rock outcrop	880.440
		Snow covered or Glacial Area	7043.520
		Total	40768.201
		Percentage of State area	13.846
13	Kerala	Undulating Upland with or without scrub	803.880
		Jhum or Forest Blank	3483.480
		Barren Hill ridge or Rock outcrop	918.720
		Snow covered or Glacial Area	114.840
		Total	5320.920
		Percentage or State area	13.690

14	Tamil Nadu	Salt affected Land	574.200
		Gullied or Ravinous Land	4210. 800
		Undulating Upland war or without scrub	6507.600
		Jhum or Forest Blank	3330.360
		Sandy Area	1914
		Barren Hill Ridge or Rock outcrop	1071.840
		Total	17610.800
		Percentage of State area	13.538
15	Odisha	Salt affected Land	153.120
		Undulating Upland with or without scrub	12326.160
		Jhum or Forest Blank	6622.440
		Sandy Area	382.800
		Total	19484.506
		Percentage or State area	12.506
16	Karnataka	Salt affected Land	267.960
		Gullied or Ravineous land	2220.240
		Undulating Upland with or without scrub	12517.560
		Jhum or forest Blank	3100.680
		Barren Hill Ridge or Rock outcrop	3828
		Total	21934.440
		Percentage of State area	11.436
17	Andhra Pradesh	Salt affected Land	880.440
		Gullied or Ravinous Land	4517.040
		Undulating Upland with or without scrub	18642.359
		Jhum or Forest Blank	880.440
		Sandy Area	1033. 560
		Barren Hill Ridge or Rock outcrop	4708.440
		Total	30662.270
		Percentage of State area	11.076
18	Madhya Pradesh	Gullied or Ravinous Land	21283.680
		Undulating Upland with or without scrub	13436. 280
		Jhum or Forest Blank	7005.240
		Barren Hill Ridge or Rock outcrop	2411.840
		Total	54137.040
		Percentage of State area	9.966

19	Maharashtra	Salt affected Land	1186.680
		Gullied or Ravinous Lard	7388.040
		Undulating Upland with or without scrub	19637. 641
		Jhum or Forest Blank	1454.640
		Barren Hill Ridge Rock outcrop	995.280
		Total	30662.281
		Percentage of State area	9.962
20	Goa, Daman, and Dlu	Undulating Upland with or without scrub	306.240
		Jhum or Forest Blank	38.280
		Total	344.520
		Percentage of State area	9.034
21	Haryana	Salt affected Land	1110.120
		Gullied or Ravinous Land	191.400
		Jhum or Forest Blank	76.560
		Sandy Area	1492.920
		Barren Hill Ridge or Rock outcrop	842.160
		Total	3713.160
		Percentage of State area	8.396
22	Bihar	Gullied or Ravinous Land	1875.720
		Undulating Upland with or without scrub	8727.840
		Jhum or Forest Blank	2373.360
		Barren Hill Ridge or Rock outcrop	2335..080
		Total	15312.000
		Percentage of State area	8.806
23	Punjab (Including Chandigarh)		
		Salt affected Land	1301.520
		Sandy Area	2603.040
		Total	3904.560
		Percentage of State area	7.730
24	West Bengal	Salt affected Land	1263.240
		Undulating Upland with or without scrub	4249.080
		Jhum or Forest Blank	535.920
		Barren Hill Ridge or Rock outcrop	306.240
		Total	6354.480
		Percentage of State area	7.232
25	Assam	Undulating Upland with or without scrub	574.200
		Jhum or Forest Blank	2373.360
		Total	2947.560
		Percentage of State area	3.752

- Moreover, both cultivated and uncultivated regions experience soil erosion and degradation. Mining is a significant activity that contributes to deforestation and land degradation. The desert is a significant natural wasteland, and any actions associated with desertification are regarded as primary contributors to the production of such wastelands.

Wasteland Management

- Nature is our ally; we must safeguard and preserve it optimally. We must do all significant measures to prevent the destruction or degradation of nature. It is our fundamental responsibility as humans to conserve nature and implement preventive measures. Numerous measures have been implemented for the management of waste, including both refuse and soil. Innovative strategies have been developed for repurposing materials and enhancing environmental cleanliness. Rapid urbanization has resulted in significant challenges regarding wasteland management for countries. Soil is a valuable and natural resource for any country. The formation of fertile soil requires thousands of years, which is vital for our ecosystem. Nature typically requires some time to yield optimal results. Our foremost priority and inherent instinct compel us to preserve the land from degradation for the advancement of our future. The soil is the fundamental source of numerous essential resources, both directly and indirectly; so, it is imperative to conserve it to meet human need for sustenance, energy, and other necessities.
- The land's ability to yield food, fodder, fuel, and industrial raw materials is influenced by biological variables—such as climate, soil, and hydrological systems—as well as social and economic factors, including land management methods and inputs like irrigation, fertilizers, and seeds.
- What is the potential productivity of wastelands, and how can it be quantified?This is contingent upon the definition of productivity, which, like many ecological words, is frequently misinterpreted in everyday language. One definition of the productivity of an ecosystem is the amount of biomass "fixed" in a given time. This is an ongoing process, quantified as the quantity of food produced per unit of time (day, month, year).
- Our nation encompasses 329 million hectares of geographic area, of which 167 million hectares are afflicted by various forms of degradation, including water erosion, wind erosion, salinity, alkalinity, and flooding. As a result of several misuses, we have experienced a degradation of 2.1 million hectares of land annually. Deforestation for lumber and fuelwood extraction, along with the following conversion of forest land to agricultural or urban areas, significantly threatens land deterioration and accelerates wasteland formation. The growing population necessitates land suited for agriculture. To do this, it is essential to comprehend the principles of Wasteland Management and strategies for its prevention.

In the wasteland reclamation process, all causative variables must be adequately addressed to get a favorable outcome. Consequently, the government implemented various measures to enhance public knowledge regarding wasteland management.

Table 3.3.11: The Efficiency and Energy Output of Specific Plant Species in India

Name of the Species	*Tonnes/ha./Year*	*G.J/ha./Year*
Sweet sorghum	20	335
Exotic forage sorghum	68	1,250
Water hyacinth	154	2,680
Sugarcane	112	2,000
Hyloid-poplar (short rotation) three year old	20	342
Coppice crop - two year old	8.3	144
Eucalyptus species	39.1	680
Algae (fresh water pond culture)	88	1,460
Casuarina species (7-10 year rotation)	200	4,950
Lucerne (legume)	100 cu. at per ha. per year	4,600

(GJ = 10° Joules, measure of energy).

Source: Ghosh and Bisada, 1981.

- Waste management is inherently difficult. The primary emphasis on management concerns is to alleviate the strain on these delicate terrestrial ecosystems of arable land to facilitate the sensible use of land resources and optimize production. The management of wasteland must cater to the needs of local populations, restore biological equilibrium, enhance environmental conditions and hydrological systems, and mitigate soil and water erosion through the implementation of appropriate procedures. Moreover, wasteland afforestation and social forestry programs significantly contribute to the rehabilitation of wasteland.
- The reclamation of wasteland has emerged as a significant problem for the country. The reclamation and development of wasteland has four primary ecological objectives:
  - To enhance the physical composition and quality of the soil,
  - To augment the availability and quality of water,
  - To mitigate soil erosion, landslides, and flooding, and
  - To preserve the biological resources of the land for sustainable utilization.
- Strategies for the restoration of wasteland should incorporate various stakeholders, including volunteers, government officials, and farmers.

In simple terms Waste Land Management refers to land that is not used due to any obstruction is considered as Waste Land. Land that can either be cultivable or non-cultivable falls under this category of wasteland. Managing these lands for proper use or re-use it for different purposes is known as ***wasteland management****. With the upcoming technology and advanced procedures, we can contribute towards utilizing the land to its full potential. There are many factors that are responsible along with technology like climate, topography and soil properties that make it suitable for land to be used or reused. It is estimated that the biomass production in the wasteland is below 20 per cent due to which it is not considered suitable for use. Usually, when a land does not return anything even when yield properly is a wasteland. A land that does not serve any purpose for cultivation falls under this category.*

Furthermore, supplementary techniques should be implemented to enhance agricultural productivity and rehabilitate wastelands through afforestation. Each district should recognize the technological factors contributing to wasteland and various reclamation methods should be imparted. A program should be established for the execution of wasteland greening, incorporating modern reclamation techniques and a framework for sowing high-quality seeds. Farmers should be incentivized to engage in various methods and techniques, such as contour ploughing, to enhance wasteland through crop cultivation.

Specific Measure for Improvement of Wasteland

- Extensive afforestation contributes to the preservation of soil fertility. The soil is deficient in moisture; hence, extensive irrigation may be implemented for its conservation.
- Regulate the river's water and restrict grazing or overgrazing.
- Soil and water conservation with a plowing technique.
- Reforestation can occur in areas devastated by fires, overgrazing, and excessive deforestation.
- To mitigate soil erosion, it is advisable to retain crop residue.
- Flowing water down the slope erodes and transports the soil. It can be mitigated through strip farming, terracing, and contour plowing.
- Leaching with additional water can be employed to mitigate salinity in damaged land. Implementing crop rotation and cultivating diverse plant species helps enhance soil fertility.
- The redevelopment of the ecosystem facilitates the reclamation of minerals from the wasteland.

Although there is a lot of geographical areas available, a small amount of it is usable. There are many places that are become barren due to a climatic condition which does not make it suitable to grow something. For example, Antarctica is the highest, darkest and coldest continent on earth. It is surrounded by ice and strong winds. Due to degradation by many agents, the land is not usable for forestation. There is a threat to Amazon Rainforest in becoming wasteland sooner or later. Each year thousand acres of land is cleared for raising cattle. As cattle has become a secondary source of eating while green food is still at the primary position. Waste lands can be a cause of fast spreading diseases in your locality. Gift yourself a good health by using this land for some positive activities that can help in conservation of environment and save your locality from getting any infections. Utilize your wasteland today so that you can contribute in some way towards maintaining a healthy society and healthy land.

Below are described several case studies of wasteland reclamation and development in India.

(a) Micro Watershed Development in Sukhomajri (Punjab)

The 455 people, predominantly impoverished Gujjars, subsisted by grazing their cattle and goats on hill slopes minimally adorned with grass and plants. A multitude of men sought employment at tractor, machine tool, and cement companies located beyond the hamlet. Sukhomajri lacked irrigation, water, and electricity. Its sole resources were of 10 hectares of land (half held by individual people) and over 400 goats and cattle.

Despite recognizing that grazing livestock in specific regions exacerbated erosion and impacted the yield of their rain-fed maize, wheat, and legume crops, the people perceived no viable options.

During 1974-75, the Central Soil and Water Conservation Research and Training Institute in Chandigarh undertook the desilting of the renowned Sukhna Lake, which had been 60 percent filled with silt from the Shivalik range, transported by the Sukhna Cho. P. R. Misra, Project Director of the Institute, examined the lake's catchment area and, while mapping the trajectory of the Sukhna Cho, encountered Sukhomajri.

Misra developed a system of check dams that formed three small reservoirs, with widths ranging from 60 to 120 meters. The substantial monsoon rainfall is directed into reservoirs for utilization in drinking and irrigation. Misra persuaded the villagers to refrain from grazing their cattle in the watershed areas by permitting them to harvest grass from those locations instead.

The cultivation of trees and grasses in the catchment area facilitated the regeneration of the watershed. The erosion at the 15-meter gorge at one end of the settlement was mitigated through the use of vegetation.

Rather than cultivating a single rain-fed crop, two or even three crops are now being produced year, resulting in a twelvefold increase in grass productivity. A distinctive approach to water distribution has been developed, whereby each family, regardless of land ownership or crop cultivation, is granted an equal-value water coupon.

Individuals who do not require water, such as the landless, may lend or sell their coupons to those in need of additional water. This is effective in Sukhomajri due to the village's homogeneity and the relative uniformity of landholdings.

Sukhomajri exemplifies a successful narrative. Nonetheless, irrigation has resulted in certain disparities. Large farmers with land in the command area have been utilizing water during the kharif season, resulting in insufficient water for small farmers whose properties are more distant from the reservoirs during the rabi season when water is limited.

Despite the substantial expenditure of Rs. 80,4750.00 until 1982—potentially exceeding the funds allocated to numerous other villages—Sukhomajri exemplifies the efficacy of soil protection within a catchment area. The expense of rehabilitating the catchment is significantly lower than addressing the downstream repercussions. For example, Rs. 27 million was allocated for the dredging of Sukhna Lake.

Individuals involved in the project noted that the locals were uncooperative throughout the beginning phases. Interest in investing in conservation measures arose only after greater irrigation led to increased crop yields.

The project spanned several years to address technical issues (1974-79), foster confidence among people, and establish the water distribution system (1980-83). All the major individuals dedicated considerable time to engaging with the villagers, acquiring knowledge from them prior to implementing new methodologies. The complete process spanned over ten years.

(b) Afforestation Programmes

In contrast to the substantial efforts dedicated to enhancing the productivity of arable land and mitigating soil erosion for agricultural advancement, far less focus has been directed towards increasing the productivity of forested areas. This occurred despite numerous policy suggestions highlighting the necessity to enhance the accessibility of firewood.

The National Planning Committee, established prior to independence with Jawaharlal Nehru as Chairman and which delivered its report in 1948, examined afforestation concerning soil conservation issues. It indicated that "the absence of forests in close proximity to the most densely populated areas" led to "the significant waste of the sole source of inexpensive fertilizer through the incineration of cow dung."

An examination of forestry expenditure reveals the little influence of policy pronouncements on program development.

During the 1951-69 period, the entire expenditure amounted to Rs. 1177 million, of which the farm forestry and fuel wood project, initiated in the Third Plan, represented merely 2.2 percent (NCA, 1976b). The National Commission on Agriculture (NCA) acknowledged the shortcomings of farm forestry programs in its interim report and proposed the notion of social forestry.

This had five objectives (Ministry of Agriculture, 1973):

1. Firewood supply to rural areas and replacement of cow dung as fuel;
2. Small timber supply;
3. Fodder supply;
4. Protection of agricultural fields against wind;
5. Recreational needs.

The NCA's suggestions were included into the Fifth Plan (1974-1979), leading to the initiation of the country's principal social forestry programs shortly thereafter. The Sixth Plan (1980-1985) initiated several new programs to address the escalating issue of wastelands and the declining resources from these areas.

A program for Rural Fuelwood Plantations commenced in 1980-81 to cultivate firewood in proximity to consumption sites.

Additional initiatives included the Tree for Every Child Programme and the establishment of eco-development task groups, primarily composed of ex-servicemen, aimed at rehabilitating degraded hill ecosystems through soil conservation and afforestation.

In acknowledgment of the magnitude and immediacy of the wasteland issue, Prime Minister Rajiv Gandhi, in a national broadcast (1985), urged the cessation of deforestation by fostering a grassroots movement for afforestation and proclaimed the establishment of a National Wastelands Development Board aimed at converting 5 million hectares of wasteland annually into firewood and fodder plantations.

The National Land Board was reformed into the National Land Use and Wastelands Development Council, chaired by the Prime Minister. There are two bodies beneath it.

The National Wastelands Development Board, situated within the Ministry of Environment, Forests and Wildlife, and the National Land Use and Conservation Board, positioned in the Ministry of Agriculture and Rural Development, serve as the national focal points for policy formulation, coordination, monitoring, and evaluation of programs.

(c) Social Forestry: Trees for People

No government program in India has been as fraught with misunderstanding or garnered as much media interest as social forestry. The initiative, designed to enhance the accessibility of fuel, fodder, and small timber for the rural impoverished, is a focal point of significant contention. The methodology and accomplishments are subjects of contention.

Tree planting for the rural impoverished is an element of the nation's National Programme, and the government asserts remarkable advancements, with objectives being met or surpassed (Ministry of Agriculture, 1984). Conversely, detractors assert that the program has been manipulated to advantage the upper classes and has strayed from its intended goals.

While Social Forestry Programs differ by state, the majority encompass one or more of four components.

1. Strip plantations along roadways, railways, and canals (government property) managed by the Forest Department;
2. Village woodlands on government and community land overseen by the Forest Department and community self-help initiatives;
3. Rehabilitation of degraded forest land conducted by forest departments;
4. Farm forestry on private land (including farm bunds) and leased government land (including wastelands), with support and saplings supplied by the Forest Department.

The initiative commenced in 1974, and during the Fifth Plan, Rs. 373 million were provided, approximately half of the amount suggested by the National Commission on Agriculture. During the Fifth Plan period, 50 percent of the 1.2 million hectares planted was allocated to the social forestry program.

In the Sixth Plan, this allocation was raised to Rs. 3,518 million, constituting 51 percent of the entire forestry budget of Rs. 6,925 million, which exceeded the combined forestry budget of the preceding 30 years, amounting to Rs. 4,820 million.

The Sixth Plan of Social Forestry Programs was executed within three categories.

1. Central Government initiative for social forestry, encompassing a Rs. 500 million rural fuel wood plantation program in 157 fuel-deficient districts;
2. Externally funded projects across nine states (with a government allocation of Rs. 1,600 million);
3. State government initiatives (Rs. 1,418 million).

During the Seventh Plan, a National Umbrella Project was proposed to expand foreign-assisted social forestry initiatives nationwide, requiring an investment of Rs. 800 million (Guhathakurta, 1984).

While the Planting Activity within the Various Components of Social Forestry has generally been satisfactory, the Program has failed to meet its broader objectives on three accounts.

1. It has provided financial support to affluent farmers;

2. It has failed to generate extensive social and environmental advantages;
3. In certain instances, it has exacerbated the plight of the impoverished.

Tree-planting can be a lucrative business opportunity in regions with a market for timber, resulting in the benefits of the initiative being seized by larger farmers. The early success of social forestry in Gujarat, Uttar Pradesh, Haryana, and Punjab was attributed to the engagement of medium farmers (possessing over 2 hectares) and large farmers (owning more than 4 hectares), alongside the increasing price of timber.

Surveys indicate that the proportion of farmers with land holdings under 2 hectares who have obtained free or low-cost seedlings ranges from 38.5 percent in Haryana and 42.5 percent in Gujarat to 64.7 percent in Jammu and Kashmir and 80.8 percent in West Bengal.

A research in Kolar district, Karnataka, underscored the inadequacy of social forestry in meeting the fundamental needs of the impoverished.

The eucalyptus plantations inside the scheme supply major rayon and paper companies; although some farmers profit from the elevated rates paid by the mills, agricultural laborers have suffered significantly. The cultivation of eucalyptus by large farmers, in lieu of traditional ragi, has resulted in diminished local food availability, leading to increased food costs, while the reduced maintenance requirements of eucalyptus have caused agricultural workers to lose their employment.

The social forestry program in Tamil Nadu has faced analogous criticism. In 1980, people in Acharapakkam, Chengalpattu district, uprooted 5,000 eucalyptus hybrid trees (CSE 1985). The state's innovative attempt to replant the exposed beds of the tanks, referred to as "foreshores," encountered difficulties.

Tamil Nadu possesses 38,000 tanks (rainwater collection reservoirs) that irrigate an aggregate area of 50,000 hectares. The tanks are shallow, featuring sloping banks, and range in size from small ponds to reservoirs exceeding 1,000 hectares. Numerous structures were constructed during the Eleventh century in the late Chola era.

The retained water is released into the fields throughout summer, revealing an expanding expanse of coastline. This acreage is currently utilized for grazing and vegetable cultivation, however the intention was to plant trees. The villagers resisted this. The poramboke, or government-owned common lands, are unavailable for planting due to encroachment for agricultural purposes.

Ultimately, the government had to concede that there was no evident social rationale for expropriating property from the impoverished to supply them with fuel at a future time.

Unlike extensive social forestry programs that struggle to assist the most disadvantaged, innovative initiatives are being devised to directly support the impoverished. In Rajasthan, a social security initiative is leasing degraded forest property to people for tree planting. The most impoverished persons have been identified and will receive 2 hectares each annually for 15 years, totaling 30 hectares.

They received 1,600 seedlings for the initial planting on two hectares. Additionally, each receives a monthly pay of Rs. 250, Rs. 500 for housing construction, and a monetary incentive for trees that survive into the second and third years. They are entitled to 20 percent of the earnings generated from the trees they cultivate.

In the Udaipur region, 435 individuals have enrolled in the program. The plan is now under revision, aiming to allocate leases of one hectare year, culminating in a maximum of ten hectares. However, 1,200 seedlings will be planted every hectare. A challenge here is that certain portions of this property are utilized for grazing.

Chapter 4

Soil Reaction and Solution

Basic Concepts

Elements and Atoms

Element

- ☆ An elementary unit of substance that cannot be further decomposed into a simpler form by physical or chemical processes, exemplified by elements like oxygen or iron. Elements constitute the fundamental components of nature.
- ☆ Elements are substances composed of a single type of atom and can exist as either individual atoms (e.g., argon) or as molecules (e.g., nitrogen). Each element is designated a symbol of one or more characters, for instance, O for oxygen and Fe for iron. Elements can be categorized into metals and non-metals.

Essential elements required by plants are

I. From water and air Carbon (C), Hydrogen (H), Oxygen (O)

II. Soil contains Nitrogen (N), Phosphorus (P), Sulfur (S), Potassium (K), Calcium (Ca), Magnesium (Mg), Copper (Cu), Iron (Fe), Manganese (Mn), Zinc (Zn), Boron (B), Molybdenum (Mo), Cobalt (Co), and Chlorine (Cl).

Atom

- ☆ An atom is the fundamental unit of an element that cannot be decomposed by chemical processes. Each atom comprises subatomic particles known as protons, neutrons, and electrons. Protons possess positive electrical charges, electrons carry negative charges, and neutrons are electrically neutral. Protons and neutrons reside in the nucleus in the center of the atom, whereas electrons orbit around the nucleus. The electrostatic force between positively charged protons and negatively charged electrons maintains atomic cohesion. Nearly all atoms include all three types of subatomic particles, with the exception of the hydrogen atom, which consists of only one proton and one electron, without a neutron.

- ☆ Atoms possess no net charge, as they contain an equal quantity of protons and electrons, which corresponds to the atomic number. The atomic number, represented by 'Z', is distinctive and unique to each element. The quantity of neutrons fluctuates and is ascertained by deducting the atomic number from the mass number (also known as the atomic weight). The atomic mass number, represented by 'A', is the total of protons and neutrons in an atom's nucleus. The atomic mass of each element is provided on the periodic table. The periodic table is a widely utilized system for organizing elements that conveys essential information regarding their chemical properties and behaviors.

Compounds, Molecules, and Atomic Bonds

- ☆ Atoms that join with one another produce more complex entities known as molecules. A collection of similar molecules composed of two or more distinct types of components is termed a compound. All compounds are classified as molecules; however, not all molecules qualify as compounds. Hydrogen gas (H_2) is classified as a molecule since it consists of a single element and cannot be designated as a compound. Water (H_2O) consists of hydrogen (H) and oxygen (O) atoms, so it is classified as a molecule or a compound.
- ☆ Molecules are denoted by the symbols of the components along by subscripts indicating the quantity of each element present. Water is denoted as H_2O, indicating the presence of two hydrogen atoms and one oxygen atom. Methane is denoted by CH_4, indicating the presence of one carbon atom and four hydrogen atoms.
- ☆ Molecules consist of clusters of atoms interconnected by chemical interactions, primarily categorized into two types: covalent and ionic connections. Atoms can unite by sharing electrons, a process known as covalent bonding. Ionic bonding occur when one atom donates at least one electron to another atom.

Ions

- ☆ Ions are charged particles created by the removal or addition of electrons. Metal atoms typically exhibit a propensity to relinquish electrons. Consequently, when there is a disparity in the number of protons and electrons in an atom, the resultant atom or molecule is termed an ion. Generally, metal atoms tend to relinquish electrons, while non-metal atoms tend to acquire electrons. Ions are typically generated when a chemical dissolves in water, such in the breakdown of salt in liquids. Ions are generated by applying direct current to a conductive solution, resulting in the dissolution of an anode through ionization.

- A cation is an ion characterized by a deficiency of electrons relative to protons, resulting in a positive charge, whereas an anion is an ion with an excess of electrons compared to protons, yielding a negative charge. The illustration below depicts calcium carbonate (left) dissolving in water (right).
- The superscript numerals denote the quantity of deficient (+) or excess (-) ions. If no numerical value is provided and only a + or - sign is present, it indicates an imbalance of a single electron.

$$CaCO_3 \rightarrow Ca^{2} + CO_3^{2-}$$

Where: $CaCO_3$ = Calcium Carbonate

Ca^{2+} = Calcium (cation)

CO_3^{2-} = Carbonate (anion)

- A molecule in solution is usually in equilibrium with its constituent ions. In other words, some molecules are breaking into ions while other ions are recombining to form molecules.

Chemical Reactions

- Chemical reactions transpire when atoms are reorganized to create new molecules or compounds. For instance, carbon dioxide and water can amalgamate to produce sugar, as observed in photosynthesis. This reaction is expressed as follows:

$$6CO_2 + 6H_2O + Energy \rightarrow C_6H_{12}O_6 + 6O_2$$

(Energy is necessary for the process to occur, and this energy is released during the breakdown of sugars. This energy-releasing process is termed respiration and occurs continuously in our bodies, as well as in various forms throughout all organisms.

- Redox reactions, or oxidation-reduction reactions, entail the transfer of electrons from a reducing agent to an oxidizing agent, and are prevalent in nature. Oxidation transpires when an element or molecule relinquishes an electron, while reduction occurs when a different element or molecule acquires the electron. The electron donor undergoes oxidation, while the electron receiver experiences reduction. The term "ox" in redox originates from its initial study in aerobic conditions, when oxygen serves as the electron acceptor, hence augmenting its quantity in the resultant molecule.

- A typical redox process transpires in the soil upon the addition of ammonia: In the presence of oxygen, ammonia (NH_3) undergoes oxidation to produce nitric acid (HNO_3, now containing oxygen) and water.

$$NH_3 + 2O_2 \rightarrow HNO_3 + H_2O$$

(ammonia + oxygen → nitric acid + water)

Adsorption vs. Absorption

- Adsorption and absorption are two analogous concepts in soil science with nearly antithetical meanings. Sorption refers to the processes of absorption and adsorption. Adsorption refers to the adherence of molecules to the surface of a substance. The accumulation of a gas or liquid on the surface of a liquid or solid is referred to as the adsorption process. In soils, this pertains to the adhesion of ions to the external surfaces of mineral and organic particles.
- Absorption refers to the process by which one substance is incorporated into the volume of another substance, exemplified by water being absorbed by a sponge or nutrients and water being assimilated by plant roots.

Soil Colloids

- **Colloid:** A particle, whether mineral or organic, with a diameter ranging from 0.1 to 0.001 μm. Due to their diminutive size, colloids remain suspended in a solution, floating for extended periods without precipitating. Clay particles and soil organic matter, recognized as the most chemically active components of soil due to their extensive surface area, are typical examples of soil colloids.
- **Importance:** Colloids possess features that are significant in soil chemistry and play a crucial role in ion exchange and the preservation of soil fertility. Due to their diminutive size, they possess a large relative surface area with a charge, enabling them to adsorb ions (retention of ions and molecules; anion and cation exchange capacity) and hence inhibit ion leaching. It also aids in preserving and enhancing soil structure by binding particles together (because to its shrinking and swelling qualities) and significantly influences soil hydraulic characteristics, such as water-holding capacity (a higher concentration of colloids correlates with an increased ability to retain water).

Soil Solution

- The soil solution is the medium in which reactions between soil particles and the solution occur. The water present in the soil is termed the soil solution, as it comprises dissolved substances (cations and anions) along with suspended colloidal particles of clay and organic matter. The soil solution serves as the reservoir of mineral nutrients for all terrestrial species and encompasses a diverse array of solutes, perhaps encompassing every element from the periodic table. Although plants primarily derive their nutrients from the soil solution, which is required for nutrient absorption by roots, this solution does not provide a enough supply of nutrients to sustain the plant throughout its lifespan. The soil solution is highly dynamic and reacts strongly to fluctuations in the nutrient supply and demand, resulting in significant variations in its composition across both temporal and spatial scales.

Typically, these nutrients are restored from the reservoir of exchangeable nutrients. A greater quantity of nutrients is included in the stable pool, which is bound in solid form as minerals or organic matter.

Cation Exchange Capacity (CEC) and Base Saturation

- Cation exchange capacity (CEC) is a critical soil attribute that assesses the soil's capability to store and deliver plant nutrients, particularly positively charged ions such as calcium, magnesium, potassium, ammonium, and various micronutrients. Plants predominantly absorb nutrients in their ionic form through their roots. A substantial number of these nutrients are absorbed as cations (positive ions). Most soils possess a certain capacity to retain cations at negatively charged sites on clay and organic particles, known as colloids, referred to as exchange sites on soil particles.
- The Cation Exchange Capacity (CEC) is described as the total quantity of cations that are loosely adsorbed to the edges or surface of soil particles by electrostatic attraction, allowing for facile exchange with similarly charged cations in the soil solution. This soil characteristic significantly affects soil pH, nutrient availability, and structural stability.

Measurement of CEC

- CEC is quantified in milliequivalents (meq) per 100 grams of soil (meq/100g) or centimoles (cmol) per kilogram (cmol+/kg). These are two distinct methods of articulating the identical numerical values. Cation exchange capacity typically varies from 5 to 50 meq/100g. Soils deficient in colloids (clay and organic matter) typically exhibit lower cation exchange capacity (CEC) compared to soils rich in colloidal particles. in sandy soils deficient in organic matter to above 20 meq/100g in finer-textured soils and those rich in organic matter. Soils with low cation exchange capacity (CEC) are more vulnerable to the leaching of cationic nutrients and may lack the ability to retain sufficient nutritional cations for an entire growing season.

Major Factors Influencing CEC

- The cation exchange capacity (CEC) of soil fluctuates based on clay concentration, clay type, organic matter quantity, and soil pH. The quantity and kind of clay: Soils with a greater clay content and reduced sand content generally exhibit a higher cation exchange capacity (CEC), while soils with diminished clay content and increased sand content typically have a lower CEC. Increased clay content in the soil correlates with elevated cation exchange capacity (CEC). Soils rich in smectites, montmorillonite, and illite clay minerals exhibit a higher cation exchange capacity (CEC) compared to soils containing kaolinite clay minerals.

- Amount of organic matter: A strong association exists between the quantity of organic matter and the cation exchange capacity (CEC) values in the soil. Increased organic matter in the soil correlates with elevated cation exchange capacity (CEC), as organic matter serves as a significant supply of negative electrostatic sites for cation exchange.
- pH-dependent CEC: Clay minerals and organic matter exhibit a cation exchange capacity that fluctuates with pH levels. As pH rises, the quantity of negative charges on clay or organic matter particles increases, thus elevating the cation exchange capacity (CEC).

Base Saturation

- Base saturation (BS) denotes the proportion of cation exchange capacity (CEC) sites filled by basic cations (often Ca^{2+}, Mg^{2+}, K^+, and Na^+) as opposed to acidic ions (H^+ or Al^{3+}). It often signifies the equilibrium between acid cations (H^+ or Al^{3+}) and base cations (commonly Ca^{2+}, Mg^{2+}, K^+, and Na^+) present in the exchangeable sites of soil.
- Base saturation is frequently represented as a percentage. The word exchangeable bases often denotes the Ca^{2+}, Mg^{2+}, K^+, and Na^+ ions adsorbed to cation exchange capacity (CEC) sites. If all cations are basic and none are acidic, there would be complete base saturation.

Significance of Base Saturation

- Base saturation is one of the most important soil properties and knowledge on BS will help us in understanding soil fertility status and manage our soil for sustainable crop production. Soils with elevated base saturations are deemed more fertile, as numerous "bases" contributing to this condition contain essential plant nutrients, making it a superior predictor of nutrient availability. The quantity of fertilizer necessary for crop cultivation and the amendments needed (such as lime and gypsum) to rectify problematic soils (including acidic and alkaline soils) can significantly fluctuate based on the soil's base saturation capacity.
- The base saturation is typically correlated with pH; it increases as soil pH rises, and when pH exceeds 7, base saturation is generally greater than 90 percent. Due to rainfall's propensity to leach bases from the soil, regions with elevated precipitation typically have lower base saturations

compared to those with reduced rainfall, unless the parent material is rich in bases, such as limestone.

Anion Exchange Capacity (AEC)

- ***Definition:*** Positively-charged cations adhere to negatively-charged sites, whereas negatively-charged anions adhere to positively-charged sites. This refers to anion exchange capacity (AEC). Nutrients typically provided by anions include nitrogen (as NO_3^-), phosphorus (as HPO_4^{2-}), sulfur (as SO_4^{2-}), chlorine (as Cl^-), boron (as $B_4O_7^{2-}$), and molybdenum (as MoO_4^{2-}). These anions exhibit distinct adsorption affinities on positively charged clay minerals: PO_4^- is strongly adsorbed, SO_4^- is adsorbed to a lesser extent, and NO_3^- is not adsorbed at all. Consequently, NO_3^- leaching is more susceptible than other anions, even in soils with elevated AEC, and can thus be readily leached from the soil system. In most soils, cation exchange capacity (CEC) typically exceeds anion exchange capacity (AEC).
- **Measurement:** Similar to CEC, AEC denotes the quantity of anionic negative charges held by 100 grams of soil, quantified in milliequivalents (meq) per 100g of soil or centimoles (cmol) per kg.
- ***pH-dependent AEC:*** Similar to cation exchange capacity (CEC), soil anion exchange capacity (AEC) is likewise reliant on pH. The majority of clay particles possess exclusively negative exchange sites, resulting in cation exchange capacity (CEC) under neutral and elevated pH circumstances, and occasionally anion exchange capacity (AEC) at low pH levels. Soil organic matter possesses both negative and positive exchange sites; it typically exhibits cation exchange capacity (CEC) and may demonstrate anion exchange capacity (AEC) in extremely low pH circumstances (2 or lower). The majority of the most fertile soils in the U.S. exhibit a pH significantly exceeding the levels required for AEC, rendering this process a negligible factor in nutrient supply in this region. Highly worn tropical soils are more prone to possess anion exchange capacity (AEC), indicating a greater affinity for attracting and retaining anions over cations. Soils exhibiting anion exchange capacity generally comprise worn kaolin minerals, iron and aluminum oxides, and amorphous substances. In heavily weathered volcanic soil, iron and aluminum hydroxides/oxides develop on the surface of clay minerals, which possess positive charges. These positively charged oxides and hydroxides specifically absorb negatively charged anions; hence, these soils exhibit a higher anion exchange capacity (AEC) than cation exchange capacity (CEC).
- ***Nutrient leaching:*** Due to the minimal adsorption of anions, many, especially nitrates, are readily leached through the soil by rainfall or excessive irrigation. This may result in groundwater contamination, which can occur in organic farming if nitrogen is not properly handled.

Oxidation and Reduction Reactions

- Soils that fluctuate between wet and dry conditions transition from high oxygen levels to low oxygen levels. The presence or lack of oxygen dictates the chemical reactions of soils.
- Oxidation refers to the loss of electrons, while reduction denotes the acquisition of electrons at the soil surface. Such interactions occur daily and are responsible for the formation of substances like rust. Soils, due to their high iron content, can oxidize, or if saturated with water, may exhibit a pale gray hue.
- This contributes to the various hues present and generates the speckles typically located deeper in the soi!.

Soil pH

- Soil pH is utilized to assess soil response. Soil pH, or soil response, indicates the soil's acidity or alkalinity and is quantified in pH units. Soil pH quantifies the activity of hydronium ions (H_3O^+ or, more often, H^+) in the soil solution. Soil pH is defined as the negative logarithm base 10 of the activity of H^+ ions (moles per liter) in the soil solution. It is expresses as: $pH = 1/\log 10[H^+]$ or $= - \text{Log}\ [H^+]$
- The pH scale ranges from 0 to 14, with a pH of 7 indicating neutrality, devoid of acidity or basicity. As the concentration of hydrogen ions in the soil rises, the soil pH diminishes, resulting in increased acidity due to the presence of ionized (or free) H^+ ions. As pH decreases from 7 to 0, the soil becomes progressively more acidic, while as pH increases from 7 to 14, the soil becomes increasingly alkaline, characterized by the presence of ionized (or free) OH^- ions.
- Soil response is significant as it forecasts nutrient availability for plant growth, regulates the chemical environment for both plants and microorganisms, and affects the solubility of harmful substances in the soil. Most plant species thrive optimally within a pH range of 5.5 to 6.5 or 7.0, however others exhibit a preference for high values. For instance, potatoes, azaleas, and blueberries thrive optimally in more acidic environments. In contrast, alfalfa and sweet clover thrive under somewhat alkaline circumstances.
- Descriptive classifications corresponding to certain ranges in soil pH are:
  - Extremely acid: < than 4.5; (lemon=2.5; vinegar=3.0; stomach acid=2.0; soda=2–4)
  - Very strongly acid: 4.5–5.0; (beer=4.5–5.0; tomatoes=4.5)
  - Strongly acid: 5.1–5.5; (carrots=5.0; asparagus=5.5; boric acid=5.2; cabbage=5.3)
  - Moderately acid: 5.6–6.0; (potatoes=5.6)

- Slightly acid: 6.1–6.5; (salmon=6.2; cow's milk=6.5)
- Neutral: 6.6–7.3; (saliva=6.6–7.3; blood=7.3; shrimp=7.0)
- Slightly alkaline: 7.4–7.8; (eggs=7.6–7.8)
- Moderately alkaline: 7.9–8.4; (sea water=8.2; sodium bicarbonate=8.4)
- Strongly alkaline: 8.5–9.0; (borax=9.0)
- Very strongly alkaline: > than 9.1; (milk of magnesia=10.5, ammonia=11.1; lime=12)

Origins of H^+ and OH^-

Strongly Acid Soils pH < 5

Adsorbed H^+ and Al_3^+ predominate in the cation exchange capacity (CEC). The presence of these cations in solution contributes to a low pH level. In highly acidic conditions, the majority of aluminum dissolves as Al_3^+ or $Al(OH)_2^+$ cations, which are preferentially adsorbed over hydrogen. The Al_3^+ ions in solution, in equilibrium with exchangeable Al, tend to undergo hydrolysis, resulting in the formation of $Al(OH)_2^+$ and a hydrogen ion upon reaction with water. Adsorbed hydrogen constitutes a negligible supply of hydrogen, as the adsorbed hydrogen ions are in equilibrium with those present in the soil solution.

Moderately Acidic Soils 5 pH < 6.5

Within this pH range, there exists a diminished proportion of adsorbed acidic species and an elevated percentage of base saturation. Adsorbed and dissolved aluminum undergoes more hydrolysis. Exchangeable H^+ and $Al(OH)_x^{(3-x)+}$ contribute to the mild acidity of the soil.

Neutral to Alkaline Soils pH 6.5

The exchangeable sites are primarily occupied by basic cations, while the soil solution or exchange sites are no longer predominantly influenced by hydrogen or aluminum ions. The origin of the hydroxyls is the substitution of hydrogen and aluminum with base cations. The residual H^+ and $Al(OH)_x^{(3-x)+}$ associated with pH-dependent exchange sites can be regarded as bound. Exchange equilibria significantly favor the adsorption of these species at these specific locations, resulting in a low concentration in solution. In the relative scarcity of acidic cations, pH may exceed 7.

Carbonate and bicarbonate salts exist at alkaline pH levels. The hydrolysis of these anions produces OH^-. For instance,

$$CaCO_3 \rightarrow Ca^{2+} + CO_3^{2-}$$

$$CO_3^{2-} + 2H_2O \rightarrow H_2CO_3 + 2OH^-$$

Classification of Soil Acidity

There are three general pools, or sources, of acidity: active, exchangeable and residual acidity.

Active acidity refers to quantity of hydrogen ions (H^+) in soil solution. Active acidity may be directly determined using a pH meter. These active pool of acidity directly affects plant growth and development. Exchangeable acidity refers to adsorbed acidic cations, H^+ and $Al(OH)_x^{(3-x)+}$ on the exchangeable sites. That is subject to displacement by addition of an excess of extracting cation such as K^+. To correct exchangeable acidity more amount of lime is required. Exchangeable acidity can be 100 times greater than active acidity. Residual is all the bound H^+ and $Al(OH)_x$ $(3\text{-}x)^+$ in soil mineral that is unextractable by unbuffered salts. Based on the removal of these acid cations from particular pools, the aluminum and hydrogen of one pool will replenish the aluminum and hydrogen of another pool. For example, if aluminum and hydrogen are removed from active pools (soil solution), the acid cations of the exchangeable acidity will replenish the soil solution. Similarly, minerals containing aluminum and hydrogen dissolve and release these cations as they are removed from the exchangeable pool. Total acidity includes active, exchangeable and residual acidity. Managing soil acidity must account for exchangeable acidity as well as active acidity.

Measuring Soil pH

- Soil pH offers numerous insights about soil characteristics and can be readily measured. The most precise technique for measuring soil pH is utilizing a pH meter with a glass electrode. The glass pH electrode is quite recent. A potential difference arises across the glass membrane when submerged in a solution with H+ activity distinct from that of the internal solution. This potential difference is adjusted according to the solution's pH.
- A second method, which is straightforward and uncomplicated however less precise than utilizing a pH meter, involves the use of certain indicator solutions or color dyes. Indicator dyes have been utilized for an extended period. The color varies based on the presence of either the undissociated or dissociated form.

Buffering

- Denotes the propensity of soils to withstand alterations in pH when acid or base is introduced.
- The introduction of acid results in enhanced adsorption of H^+ or the consumption of H^+ by its interaction with $Al(OH)_3$, yielding soluble Ahydroxy ions, $Al(OH)_x^{(3-x)+}$. At elevated soil pH, carbonates or bicarbonates will interact with introduced acid.

$$CO_3^{2-} + 2H^+ \rightarrow H_2CO_3$$

$$H_2CO_3 \rightarrow H_2O + CO_2$$

- Conversely, the introduction of OH^- results in its consumption through reaction with H+ or $Al(OH)_x^{(3-x)+}$.

- The magnitude of buffering is quantified in the buffer curve of a soil, which correlates solution pH with % base saturation. The buffer curves of soil vary based on the types and quantities of colloids present.

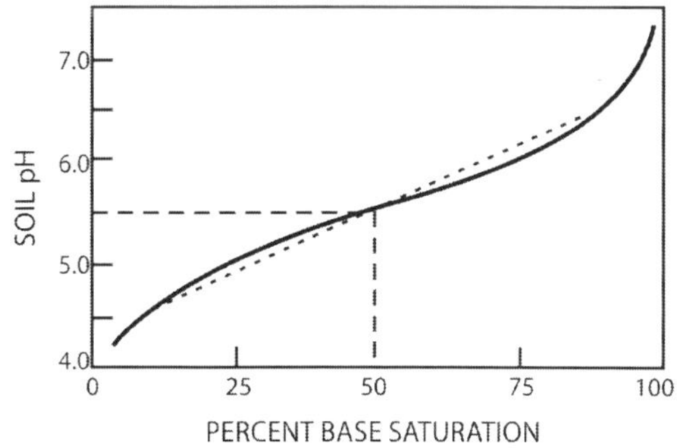

- Illustration of a buffer curve demonstrating the elevation in pH corresponding to an increase in percentage base saturation.
- Percentage base saturation is contingent upon cation exchange capacity (CEC) and the quantity of adsorbed basic cations; thus, a higher CEC necessitates a bigger volume of acid or base to induce a variation in soil pH.

The pH Influences Nutrients, Minerals, and Growth

- Soil pH significantly influences the availability of certain plant nutrients by altering their chemical form in the soil. The influence of soil pH significantly affects the solubility of minerals and nutrients. Fourteen of the seventeen vital plant nutrients are sourced from the soil. Prior to use by plants, a nutrient must be solubilized in the soil solution. The majority of minerals and nutrients exhibit greater solubility or availability in acidic soils compared to neutral or mildly alkaline soils. Generally, low pH diminishes the availability of macronutrients and secondary nutrients, whereas high pH decreases the availability of most micronutrients.
- Phosphorus is not easily soluble in soil but is most accessible in soil with a pH of 6.5. Soils with a pH of 4.0-5.0, characterized as very and strongly acidic, may have elevated levels of soluble aluminum, iron, and manganese, which can be detrimental to the growth of certain plants, while the availability of calcium, phosphorus, and magnesium is diminished. An adverse consequence of excessive soluble aluminum is the inhibition or cessation of root development. The optimal soil pH range for ideal plant growth differs across various crops. Soil pH levels between 6.0 to 7.0 are optimal for the majority of plants, as most nutrients are accessible within this pH range.
- However, many plants, like azaleas, rhododendrons, blueberries, white potatoes, and coniferous trees, endure highly acidic soils and thrive

effectively. Additionally, several plants thrive exclusively in slightly acidic to moderately alkaline soils. Nonetheless, soil with a slightly alkaline pH (7.4-7.8) or higher can impede iron availability for pin oak and several other tree species in Central New York, resulting in leaf chlorosis that induces stress, ultimately leading to tree decline and even mortality.

- The soil pH can affect plant growth by influencing the activity of helpful microbes. Bacteria that breakdown soil organic materials are impeded in strongly acidic soils. This inhibits the decomposition of organic matter, leading to its buildup and the sequestration of nutrients, especially nitrogen, contained within the organic matter. Moreover, a low pH in topsoils may influence microbial activity, particularly reducing Rhizobia bacteria and legume nodulation. Moreover, low soil pH inhibits the biological conversion of ammonium to nitrate.

Natural Acidification of Soil pH

- Soils typically become acidic due to:
    - Rainwater possesses a mildly acidic pH (about 5.7) resulting from an interaction with atmospheric CO, which generates carbonic acid. As rainwater percolates through soil, it extracts essential cations (calcium, magnesium, potassium, and sodium);
    - Carbon dioxide produced by the decomposition of organic matter and root respiration dissolves in soil water, forming a weak organic acid (carbonic acid), which subsequently leads to the leaching of basic cations from the soil.
    - Carbonic acid, organic acids, and H^+ produced during the oxidation of reduced nitrogen and sulfur in organic matter displace basic cations on soil colloids. Basic cations are prone to leaching loss under humid circumstances. As a result, the percentage base saturation diminishes and the pH drops.
    - The combustion of fossil fuels emits sulfur and nitrogen oxides into the atmosphere. These substances react with atmospheric water to produce sulfuric and nitric acid in precipitation (acid rain), hence expediting the natural leaching acidification process.

How a Soil Management also Affects its pH

Acidifying Fertilizers

Generation of potent inorganic acids, including nitric and sulfuric acid, through the oxidation of ammonium and sulfur fertilizers. Ammonium fertilizers undergo nitrification in the soil, resulting in the formation of nitrate and the release of hydrogen ions. These inorganic acids result in the formation of highly acidic soils. The microbial oxidation of reduced nitrogen serves as a source of hydrogen ions.

$$NH_4^+ + 2O_2 \rightarrow NO_3^- + 2H^+ + H_2O$$

Disposal of Organic Wastes

Results in an increased rate of acidification caused by organic and mineral acids generated through microbial oxidation. Acidification can be prevented by the addition of lime to organic material.

Drainage of Certain Coastal Wetlands that Contain High Levels of Reduced S

The oxidation of sulfur or sulfide in the drained material generates sulfuric acid. A significantly low pH may occur.

Irrigation Water Quality

The accumulation of elevated Na^+ concentrations will likely increase soil pH to high levels. Additional adverse impacts encompass the dispersion of soil colloids, leading to diminished hydraulic conductivity, and salinization.

Measure to Correct Soil pH

Depending on the plants to be cultivated (or other objectives), it may be necessary to adjust the soil pH upward or downward. The pH of alkaline soil can be lowered by incorporating acidifying chemicals or acidic organic substances. Elemental sulfur (90–99 percent S) is utilized to decrease soil pH as it gradually oxidizes in the soil to produce sulfuric acid.

$$\mathbf{2S + 3O_2 + 2H_2O \rightarrow 2H_2SO_4}$$

Acidifying fertilizers, including ammonium sulfate, ammonium nitrate, and urea, can lower soil pH as ammonium oxidizes to produce nitric acid.

Typically, lime is employed to increase soil pH. Liming is the most cost-effective technique for mitigating soil acidity. Calcium and magnesium carbonates, oxides, and hydroxides constitute agricultural limes. Among these, pulverized limestone, referred to as calcite ($CaCO_3$) or dolomite ($CaMg(CO_3)_2$), is the most prevalent. Quick lime (CaO) and hydrated lime ($Ca(OH)_2$) are utilized as liming materials.

When lime is added to an acid soil, bicarbonate is formed by the following reaction.

$$CaO + 2H_2CO_3 \rightarrow Ca(HCO_3)_2 + H_2O$$

$$Ca(OH)_2 + 2H_2CO_3 \rightarrow Ca(HCO_3)_2 + 2H_2O$$

$$CaCO_3 + H_2CO_3 \rightarrow Ca(HCO_3)_2$$

Bicarbonate reacts with H^+

$$H^+ + HCO_3^- \rightarrow H_2CO_3$$

$$H_2CO_3 \rightarrow H_2O + CO_2$$

Reduction of solution H^+ drives the desorption of adsorbed H^+ and acidic hydrolysis of $Al(OH)_x^{(3-x)+}$ to $Al(OH)_3$.

Despite the elevation of soil pH with lime application, soil acidification persists due to the leaching of newly introduced basic cations, decomposition of organic matter, and other factors. Consequently, regular lime application may ultimately be necessary.

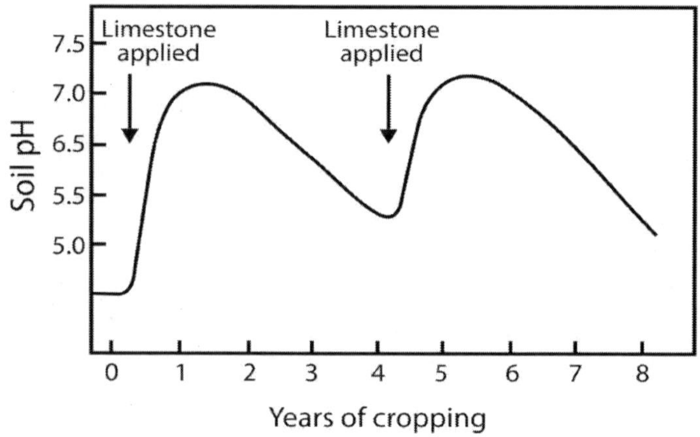

Lime Requirement

- Lime is typically applied to acidic soils to elevate soil pH. The incorporation of lime substitutes hydrogen ions and elevates soil pH, hence resolving most significant issues related to acidic soils, while also supplying two essential nutrients, calcium and magnesium, to the soil. Lime enhances the availability of phosphorus in the soil for plant growth and accelerates the breakdown of organic materials, hence increasing nitrogen availability. Liming materials are cost-effective, easy to manage, and do not leave undesirable residues in the soil.

The quantity of liming necessary is contingent upon: 1) desired pH alteration, 2) soil buffer capacity, 3) kind and purity of lime material, and 4) lime fineness.

- The principal determinants of lime quality are neutralizing value and particle size. The neutralizing capacity of lime is represented as a percentage of pure calcium carbonate, assigned a value of 100 percent. A higher neutralising value allows for reduced lime usage or increased area treatment for the same pH alteration. Oxides and hydroxides react more rapidly than limestones; nonetheless, they are costlier.
- Lime containing a greater ratio of fine particles will react more rapidly to neutralize soil acidity, which is advantageous for ameliorating acidic soil conditions. If at least 50 percent of the particles pass through a 60 mesh screen (0.25 mm), the limestone will sufficiently mitigate soil acidity in a timely manner.

Subsoil Acidity Correction

- Given the limited solubility of limestone, its incorporation is most efficacious. This presents no issue for rectifying topsoil acidity;

nevertheless, the deep inclusion of lime to mitigate subterranean acidity is challenging and costly.

- Gypsum, $CaSO_4$ $2H_2O$, serves as a more soluble alternative. Gypsum is a soil conditioner utilized to rectify aluminum issues in the underlying soil strata. It is essential to differentiate between lime and gypsum. Both lime and gypsum serve as calcium supplies; however, lime directly elevates soil pH, whereas gypsum does so indirectly.
- The Ca^{2+} replaces certain adsorbed Al_3^+, rendering the Al_3^+ susceptible to leaching. The sulfate ion (SO_4^{2-}) may precipitate aluminum ions (Al_3^+). Furthermore, SO_4^{2-} is preferentially adsorbed onto soil colloids, displacing OH^- in the process.

Soil as a Substrate for Vegetative Development

- We are specifically focused on the physical aspect of soil fertility within the soil ecosystem.
- For a nutrient to be accessible for plant absorption, it must satisfy two criteria:
  - It must be in the appropriate chemical form to traverse the root membrane; and
  - It must be accessible at the surface level.
  - Seventeen nutrients are necessary for plant growth and development. Thirteen are assimilated by the roots. Once absorbed by the roots, the nutrients are transported into the plant and utilized for various activities and processes. Generally, there are three methods via which nutrients access the surface of root hairs;
    - root interception
    - mass flow
    - diffusion

Root Interception

- Root interception is the mechanism by which roots penetrate the soil profile to access nutrients. Plant roots continuously expand, unblocking obstructed pores, and grow from nutrient-depleted places to regions with higher nutrient concentrations. Compaction can significantly restrict root development and nutrient absorption throughout the soil profile.
- Despite possessing vast root systems, many plants, including cereals and other grasses, interact with less than 5 percent of the soil volume. The root interception mechanism is highly beneficial, as root growth can extend into regions where mass flow and diffusion subsequently dominate.
- A root may develop within a few millimeters of a soil phosphorus hotspot. Although the root does not physically contact the nutrient, it is sufficiently proximal for diffusion to facilitate the transfer of phosphorus into it.

- In certain instances, the presence of mycorrhizal fungus enhances the nutrient-absorption capability of root systems. Root interception facilitates the absorption of certain calcium, magnesium, zinc, and manganese.

Mass Flow

- Vegetative plants consistently absorb water from the soil profile, a process facilitated by transpiration (the loss of water from the plant through stomata on the leaves). Plant nutrients exist as soluble nutrients dissolved in soil water. The nutrients are conveyed with the water to the root surface.
- Nutrients, including nitrogen in the form of nitrate and sulfur as sulfate, are loosely bound to soil particles and can migrate with water. The principal ions transported predominantly by mass flow are the nitrate and chloride anions, with the sulfate anion being moderately involved. Nutrients, such as phosphorus in the form of orthophosphate, that are heavily adsorbed on soil particles cannot access roots via mass flow. Mass flow facilitates the absorption of additional plant nutrients such as calcium, magnesium, sulfur, copper, boron, manganese, and molybdenum. Nonetheless, its absorption via mass flow is contingent upon the soil's physical and chemical properties, as well as the concentration and attributes of that element within the soil.

Diffusion

- Diffusion is the process by which ions migrate down a gradient from an area of high concentration to one of lower concentration, achieving uniform distribution. Consider a tank of water containing a detachable partition at its center. Ink is poured on one side of the barrier, while the opposite side remains clean water. Gradually removing the barrier will allow you to observe the ink and water amalgamate as the molecules transition from a region of high concentration (the inky side) to a region of low concentration (the pure water side).
- Nutrients migrate from regions of elevated concentration in the soil solution to regions of diminished concentration. This process is notably sluggish; yet, it serves as the primary mechanism for the transportation of phosphorus and potassium, which are highly adsorbed in soils and exist in low quantities within the soil solution.

Preparation and Standardization of Reagent Solutions

- Chemical reagents are produced and sold in various purity grades. Typically, the most refined reagents are sold as "analytical reagent" or AR-grade. Additional labels include “LR,” denoting laboratory reagent, and “CP,” signifying chemically pure.

- The potency of substances is denoted as normalcy or molarity. Consequently, it is beneficial to possess knowledge regarding the potency of the acids and alkalis frequently utilized in chemical laboratories.
- Below are definitions of several key phrases frequently utilized in a laboratory for chemical analysis.

Molarity

Molarity is defined as the number of moles of solute per liter of solution. A one-molar (M) solution comprises one mole or one molecular weight in grams of a material per liter of the solution. The molar technique of quantifying concentration is advantageous as equal volumes of equimolar solutions contain an identical number of molecules.

Normality

The normalcy of a solution is defined as the number of gram equivalents of the solute per liter of the solution. Similar to molarity, normality connects the quantity of solute to the overall volume of solution; however, normality is exclusively applicable to acids and bases. It is mostly utilized as an indicator of reactive species in a solution and during titration reactions, particularly in contexts requiring acid-base chemistry. It is typically denoted by the letter N. Semi-normal, pentinormal, desi-normal, centi-normal, and milli-normal solutions are typically denoted as 0.5N, 0.2N, 0.1N, 0.01N, and 0.001N, respectively. Nevertheless, molar expression is favored due to the awkward representation of "odd" normalities, such as 0.121N, in fractional form.

Normality is defined as the equivalent concentration of a solution. The definition of a normal solution include the term "equivalent weight." This quantity fluctuates depending on the reaction type. Consequently, providing a definitive definition of equivalent weight that encompasses all reactions is challenging. It frequently occurs that a same compound exhibits varying equivalent weights in distinct chemical processes. A scenario may occur in which a solution possesses a standard concentration for one application and a distinct normalcy for another chemical reaction. Therefore, the molarity system is favored.

Table 4.1: Strength of Commonly Used Acids and Alkalis

Reagent/ Chemical	*Normality (Approx.)*	*Molarity (Approx.)*	*Formula Weight*	*Per cent by Weight (Approx.)*	*Specific Gravity (Approx.)*	*Milliliters Required for 1N/litre Solution (Approx.)*	*Milliliters Required for 1M/litre Solution (Approx.)*
Nitric acid	16.0	16.0	63.0	70	1.42	63.7	63.7
Sulphuric acid	35.0	17.5	98.0	98	1.84	28.0	56.0

Hydrochloric acid	11.6	11.6	36.5	37	1.19	82.6	82.6
Phosphoric acid	45.0	15.0	98.0	85	1.71	22.7	68.1
Perchloric acid	10.5	10.5	100.5	65	1.60	108.7	108.7
Ammonium hydroxide	15.0	15.0	35.0	28	0.90	67.6	67.6

Equivalent Weight

The equivalent weight (Eq W), commonly referred to as gram equivalent weight, is the mass in grams that corresponds to a gram atom of hydrogen or hydroxyl, half a gram atom of oxygen, or a gram atom of a univalent ion in a chemical process. In contrast to atomic weight, which is dimensionless, comparable weight possesses dimensions and units of mass. In acid-base processes, the equivalent weight of an acid or base is the mass that provides or interacts with one mole of hydrogen ions (H^+). In redox reactions, the equivalent weight of each reactant either donates or accepts one mole (6.023×10^{23}) of electrons (e^-). The normalcy of a solution is defined as one equivalent weight of a substance dissolved in one liter of solution (1N). Annex 10 provides the equivalent and molecular weights of some significant substances.

Milli-equivalent Weight

Equivalent weight, typically denoted as milli-equivalent weight (mEq W), refers to the equivalent weight in grams divided by 1000. It is frequently articulated as "me." This value is the most useful as it represents the weight of a substance contained in or equivalent to 1 ml of a 1N solution. Consequently, it serves as a unit applicable to both volumes and weights, facilitating the conversion of a solution's volume to its corresponding weight, and a substance's weight to its equivalent solution volume: number of mEq = volume × normalcy.

Buffer Solutions

- A buffer solution is an aqueous solution including either a weak acid and its corresponding salt or a weak base and its salt (e.g., CH_3COOH + CH_3COONa and NH_4OH + NH_4Cl) that exhibits the virtue of resisting pH alterations with the addition of an acid or base. Buffer solutions maintain their resistance to pH fluctuations due to the equilibrium established between the weak acid HA and its conjugate base A^-.

$$HA \rightleftharpoons H^+ + A^-$$

- Buffer solutions are widely categorized into two basic types: acidic and alkaline buffers.
    - An acidic buffer solution is defined as one with a pH below 7. Acidic buffer solutions with a pH of 4.7 are often composed of a weak acid

and its corresponding salt, frequently a sodium salt, such as a mixture of ethanoic acid and sodium ethanoate in solution.

- An alkaline buffer solution possesses a pH exceeding 7. Alkaline buffer solutions with a pH of 9.2 are often composed of a weak base and its corresponding salt, such as a mixture of ammonia solution and ammonium chloride solution.

- The essential characteristics of a buffer solution are:
  - It possesses a specific pH value.
  - The pH value remains unchanged throughout prolonged storage.
  - The pH value is only marginally affected when a strong base or strong acid is introduced.
- Due to this property, conveniently manufactured buffer solutions with known pH are utilized to verify the accuracy of laboratory pH meters.

Titrations

Titration is a prevalent laboratory technique for quantitative chemical analysis employed to ascertain the unknown concentration of a specified analyte by precisely measuring the volume of analyte that reacts with a standard solution. The solution of known concentration utilized in the titration is referred to as the titrant. The analyte in the solution is referred to as the titrate.

Table 4.2: Equivalent and Molecular Weights of Compounds

Compound	*Formula*	*Molecular Weight (g)*	*Equivalent Weight (g)*
Ammonium acetate	CH_3COONH_4	77.08	77.08
Ammonium chloride	NH_4CI	53.49	53.49
Ammonium fluoride	NF_4U	37.04	37.04
Ammonium nitrate	NH_4NO_3	80.04	80.04
Barium chloride	$BaCl_2.2H_2O$	244.28	122.14
Boric acid	H_3BO_3	61.83	20.61
Calcium acetate	$(CH_3COO)_2Ca$	158.00	79.00
Calcium carbonate	$CaCO_3$	100.09	50.05
Calcium chloride (dehydrate)	$CaCl_2.2H_2O$	147.02	73.51
Calcium hydroxide	$Ca(OH)_2$	74.00	37.00
Calcium nitrate	$Ca(NO_3)_2$	164.00	82.00
Calcium sulphate	$CaSO_4.2H_2O$	172.17	86.08
Ferrous ammonium sulphate	$FeSO_4(N1_4)_2\ SO_4H_2O$	392.13	192.13
Ferrous sulphate	$FeSO_42H_2O$	278.01	139.00
Magnesium chloride	$MgC1_2$ 6HzO	203.30	101.65

Magnesium nitrate	$Mg(NO_3)_2 6H_2O$	256.41	128.20
Potassium chloride	KCI	74.55	74.55
Potassium dichromate	$K_2Cr_2O_2$	294.19	49.04
Potassium hydroxide	KOH	56.10	56.10
Potassium permanganate	$KMnO_4$	158.03	31.60
Potassium nitrate	KNO_3	101.10	101.10
Potassium sulphate	K_2SO_4	174.27	87.13
Potassium hydrogen phthalate	$COOH\ C_5H_4,\ COOK$	204.22	204.22
Oxalic acid	$C_2H_2O_4.2H_2O$	126.00	63.00
Silver nitrate	$AgNO_3$	169.87	169.87
Sodium acetate (anhydrous)	CH_3COONa	82.04	82.04
Sodium bicarbonate	$NaHCO_3$	84.01	84.01
Sodium carbonate	Na_2CO_3	106.00	53.00
Sodium chloride	NaCI	58.45	58.45
Sodium hydroxide	NaOH	40.00	40.00
Sodium nitrate	$NaNO_3$	84.99	84.99
Sodium oxalate	$Na_2C_2O_4$	134.00	67.00
Sodium sulphate	Na_2SO_4	142.04	71.02
Sodium thiosulphate	$Na_2S_2O_3.5H_2O$	248.18	248.18

The reaction's completeness is evaluated using a suitable indication.

Indicators

An indicator is a chemical that provides a visible sign, typically by a color change, at the endpoint upon completion of a reaction. Indicators may also demonstrate alterations in various physical features; for instance, olfactory indicators exhibit changes in their scent. The predominant indicators utilized in volumetric analysis include:

1. Internal indicators,
2. External indicators,
3. Self-indicators.

Internal Indicators

- ✰ Internal indications, including methyl red, methyl orange, bromothymol blue, phenolphthalein, and diphenylamine, utilized in acid-base neutralization reactions, are referred to as acid-base internal indicators. In simpler terms, indicators utilized within the solution are referred to as internal indicators. Upon the completion of the titrant's reaction with the titrate, a color shift occurs, contingent upon the solution's acidity or basicity, facilitated by the indicator, which also signifies the conclusion of the titration.

- The indicator employed in precipitation reactions, such as the titration of a neutral NaCl solution (or chloride ion) with silver nitrate ($AgNO_3$) solution, is K_2CrO_4. Upon the conclusion of the titration process between $AgNO_3$ and Cl, when no additional Cl is present to react with silver ions to produce AgCl, the chromium ions react with Ag_2^+ ions to yield the sparingly soluble Ag_2CrO_4, characterized by a brick red hue. This signifies that Cl has been fully titrated and that the endpoint has been reached.
- Redox indicators are frequently employed. Oxidation-reduction indicators are compounds that exhibit a reversible color change between their oxidized and reduced states. Diphenylamine exhibits a blue-violet hue in its oxidized state and is colorless when reduced. Ferrocin has a blue hue in its oxidized state and a red hue in its reduced form.

External Indicators

Indicators utilized outside the titration mixture, rather than being incorporated into the solution, are referred to as external indicators. Potassium ferricyanide $[K_3[Fe(CN)_6]$ serves as an external indicator in the titration of potassium dichromate ($K_2Cr_2O_7$) and ferrous sulfate ($FeSO_4$) in an acidic media. During this titration, many drops of indicator are applied on a white porcelain tile. A glass rod immersed in the titration solution is removed and placed in contact with the indicator droplets on the tile. A strong blue hue is observed, which transitions to a greenish tint upon the conclusion of the titration.

Self-Indicators

A substance is classified as a self-indicator if it functions as an indicator during titration. Upon conclusion of the reaction, the titrant relinquishes its color due to a marginal surplus in minute quantities. During the titration of $KMnO_4$ with ferrous sulfate, the introduction of $KMnO_4$ initiates a reaction with the colorless $FeSO_4$. Upon the conclusion of titration, the minimal surplus of $KMnO_4$ imparts a pink hue to the solution, serving as a self-indicator that signifies the completion of the titration.

Standard Solutions

- A standard solution is a solution that contains a precisely known concentration (strength) of an element or chemical. A known mass of solute is dissolved to create a given volume. It is typically given as grams equivalent or grams mole per liter of solution. A solution is classified as 1N if it contains 1 gram equivalent weight of a chemical or compound. If it contains 2 g equivalent weights of the component, it is classified as 2N.
- All titrimetric procedures rely on standard solutions containing known quantities of reagents in a unit volume of the solution. A solution is formulated with about the necessary concentration. This solution is subsequently standardized through titration with another drug available in a highly pure state. Consequently, potassium permanganate solution can be standardized using sodium oxalate, which is readily obtainable

in high purity due to its ease of drying and non-hygroscopic nature. A material with steady weight and purity is referred to as a "primary standard."

- A primary standard must possess the following attributes.:
  - The concentration must stay consistent at all times to eliminate the necessity for restandardization.
  - It must be available in a pure form or in a condition of established purity.
  - It must exhibit a singular reaction during titration (very stable/low reactivity) with no occurrence of side reactions.
  - It must be non-hygroscopic. Salt hydrates are typically unsuitable as primary standards. It must be easily accessible and cost-effective.
  - Typically, it should possess a substantial equivalent weight to minimize weighing errors.
  - An acid or base should ideally be strong, possessing a high dissociation constant for use as standards.
  - A primary standard solution is one that can be directly prepared by weighing the substance, allowing for the titration and standardization of other solutions of approximate strength.

The following are primary standards:

Acid-Base Titration

- Potassium hydrogen phthalate
- Anhydrous sodium carbonate
- Borax

Oxidation-Reduction Titrations (Redox titrations)

- Potassium dichromate
- Potassium bromate
- Sodium oxalate
- Potassium ferrocyanide
- Potassium iodate
- Arsenic trioxide

Precipitation Titrations

- Sodium chloride
- Potassium chloride

Non-Aqueous Titration

- Benzoic Acid

Secondary standard solutions are made by dissolving slightly more than the gram equivalent weight of a chemical per liter of solution, followed by precise standardization with a primary standard solution. A secondary standard solution is one in which the concentration of the dissolved solute is not established by the weight of the chemical dissolved, but rather through the titration of a specified amount of a main standard solution. Secondary standards are frequently employed to calibrate analytical techniques. A secondary standard is a substance that cannot serve as a primary standard for one or more reasons. Sodium hydroxide cannot serve as a primary standard because it absorbs moisture and carbon dioxide from the atmosphere, resulting in significant fluctuations in the composition of its solution over time. Sodium thiosulphate similarly takes CO_2 from the atmosphere and decomposes, resulting in the deposition of sulfur at the bottom. The compositions of solutions of numerous chemicals, such as mineral acids, potassium permanganate, and iodine, also exhibit variability over time. Consequently, these cannot serve as primary criteria. Commonly utilized secondary standards include sulfuric acid, hydrochloric acid, and sodium hydroxide.

Standard solutions of all reagents necessary for laboratory analysis must be prepared and readily available prior to initiating any examination. Nevertheless, their potency must be assessed periodically, or new reagents should be synthesized prior to analysis.

All titrations pertaining to acidimetry and alkalimetry necessitate standard solutions. These may be created either from standard substances through direct weighing or by standardizing a solution of near normalcy by titrating against a prepared standard. The procedures for creating standard solutions of certain non-primary standard compounds frequently utilized are outlined below.

Standardization of Hydrochloric Acid (HCl)

Concentrated hydrochloric acid (HCl) has a normality of around 11N. Consequently, to make a standard solution, namely a decinormal (0.1N) acid solution, it is diluted approximately 100-fold. Measure 10 ml of acid and dilute it with distilled water to achieve a total volume of approximately 1 liter. Conduct a titration of this acid with 0.1N Na_2CO_3 (primary standard) utilizing methyl orange as the indicator. The hue transitions from pink to yellow upon neutralization of the acid. If 10 ml of acid and 12 ml of Na_2CO_3 are utilized in the titration, then:

Acid		*Alkali*
V1 × N1	=	V2 × N2
10 × N1	=	12 × 0.1
10 N1	=	1.2
N1	=	0.12

Thus, the normality of the acid is 0.12.

Standardization of Sulphuric Acid

The normality of sulfuric acid (H_2SO_4) can be ascertained in a manner analogous to the aforementioned. It requires a dilution of around 350 times to achieve a normalcy of about 0.1N, given its initial normality of approximately 35. Subsequently, titrate with standard Na_2CO_3 to ascertain the precise normalcy of H_2SO_4.

Standardization of Sodium Hydroxide

As per the above method, the normality of HCl/H_2SO_4 has been fixed. Therefore, to determine the normality of sodium hydroxide (NaOH), titration is carried out by using either of these standard acids. To determine the molarity, molar standard solutions are used.

Potassium hydrogen phthalate can serve as a primary standard for the standardization of NaOH or any other alkali, rather than use secondary standards for titration.

Quality Check in Analysis

Use of Blank

- ☆ A blank determination is an analysis conducted without the analyte or attribute; in essence, it is an analysis performed without a sample, executing all procedural steps using only the reagents. A blank is utilized to account for any contamination in the substances employed during the actual analysis. The estimate of the blank is deducted from the sample estimates. "Sequence control" samples are utilized in extended batches during automated analysis.
- ☆ Typically, two samples are utilized as standards to assess analytical accuracy: one containing a low concentration and the other a significantly high concentration of a known analyte, both within the method's operating range.

Blind Sample

A blind sample has a known quantity of analyte. The laboratory director introduces this sample in batches and at unpredictable intervals to the analyzer. Diverse categories of sample material can function as blind samples, including control samples or adequately sized remnants of test samples that have undergone multiple analyses. The analyst must be cognizant of the potential existence of a blind sample without being able to identify the material as such.

Chapter 5

Soil Properties

Introduction

- The characteristics of soil are predominantly influenced by its parent material and the weathering processes that occur throughout its creation. The growth of plants is affected by the physical, chemical, and biological characteristics of the soil.
- The qualities of soil are interrelated. Comprehending soil characteristics is crucial for nutrient management and land-use planning.

Physical Properties

The physical qualities of soil that influence a plant's growth potential include:

- Soil texture influences the soil's capacity for nutrient retention (cation exchange capacity) and water retention. Texture denotes the relative proportions of various particle sizes (sand, silt, and clay) within the soil. It is a consistent characteristic of soils and, thus, is utilized in soil classification and description.
- Soil structure influences aeration, water retention, drainage, and root and seedling penetration. Soil structure denotes the configuration of soil particles into aggregates (or peds) and the spatial distribution of interstitial pores. It is an unstable attribute significantly affected by soil management techniques.

Soil Texture

- Soil texture, defined as the tactile quality of soil, is dictated by the relative quantities of sand, silt, and clay present within it. When moist, sandy soils have a gritty texture, silty soils possess a smooth and silky feel, and clayey soils are characterized by a sticky and plastic consistency, allowing for molding.

- Soils with a significant sand content are termed 'light', whereas those with a substantial clay content are termed 'heavy'.

Soil Textural Classes

- The nomenclature of soil textural classes aims to convey information on their textural composition and physical characteristics. The three fundamental categories are sands, clays, and loams.
- A soil classified as sand comprises a minimum of 70 percent by weight of sand. A clay soil must comprise a minimum of 35 percent clay, and typically no less than 40 percent.
- A loam soil is, ideally, a composite of sand, silt, and clay particles that demonstrate both light and heavy characteristics in approximately equal proportions; thus, a soil classified as loam will contain varying quantities of sand, silt, or clay.
- Supplementary texture class designations are derived from these three fundamental categories. The fundamental group designation consistently appears at the end of the class name. Consequently, loamy sand belongs to the sand category, while sandy loam is classified under the loam category.

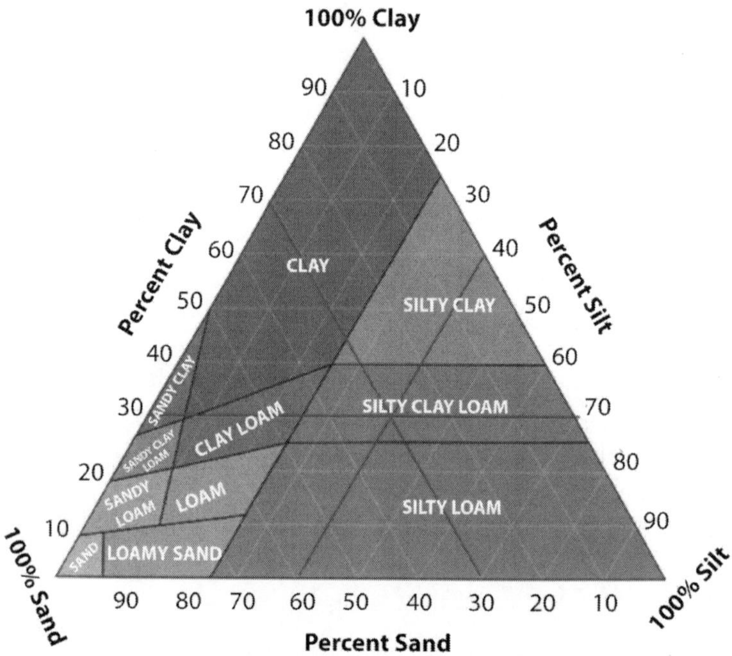

Figure 5.1: Soil Texture Triangle
(*Source*: Image adapted from Hunt and Gilkes, 1992).

Laboratory Assessment of Soil Texture

- Particle size distribution can be ascertained through laboratory analysis, with results presented as percentages. The texture is established by extending lines from the percentage point on the pertinent axis parallel to the triangle's side at the zero end of the same axis; the intersection of the three lines denotes the soil texture. A soil with 40 percent silt, 30 percent clay, and 30 percent sand is classified as silty clay loam, as indicated by the red lines in the aforementioned figure.
- Soil texture affects various physical aspects of soil, including water retention and drainage capabilities. Coarse-textured sandy soils typically exhibit elevated infiltration rates but possess limited water retention capacity. Silt particles are far smaller than sand, possess a larger surface area, and are typically highly fertile. Silty soils retain less moisture than clay soils; yet, a greater proportion of that moisture is accessible to plants. Clay soil with a fine texture typically exhibits a reduced penetration rate while possessing a substantial water retention capacity. Clay particles absorb water and undergo hydration, resulting in soil expansion when wet and contraction when dry.
- Soil texture also affects the soil's intrinsic fertility. A gram of clay particles can adsorb more nutrients than a gram of sand or silt particles due to the significantly larger surface area available for adsorption. Clay constitutes the active component of the soil. It is the location where soil nutrients are retained and primarily from which they are exchanged. The clay fraction significantly influences soil structural stability and, consequently, erosion risk.
- The comparatively inert sand and silt components can be termed the soil "skeleton," whereas the clay can be considered the "flesh" of the soil. The three portions of the solid phase collectively form the soil matrix.PThe soil texture can be readily assessed in the field utilizing the soil texture key – Refer to Table 5.1. Initially, mold a small quantity of dirt into a sphere approximately 4 cm in diameter, ensuring the removal of any stones and organic matter. Gradually moisten the soil and shape or compress it to create a ribbon between your thumb and forefinger. The ribbon's length and the ball's characteristics enable the estimation of the soil's texture class.

Soil Structure

- Soil structure denotes the configuration of soil particles (sand, silt, and clay) and pores, as well as the capacity of these particles to form aggregates. Aggregates are clusters of soil particles bound by organic matter or chemical forces. Pores are the voids within the soil.
- The interstices among the aggregates are typically substantial (macropores). Their substantial size facilitates effective aeration, swift

water infiltration, effortless plant root penetration, enough water drainage, and optimal conditions for soil microorganisms to flourish. The diminutive pores inside aggregates or between soil particles (micropores) retain water against gravitational forces (capillary action), yet not so firmly that plants are unable to draw the water.

Table 5.1: Soil Attributes Reflective of Soil Texture

Soil Texture	*Description*	*Length of Soil Ribbon (mm)*
Sand	The soil stays loose and separated, and can only be accumulated in the form of a pyramid.	<15
Sandy loam	The soil contains enough silt and clay to become sticky and can be made into the shape of a fragile ball.	15 – 25
Silty loam	Similar to the sandy loam, but the soil can be shaped by rolling into a small, short cylinder. Soil has a 'silky' feel.	25
Loam	Contains almost the same amount of sand, silt and clay. Can be rolled into a 15cm long (approximately) cylinder that breaks when bent.	25
Clay loam	Similar to loam, although the cylinder can be bent into a U shape (without forcing it) and does not break.	40-50
Fine clay	The soil cylinder can be made into the shape of a circle but shows some cracks.	50-75
Heavy clay	The soil cylinder can be shaped into a circle without showing any cracks.	>75

Source: Euroconsult 1989, McDonald *et al.,* 1990 cited in Moody and Cong 2008.

- ✰ A well-structured soil produces stable aggregates and has numerous pores of diverse sizes, as illustrated in Figure 5.2a. Such soil is friable, readily manipulated, and facilitates the emergence of germinating seedlings, enabling them to rapidly develop a robust root system.
- ✰ An inadequately organized soil possesses either a limited number of unstable aggregates or few pore spaces – Refer to Figure 5.2b. An inadequately constructed soil can lead to unproductive, compacted, or soggy conditions characterized by deficient drainage and aeration. Inadequately constructed soil is more susceptible to slaking and erosion.

Pore Spaces

- ✰ The voids between soil particles (clay, silt, and sand) and among and within aggregates (clusters of soil particles) are referred to as pore spaces. They are the portion of the soil occupied by air and water. Water displaces air in the soil; thus, the air content of soil is inversely correlated with its water content.
- ✰ Elevated moisture levels in soils indicate a reduced presence of air inside the substrate. This leads to elevated carbon dioxide levels and diminished

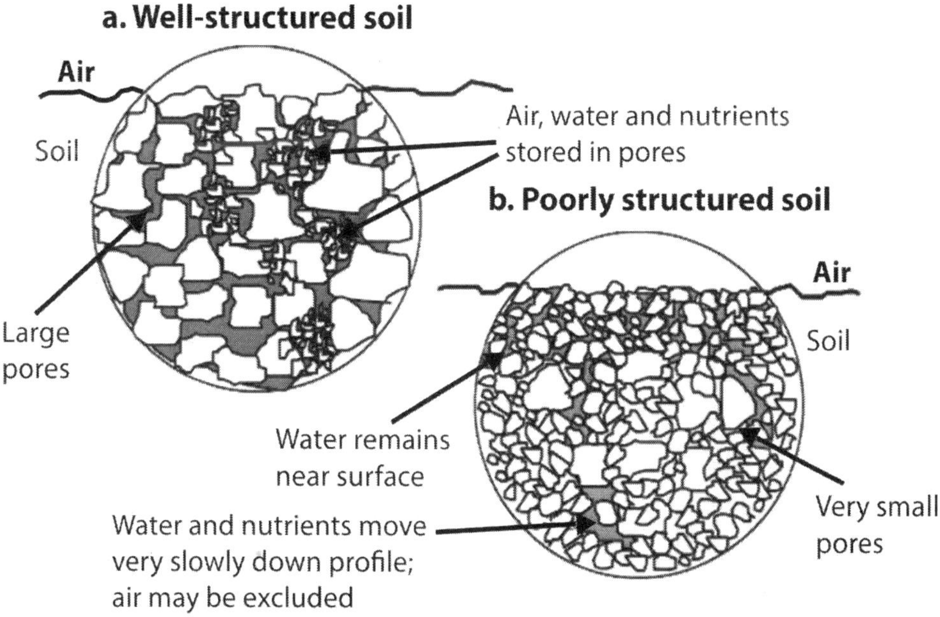

Figure 5.2a-b: Soil Structure.

oxygen levels in the soil, which is detrimental to plant growth. These circumstances also promote denitrification, the biological process that transforms nitrate-nitrogen into the greenhouse gas nitrous oxide.

- ✰ Soil air varies from atmospheric air due to its more changeable composition, increased humidity, higher carbon dioxide levels, and lower oxygen concentration compared to the atmosphere.
- ✰ The quantity and dimensions of pore spaces are influenced by the size of soil particles (soil texture) and the configuration of these particles into aggregates (soil structure). The macropores facilitate the unobstructed movement of air and percolating water through the soil. The diminutive pores (micropores) impede air flow and significantly restrict water transport.
- ✰ Soil biology contributes to the cohesion of soil. A sandy soil may lack adequate organic matter to consolidate the sand particles into bigger aggregates. In this instance, the soil will possess numerous huge pore spaces and a minimal number of tiny pores.
- ✰ The plant roots will receive ample aeration, while water will drain efficiently through the soil with minimal retention. Conversely, a dense, compacted clay soil will possess numerous small pores and a scarcity of large pores. Plants experience distress as water is excessively bound within the minuscule pores, rendering plant roots unable of extracting it from the soil. The soil has inadequate aeration and insufficient drainage. As a result, the oxygen is depleted.

Soil Water

- Soil moisture significantly affects plant development and the biological activity of the soil. It serves as a medium for compounds to dissolve, including nutrient components, so making them accessible to plant roots.
- Water facilitates the transportation of nutrients away from the farm and adds to erosion and weathering processes. Soil texture affects water retention and the rate of water infiltration into the soil.

Too Much Water

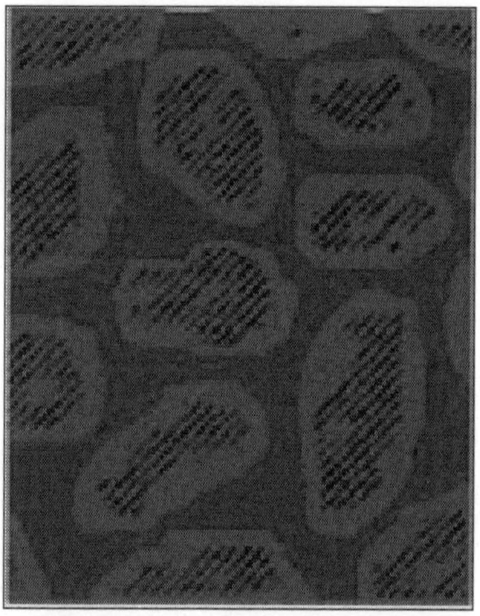

- When all soil pores are filled with water due to precipitation or irrigation, the soil may become saturated or waterlogged. Plants necessitate both air and water in the soil. Prolonged waterlogging of soil, particularly over a few days, can adversely affect plants.
- Plants demand oxygen for respiration and energy production; without it, they cannot thrive. Fertilizer application should be avoided when soils are wet.

Too Little Water

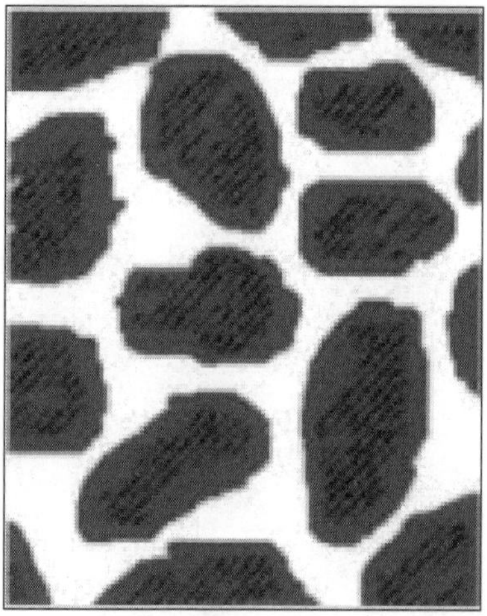

- As the soil desiccates, soil particles, especially clay, retain water more firmly than the plant can extract it. Consequently, water is retained in the soil with increasing force as the soil desiccates. At this juncture, when the plant fails to absorb sufficient water, it wilts and does not recuperate. This is referred to as the wilting point or the lower extractable limit.

The Right Balance of Air and Water

- Once the soil becomes saturated and begins to drain, the enormous pore spaces regain air, providing sufficient water for plants. This is the state when the soil reaches field capacity. Field capacity fluctuates with on soil texture, ranging from -1/10 bar for coarse-textured soils to -1/3 bar for medium to fine-textured soils.
- Once plants have depleted the readily available water, the soil attains its replenishment point. The soil moisture level between the refill point and

field capacity is referred to as readily accessible water (RAW).

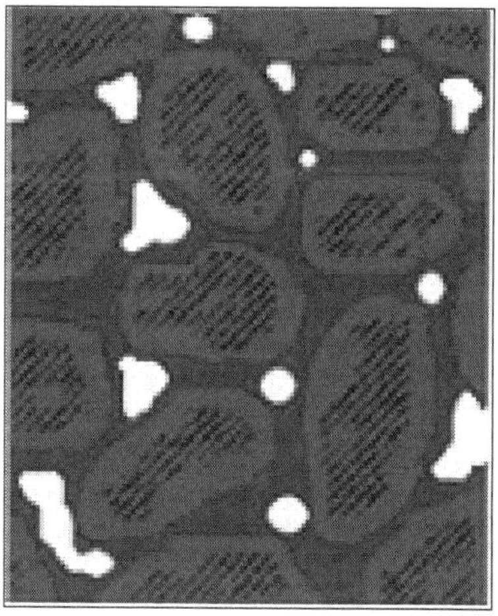

- RAW refers to the water that plants can readily absorb from the soil, and it is the target level that irrigators want to sustain, unless they are deliberately inducing stress in the plants. The figure on the right illustrates that sandy soils necessitate less water prior to its availability to plants, in contrast to clay soils, which require saturation before water becomes accessible to plants.

Chemical Properties

The chemical characteristics of soils crucial for plant growth are:

- Nutrient availability and cation exchange capacity influence the soil's intrinsic fertility and its capability to retain nutrient cations, including calcium, potassium, and magnesium.
- The chemical properties of the soil solution that influence pH and salinity.
- The sodicity of the soil influences soil stability and the availability of nutrient cations.

Nutrient Availability and Cation Exchange Capacity

- A significant proportion of plant nutrients in the soil is bound in complex compounds that are inaccessible to plants. The lesser fraction exists in more straightforward, soluble forms that are accessible to plants.
- The intricate molecules are progressively transformed into simpler compounds by chemical weathering and biological processes. The chemical fertility of soil is contingent upon the accessibility of nutrients in a form that plants can utilize, known as nutrient availability.
- The availability of soil nutrients is contingent upon various parameters, including soil pH, soil solution, soil type, and the age, type, and root system of the plant.
- Plant nutrients consist of individual elements (e.g., potassium (K)) or compounds of elements (e.g., ammonium nitrate (NH_4NO_3)). In every instance, nutrients consist entirely of atoms.
- Plants take mineral nutrients from the soil solution in the form of ions. An ion is a charged particle created by the loss or gain of electrons from an atom or molecule. A positively charged ion is referred to as a cation. An ion possessing a negative electrical charge is referred to as an anion. Cations comprise sodium (Na^+), potassium (K^+), calcium (Ca^{++}),

Cation-exchange capacity (CEC) is the maximum quantity of total cations that a soil is capable of holding, at a given pH value, available for exchange with the soil solution. It is the sum total of exchangeable cations that can be adsorbed by a soil colloidal surface. CEC is used as a measure of fertility, nutrient retention capacity, and the capacity to protect groundwater from cation contamination. It is expressed as cmol(p⁺)kg⁻¹ or meq. per 100 g, where (p⁺) denotes proton.

magnesium (Mg^{++}), and aluminum (Al^{+++}). Anions comprise chloride (Cl^-), nitrate (NO_3^-), sulfate (SO_4^{2-}), carbonate (CO_3^{2-}), phosphate (HPO_4^{3-}), and boric acid (H_3BO_3).

- A single plus sign or minus sign indicates that an ion possesses one positive or negative electrical charge, respectively. Multiple plus or minus signs indicate that an ion possesses two or more positive or negative charges.
- The substance that is adsorbed is referred to as the adsorbate, while the medium on which adsorption takes place is known as the adsorbent.
- The availability of phosphorus is significantly affected by adsorption reactions involving calcium, aluminium, iron, manganese, and the reactive surfaces of specific clay minerals. These processes can 'sequester' phosphorus, rendering it less accessible to plants. The extent of fixation is contingent upon pH levels. In alkaline soils, phosphorus binds with calcium, while in acidic soils, it associates with iron and aluminum, resulting in reduced phosphorus availability for plants.
- Cations and anions are not retained equally by soil particles. An increase in positive charges enhances the capacity to bind with a negatively charged surface. An increase in negative charges enhances the capacity to bind with a positively charged surface. The order of *Lyotropic series* of adsorption is; $Al^3z > Hz > Ca^2z > Mg^2z > Kz = NH_{„}^{+} >$ Naz. For instance, plant root cells can excrete Hz ions that can displace less potent ions such as Kz, making them available for plant absorption.
- The ions may be cations or anions::
  - Uptaken by plant roots.
  - Extracted from the soil through soil water.
  - Adsorbed onto the surfaces of negatively and positively charged soil particles.
  - The ability of soil to absorb nutrients as cations is referred to as its cation exchange capacity. Cation exchange capability is assessed using a soil analysis.

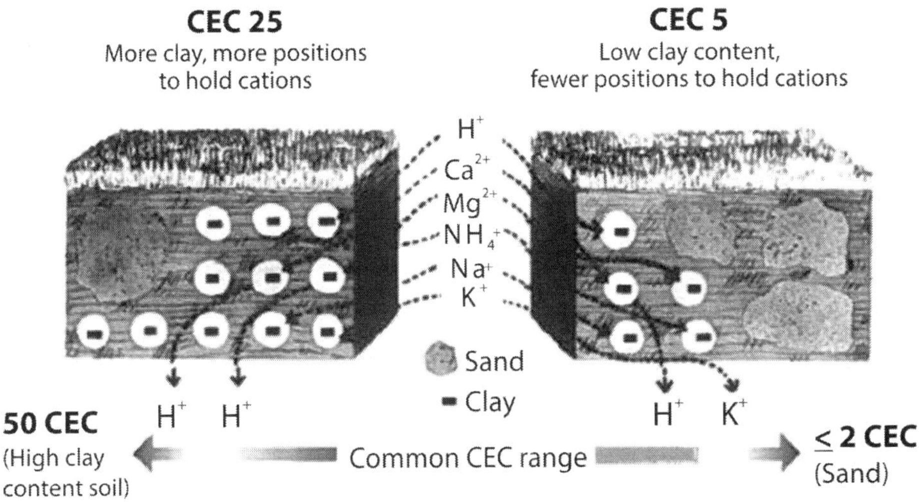

Figure 5.3: Illustration of Cation Exchange Processes (*Source*: Fertiliser Industry Federation of Australia 2006, pg. 6).

Cations are retained on the surfaces of soil minerals and organic materials, as well as inside the crystalline structures of certain clay minerals. A larger surface area for cation adsorption correlates with increased soil fertility. Consequently, soil texture influences soil fertility due to the particle sizes comprising different soil texture classes, as well as the quantity of organic matter present. Some clay minerals possess a positive charge at their edges, which attracts and retains anions.

Table 5.2: Surface Area of Soil Particles and Organic Matter

Particle	*Square cm per gram*
Coarse sand	23
Very fine sand	90
Silt	230
Clay	around 8,000,000
Organic matter (humus)	around 8,000,000

Source: CSIRO (1979).

- ☆ As long as nutritional cations and anions are adsorbed onto soil particles, they cannot be assimilated by plants or leached from the soil, unless the entire clay particle is removed through erosion. Nevertheless, they are not firmly bound and can be substituted with other ions with a similar charge present in the soil solution.
- ☆ In these exchanges, certain cations, such as Ca^{2+}, are retained more firmly than others, including Na^{+} and Mg^{2+}. Once the nutrients are present in

the soil solution, they may be assimilated by the plant's roots, utilized by soil microorganisms, or lost by leaching.

Soil Water Solution

- Soil water refers to the water retained within the pores of the soil. The soil solution comprises soil water and its dissolved salts, including cations and anions.
- The soil solution serves as the medium via which the majority of soil nutrients are delivered to growing plants. It also influences soil salinity and pH levels.

Soil Salinity

- Soil salinity refers to the elevated concentration of salts within the soil solution. As soil moisture diminishes, particularly due to evaporation, the concentration of soluble salts, including sodium, calcium, magnesium, and potassium chlorides and sulfates, in the soil solution escalates. These salts may already exist in the soil solution or may be transported upward from the groundwater by capillary action if the water table ascends.
- The concentration of soluble salts may reach levels that hinder plant growth. Soils with a salt concentration in the plant root zone that significantly disrupts plant growth are termed saline soils. These soils have a white salt encrustation on the surface. In India, these soils are referred to as 'reh' in certain regions and 'thur' in others.
- Salinity may manifest in both dryland and irrigated agricultural systems. The salinity remains consistent in all scenarios; only the initiating factors and management strategies may alter.

Soil pH

- The soil solution may be neutral, acidic, or alkaline. This is referred to as soil pH. The pH quantifies the quantity of hydrogen ions (H^+) in the soil solution on a logarithmic scale from 0 to 14. When a soil solution has a higher concentration of H^+ ions, it is classified as acidic. When there is a reduced concentration of H^+ ions, indicating an increased presence of hydroxyl (OH^-) ions, the soil solution is classified as alkaline.

Soil reactivity is expressed in terms of pH and is a measure of the acidity or alkalinity of the soil. More precisely, it is a measure of hydrogen ion concentration in an aqueous solution and ranges in soils from 3.5 (very acid) to 9.5 (very alkaline). The effect of pH is to remove from the soil or to make available certain ions. Soils with high acidity (<5.5) tend to have toxic amounts of aluminium and manganese. Soils with high alkalinity (>8.5) tend to disperse. Soil organisms are hindered by high acidity, and most agricultural crops do best with mineral soils of pH 6.5.

- The pH of soil influences nutrient availability and microbial activity. It may also influence the concentration of exchangeable aluminium.

Sodicity

- The sodicity of soil pertains to the ratio of exchangeable sodium cations to other cations adsorbed in the soil. A soil containing 6 percent or more of its exchangeable cations as sodium is referred to as sodic soil.PSodic soils have inadequate structure and rapidly disperse upon wetting. Seedlings struggle to penetrate a desiccating scattered surface, resulting in inadequate germination and survival rates.
- Dispersion occurs due to weak positive charges, such as sodium, and is mitigated by gypsum application, which substitutes sodium ions with calcium ions.

Sodicity refers to an excess of exchangeable sodium in the soil. Sodic soils tend to occur within arid to semiarid regions and are often unstable, exhibiting poor physical and chemical properties, which impede water infiltration, water availability, and ultimately plant growth.

Calcium Carbonate Content

- Calcium carbonate ($CaCO_3$) is a poorly soluble salt that exists in various forms and concentrations in soils.
- Calcium carbonate, in moderate quantities, benefits soil structure and is commonly employed to amend the pH of acidic soils; however, when calcium levels surpass the soil's absorption capacity, it interacts with other elements, resulting in insoluble compounds that are challenging for plants to assimilate.
- Excessive calcium levels may inhibit the availability of phosphorus, boron, and iron to plants.

Calcium Sulfate (Gypsum) Content

- In soils, significant secondary deposit of gypsum ($CaSO_4.2H_2O$) may take place, especially in arid regions. The soils influenced by gypsum are mostly formed in unconsolidated alluvial, colluvial, and aeolian deposits of base-rich weathered material.
- The natural vegetation on gypsum-rich soils is limited and primarily consists of xerophytic shrubs, trees, and/or ephemeral grasses.

Biological Properties

Living Organisms

- A diverse array of living species inhabits healthy soil, ranging from big entities like earthworms to microscopic microbes. Soil organisms facilitate the decomposition of organic materials. The burrowing behavior of bigger species integrates organic materials into the soil and generates substantial pore spaces that enhance soil aeration and facilitate rapid water infiltration. Millions of creatures, varying from sub-microscopic to macroscopic sizes, develop within these pore spaces.

- Microorganisms, including bacteria, actinomycetes, fungus, yeasts, algae, and protozoa, further degrade organic materials, so releasing nutrients in a form accessible to plants.
- Living organisms constitute a significant component of the soil. Their existence is promoted by elevated organic matter levels, sufficient soil moisture, and optimal drainage and aeration.
- Soil's nutritional needs are fulfilled by soil organic matter, with reduced compounds being continually incorporated. When oxidized by the heterotrophic soil population, these chemicals provide energy for their growth. Soil comprises all the major and minor nutritional elements necessary for the growth and proliferation of soil organisms.PIn healthy soil, the biomass of creatures residing underground significantly surpasses the biomass of terrestrial domestic animals.

Organic Matter

- Organic matter encompasses all living entities or the remnants of such entities. In the context of soil composition, organic matter refers to the accumulation of decomposed plant and animal wastes in the soil. It serves as the nutritional substrate for soil microbes, without which the soil would be profoundly barren.
- Organic soils, including peats, comprise between 20 percent and 95 percent organic matter. Mineral soils comprise between a trace and 15 to 20 percent organic matter. Organic matter consists of approximately 57 percent organic carbon.
- The impact of organic matter on soil characteristics, and subsequently on plant development, is significantly more substantial than this minor fraction suggests.
- Organic matter in soil enhances soil structure, porosity, drainage, and increases nutrient availability and water retention capacity.
- Organic matter serves as a nutritional source for soil microorganisms. Soils with less organic matter may exhibit inadequate structure, retain limited moisture, and are susceptible to erosion or nutrient leaching. The exception pertains to cracking clay soils, wherein clay minerals predominantly influence the structure.
- Soils rich in organic matter exhibit superior structure, enhanced water retention, and less erosion and nutrient leaching.

Organic matter is essential for nutrient cycling in the soil. Humus is produced when organic matter is completely decomposed. Humus varies in color from brown to black, with its color intensity being affected by climate (precipitation and temperature) rather than the quantity of organic matter in the soil.

Humus possesses beneficial properties, such as nutrient adsorption, superior water retention compared to clay, and enhancement of soil structure due to its limited flexibility and excellent cohesiveness. Consequently, organic matter is crucial for preserving a loose, friable soil structure.

Soil Biota

- The soil harbors a diverse spectrum of life forms, from submicroscopic entities like viruses to earthworms and larger burrowing mammals such as gophers and ground squirrels. Microscopic organisms in the soil are typically referred to as "soil microflora" (plant kingdom), whereas larger organisms are designated as macrofauna (animal forms).
- Soil fauna, particularly earthworms, isopterans, isopods, and coleopterans, together with certain insects, positively influence the soil by their burrowing and feeding behaviors, enhancing aeration and drainage via structural alterations of the soil profile. Generally, they exert a minimal influence on soil chemical characteristics; nonetheless, their actions indirectly promote microbial activities by fostering a more conducive soil environment. They facilitate the breakdown of organic wastes in soil through mixing, churning, or fragmentation while consuming plant material.

Soil Microorganisms

- Soil microorganisms exist in vast quantities and exhibit a remarkable diversity of shapes and functions. The predominant microbiological groups in soil comprise bacteria (such as Pseudomonas, Clostridium, Bacillus), actinomycetes (including Streptomyces, Nocardia), fungus, algae (including cyanobacteria), protozoa, nematodes, and viruses.
- Due to their diminutive cell size (ranging from one to several micrometers), vast quantities of soil microbes can inhabit a very compact volume, so space is never a limitation for soil microbes. Soil bacteria can number in the millions or more per gram of fertile soil, roughly equivalent to the amount of a red kidney bean.
- It is evident that bacteria constitute the predominant population among soil microorganisms. The microbial biomass contributed by the distinct groups may be more significant than the numerical counts of diverse soil microorganisms. Soil fungi contribute the greatest biomass among microbial groups. Their significant contribution to biomass is why they are typically considered the predominant decomposer microorganisms in the soil. It is astonishing that there are numerous bacteria beneath

your feet as you traverse a grassland. Actinomycetes are analogous to bacteria, exhibiting aerial hyphae similar to fungus. These organisms exhibit traits of both bacteria (cellular dimensions, architecture, and reproductive methods) and fungal (branching morphology). Primarily located in neutral to mildly alkaline soils.

- A fungus, *Armillaria bulbosa*, identified in Michigan, USA, may be the largest organism on Earth or among the largest. Researchers identified the fungus proliferating amid the roots of deciduous trees in a woodland. The microscopic, branched filaments known as hyphae of the fungus encompass a 14.8-hectare (37-acre) expanse of land. Meticulous genetic study has demonstrated that the filaments comprise a singular creature. Fungi typically expand radially in a circular formation as they proliferate through the earth.
- The fairy rings of mushrooms, named for the belief of ancient peoples that they symbolized the pathways of fairies dancing at night, are commonly observed in lawns or on golf courses and signify the outer perimeter of a growing fungus. Researchers estimate that the identifiable component of the Michigan fungus may weigh approximately 100 tons, marginally less than a blue whale.
- The importance of substantial microbial biomass in the soil is rooted not only in its extensive metabolic capacity but also in the remarkable diversity of biochemical processes associated with the soil microbial community. Soil microorganisms interact not just with their own group but also with other microbial communities. Degradation of plant components happens far more rapidly in the presence of a mixed soil microbial population than when one or more types of soil microorganisms are absent from the system.
- Soil biota can be categorized into trophic levels. The soil microbial population, situated at the foundation of the trophic levels, decomposes plant, animal, and microbial remains, while also providing sustenance for certain higher levels. Soil protozoa ingest substantial quantities of bacteria and certain fungus spores. These are subsequently ingested by larger soil organisms (nematodes, mites, etc.), which are then preyed upon by even larger creatures (e.g., worms and insects).
- Nutrients circulate within this microbial food web, which is essential for regulating soil fertility and plant productivity without external inputs like fertilizers. The significance of soil bacteria in decomposing organic waste and subsequently replenishing carbon dioxide for plants is arguably their most crucial global function.

Nutrient Cycling by Soil Microorganisms

- Soil bacteria significantly alter the quantity and shapes of many chemical components present in soil. The most significant cycles are those of carbon, nitrogen, sulfur, and phosphorus, as each of these components is crucial for soil fertility.

- The mineralization of organic molecules by soil bacteria releases carbon dioxide, ammonium (which is swiftly changed to nitrate by soil microbes), sulfate, phosphate, and inorganic forms of other elements. This constitutes the foundation of nitrogen cycling in all principal ecosystems globally.
- In global agricultural soils, which are the foundation of our food supply, the mineralization of nitrogen by soil bacteria is a critical process. In soils devoid of external nitrogen fertilizer inputs, such as most wooded areas and most grasslands, the release of ammonium from organic waste facilitates the sustained growth of new vegetation. Consequently, the soil microbial population governs the productivity of these soils, provided that other environmental parameters (moisture, temperature) are conducive.
- Fertilization of soil signifies our endeavor to equilibrate the conflict between plants and soil microorganisms for accessible nitrogen. Nitrogen sequestered in microbial cells is unavailable to plants or other bacteria until the tissue undergoes decomposition by different microbes. The nitrogen present in tissues is described as immobilized. Microorganisms are essential for the remobilization of these nutrients. These mineralization and immobilization phenomena are universal among all elements, however they are primarily significant in agriculture for macronutrients like nitrogen, phosphorus, and sulfur.
- In addition to regulating the production rates of inorganic nitrogen and sulfur, soil microbes, especially soil bacteria, can influence the ionic forms of these nutrients. Ammonium (NH_4^+) in the soil is typically swiftly reduced by bacteria to nitrite and subsequently to nitrate (NO_3^-), which can easily leak through the soil due to the actions of Nitrosomonas and Nitrobacter, respectively. Consequently, bacteria can affect the structure and, consequently, the retention of nitrogen in the soil.
- Likewise, diminished sulfur compounds like thiosulfate, elemental sulfur, and iron pyrite (FeS_2, known as "Fool's Gold") can be converted to sulfuric acid by soil microorganisms. The bacteria that facilitate the oxidation of reduced nitrogen and sulfur compounds utilize these substances as energy sources to propel their metabolism. In contrast to decomposer microorganisms that derive energy and cellular material from organic carbon molecules, specialist bacteria known as chemoautotrophs source their carbon for cellular synthesis from carbon dioxide or dissolved carbonate.
- Numerous genera of bacteria are capable of oxidizing reduced sulfur compounds. Nonetheless, a significant portion of this activity, particularly the oxidation of sulfur and pyrite, can be ascribed to bacteria belonging to the genus Thiobacillus (thio = sulfur; bacillus = rod-shaped bacterium).

Thiobacillus thiooxidans is capable of oxidizing elemental sulfur to produce sulfuric acid. Consequently, sulfur can be utilized to lower the pH of alkaline soil. Thiobacillus ferrooxidans metabolizes both iron and sulfur in iron pyrite, resulting in the production of sulfuric acid and dissolved iron. This also underpins acid mine drainage linked to global coal mining activities. The prolonged use of ammonium-based fertilizers can lead to the acidification of agricultural soils by bacterial nitrification, which involves the transformation of ammonium into nitrate accompanied by the generation of acidity. Consequently, it is evident that certain environmental issues may emerge from the actions of these chemoautotrophic soil microorganisms.

- An additional significant facet of nutrient cycling is that, under specific conditions, nitrogen and sulfur can be transformed into gaseous forms (volatilized) and then released into the atmosphere. Nitrate nitrogen can be transformed into gases like nitrous oxide (N_2O) and dinitrogen (N_2) by denitrification, a process involving the bacterial reduction of NO_3^- to N_2O or N_2 by soil bacteria in anaerobic environments. Denitrification results in the loss of nitrogen, an essential ingredient for plants, from the soil. This technique effectively eliminates surplus nitrate from wastewater.
- Sulfur, as sulfate (SO_4^{-2}), is utilized by anaerobic bacteria such as Desulfovibrio and Desulfomataculum, which convert it into hydrogen sulfide gas (H_2S). Hydrogen sulfide interacts with metal ions to produce very insoluble metallic sulfides, such as pyrite (FeS_2). It is likely that the pyrites linked to coal seams were accumulated by the activity of these bacteria in the past. The dark hue of salt marsh soils and the accompanying odor reminiscent of rotten eggs are attributable to the actions of sulfate-reducing bacteria in these environments. They confirm the presence of anaerobic conditions. Sulfur volatilization from soil signifies the depletion of a plant nutrient and contributes atmospheric sulfur, perhaps exacerbating acid precipitation.
- Nitrogen can be depleted from agricultural soils and other ecosystems. This "leak" in the terrestrial nitrogen cycle can be partially compensated for by another significant biological process known as biological nitrogen fixation. This process, exclusive to bacteria and certain other microbes, particularly cyanobacteria, involves the uptake and conversion of atmospheric dinitrogen (N_2) into forms accessible to plants. Biological nitrogen fixation is performed by free-living bacteria, cyanobacteria, and symbiotic microorganisms in diverse mutualistic partnerships with higher plants.
- The most beneficial and likely the most renowned instance of symbiotic nitrogen fixation is the Rhizobium-legume root-nodule symbiosis. Bacteria from the genera Rhizobium and Bradyrhizobium, among others, can stimulate nodule growth on the roots of some legumes, such as peas,

beans, peanuts, soybeans, and alfalfa, facilitating significant nitrogen fixation within these structures. Within the nodule, bacteria receive carbon sources (photosynthate from the plant) essential for nitrogen fixation. In exchange for this carbon, the bacteria assimilate ambient nitrogen, which is transformed into amino acids utilized by the plant for growth. This distinctive plant-microbe symbiosis enables many legumes to be self-sufficient in nitrogen, rendering them mostly independent of soil nitrogen supplies. It is unsurprising that these plants are farmed globally as supplies of sustenance, fiber, and forage.

- Approximately two-thirds of the global nitrogen supply originates from biological nitrogen fixing. Legumes have been utilized from the inception of recorded history as "soil-enhancing" crops referred to as "green manures." Green manuring is cultivating a leguminous species specifically to incorporate it back into the soil, so providing nitrogen for subsequent crops.

Nitrogen Cycle

- Soil biology is crucial in influencing various soil features; nonetheless, as a relatively nascent field, significant knowledge gaps persist about soil biology and its impact on soil properties.
- Soil organisms decompose organic debris, so rendering nutrients accessible for plant absorption. The nutrients contained within soil organisms' bodies mitigate nutrient loss through leaching.
- Microorganisms preserve soil structure, whereas earthworms play a crucial role in bioturbation inside the soil. Bacteria are essential to the Nitrogen cycle.
- Approximately 95 to 98 percent of soil nitrogen exists in organic form and is thus degraded by soil organisms. Soil mineral nitrogen is highly prone to losses.
- Mineralization is characterized as the infusion of ammonia or an ammonia compound. This is the process whereby pure nitrogen forms are transformed into ammonium by decomposers or bacteria. Upon the death of a plant or animal, or the excretion of waste by an animal, the primary type of nitrogen is organic. Bacteria, or fungi in certain instances, transform the organic nitrogen in the remains into ammonium (NH_4^+), a process known as ammonification or mineralization.The enzymatic hydrolysis of organic nitrogenous substances to the level of NH_4^+ is termed ammonification.

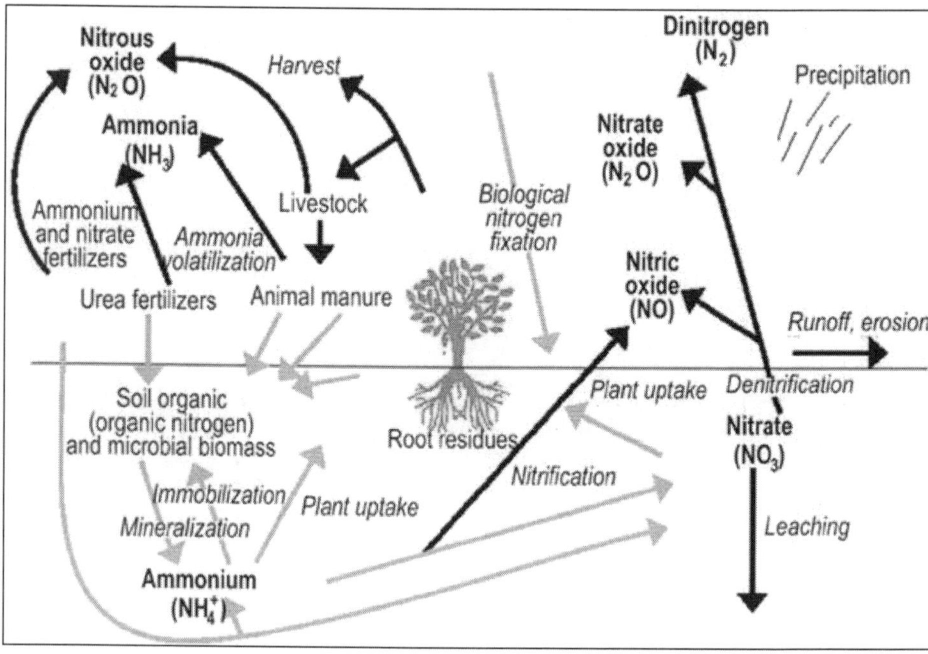

Nitrogen Cycle

- Immobilization is the process of transforming the mineral form of an element into its organic form, serving as the reverse of mineralization. During this process, ammonium ions are integrated into amino acids through a mechanism known as amination.
- Nitrification is the process by which bacteria convert nitrogen in the form of ammonium, derived from protein degradation, into nitrates, which are accessible to growing plants. It is a bifurcated process executed by two distinct categories of microorganisms. Organisms that oxidize NH_4^+ to NO_2^- include *Nitrosomonas europaea* and *Nitrosospira briensis*, whereas those that oxidize NO_2^- to NO_3^- comprise *Nitrobacter winogradskyi* and *Nitrospira gracilis*.
- Denitrification reintroduces nitrogen into the atmosphere. Denitrifying bacteria are predominantly anaerobic, including Achromobacter and Pseudomonas. The purification process under anoxic circumstances transforms nitrates and nitrites in soil into nitrogen gas or gaseous chemicals like nitrous oxide or nitric oxide. Excessive denitrification can result in significant reductions of available soil nitrogen and a consequent decline in soil fertility.
- Nitrogen fixation is performed by free-living nitrogen-fixing bacteria in soil or water, such as Azotobacter, or by those in intimate symbiosis with leguminous plants, such as rhizobia. These bacteria establish colonies

within nodules they produce on the roots of peas, beans, and analogous species. These can transform atmospheric nitrogen into nitrogenous organic compounds.

Carbon Cycle

- The carbon cycle graphic illustrates the flow of the element carbon among Earth's biosphere, pedosphere, geosphere, hydrosphere, and atmosphere. This is the paramount process on Earth since it facilitates the recycling and reutilization of its most prevalent element.
- The yearly carbon fluxes and exchanges among reservoirs transpire due to many chemical, physical, geological, and biological processes in the soil.

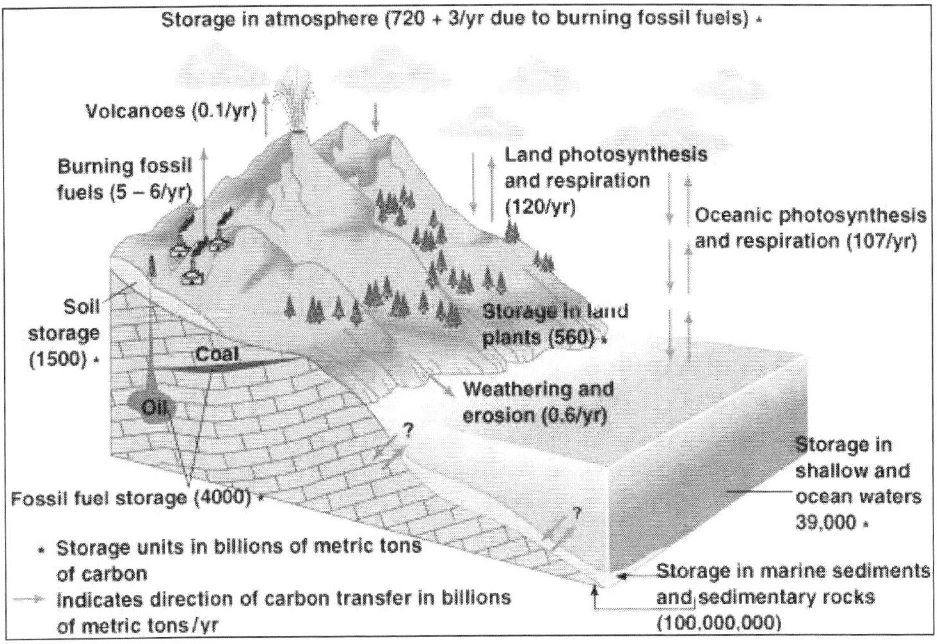

Carbon Cycle

- Soil-dwelling microorganisms recycle nutrients, including carbon and nitrogen, within the soil ecosystem. A significant portion of the organic matter introduced to the litter or root zone annually is nearly entirely decomposed by microorganisms. As a result, there exists a carbon reservoir with a rapid turnover period of approximately 1 to 3 years in several instances. The by-products of microbial consumption include CO_2, H_2O, and various other chemicals, together referred to as humus.
- Humus is less appealing to bacteria and hence decomposes at a slower rate. Once generated at superficial soil levels, a portion may migrate downhill as a clay-humus complex.

- The lowest soil layers exhibit reduced oxygen availability, which hinders microbial activity in the decomposition of humus. Ultimately, through various soil processes, this humus ascends to regions with higher oxygen levels, where bacteria subsequently decompose the humus, releasing further CO_2. This humus serves as an additional, more enduring carbon storage in the soil, with ages ranging from several hundred to a thousand years. The rapid and gradual decomposition of humus, both facilitated by microbial activity, results in an average residence time of approximately 20 to 30 years for the majority of soils.
- Soil microbes, in relation to their respiratory output, exhibit heightened sensitivity to the organic carbon content, temperature, and moisture levels; they respire more rapidly under elevated carbon concentrations, increased temperatures, and moist circumstances.

Soil Bioremediation

- We have addressed the extraordinary metabolic diversity and capability of the soil microorganisms. This capability is progressively being utilized effectively by humans. An advantageous outcome of our comprehension of soil microbial metabolism has been the creation of techniques for the bioremediation of soils tainted by hazardous waste or spilled petroleum products, both terrestrially and in aquatic environments.
- Bioremediation is the regulated application of microorganisms for the degradation of chemical contaminants. A multitude of procedures has been established to manage diverse wastes and facilitate the remediation of spilled organic substances. The core of these methods is the assertion that the metabolic functions of bacteria or fungus can be employed to decompose various organic commercial compounds (solvents, insecticides, hydrocarbon fuels, etc.).
- The two types of bioremediation are often utilized:
    - Biostimulation involves modifying the environment where the material has been released to promote the swift proliferation of bacteria. This technique often entails the use of adequate nitrogen and phosphorus fertilizers to mitigate nutritional constraints on microbial proliferation, along with implementing a method to enhance system aeration. These activities promote the proliferation of the indigenous microbial community, which often include microorganisms capable of degrading the target chemicals.
    - Bioaugmentation involves the introduction of an external microbial population to accelerate the degradation process. A multitude of microorganisms has been engineered for these objectives. Nevertheless, the complete extent of the utility of these microbial

products remains undetermined. Certain inoculants have purportedly improved the cleanup process, while others have demonstrated minimal or no impact on it. It is likely that, eventually, beneficial microbial products or procedures will be created for the remediation of oil or other chemical spills. Effective bioremediation necessitates comprehensive understanding of the characteristics that render certain bacteria more competitive than others in a certain environment. We will only understand how to apply sound ecological principles to introduce bacteria into these intricate habitats after these aspects are confirmed, ensuring their establishment and efficacy in the remediation process.

Multipurpose Tree (MPT) Species

Multipurpose Tree Species (MPTS) are a diverse group of tree species that offer multiple benefits to both human communities and the environment. These species are valued not only for their ability to provide essential resources such as timber, fuelwood, fruits, nuts, medicinal products, and fodder, but also for their ecological roles in improving soil fertility, conserving biodiversity, stabilizing landscapes, and mitigating environmental degradation. MPTS are typically integrated into agroforestry systems, reforestation projects, and landscape restoration efforts, where they help to enhance productivity and sustainability across different land-use practices. These trees are recognized for their remarkable adaptability to a variety of environmental conditions, from arid regions to tropical forests, making them a key component in efforts to combat soil erosion, desertification, and land degradation. In addition to their tangible economic and environmental benefits, MPTS also contribute to climate change mitigation through carbon sequestration, as they capture and store atmospheric carbon dioxide in their biomass and root systems. Given their versatile nature, MPTS are increasingly being utilized as a critical tool for environmental restoration and sustainable development, with particular focus on their role in bioremediation, where they help restore contaminated or degraded soils and waters. Through their multifaceted contributions, MPTS are not only enhancing food security, livelihoods, and environmental resilience but are also becoming central to the global agenda for ecosystem restoration and sustainable land management.

Some Common Multipurpose Trees of the Tropics

Tropical regions are home to a rich diversity of plant species, many of which play vital roles in the ecosystem and support local communities in a variety of ways. Among these, multipurpose trees stand out due to their ability to provide a wide range of benefits, from food and medicine to timber, fuel, and environmental services. These trees are often integral to both the livelihoods of people and the health of tropical ecosystems, offering sustainability and resilience in the face of environmental challenges. In this context, exploring some common multipurpose trees of the tropics helps to understand their diverse contributions to both human societies and biodiversity.

Common multipurpose trees in the tropics comprise:

- *Gliricidia sepium* the predominant tree utilized for living fences in Central America, as well as for fuel, fodder, and nitrogen fixation in the soil.
- Moringa (*Moringa oleifera*) consumable leaves, pods, and seeds, frequently utilized for livestock fodder and shade (it does not fix nitrogen as is sometimes assumed).
- Coconut palm utilized for sustenance, potable water (juice from the coconut inside), roofing material, fuel, and shade.
- Neem (*Azadirachta indica*) restricted use as an insect repellant, antibiotic, nitrogen enhancer for soil, windbreaks, and biomass generation for mulch and fuel.

Bioremediation of Problematic Soils through Multipurpose Tree Species (MPTS)

Multipurpose Tree Species (MPTS) are trees that provide a variety of ecosystem services and products for humans, such as food, timber, fuelwood, medicinal resources, and environmental benefits like soil fertility restoration and biodiversity conservation. These species are often used in agroforestry systems, reforestation efforts, and for landscape rehabilitation. MPTS also play a significant role in bioremediation—using living organisms to clean, restore, or detoxify polluted environments, particularly soils contaminated with heavy metals, organic pollutants, or excess nutrients. Bioremediation is a sustainable, low-cost, and effective method to mitigate soil degradation, pollution, and contamination through the use of plants, fungi, or microbes to break down harmful substances into less toxic forms.

Role of MPTS in Bioremediation of Problematic Soils

Problematic soils refer to those that are degraded or contaminated by factors such as erosion, salinization, acidification, heavy metal toxicity, organic contamination, or nutrient imbalance. MPTS can play a vital role in addressing these issues through different mechanisms:

- **Phytoremediation:** The use of plants to absorb, degrade, or stabilize contaminants from the soil and water.
- **Soil Enrichment:** The ability of certain tree species to improve soil structure, fertility, and overall health.
- **Ecosystem Restoration:** Rebuilding ecosystems and improving biodiversity by stabilizing soils and rehabilitating disturbed landscapes.

Mechanisms of Bioremediation Using MPTS

MPTS contribute to bioremediation of problematic soils via the following mechanisms:

- ☆ **Phytodegradation:** Some MPTS have the ability to break down toxic substances through biological processes. Certain plants, including trees, can metabolize harmful organic pollutants (such as pesticides, herbicides, and industrial chemicals) in their tissues, transforming them into less toxic compounds. This process often involves enzymes that decompose organic pollutants. Example: Poplar trees (Populus spp.) have been used in phytoremediation for their ability to degrade organic pollutants like petroleum hydrocarbons and chlorinated solvents.
- ☆ **Phytoextraction:** Phytoextraction is a process in which plants take up toxic substances, especially heavy metals, from the soil into their roots, stems, and leaves. This mechanism is particularly useful for the removal of metals such as cadmium, lead, arsenic, mercury, and chromium from contaminated soils. Example: Willow trees (*Salix* spp.) and Pines (*Pinus* spp.) are well-known for their ability to absorb and concentrate heavy metals, which can then be harvested and disposed of safely.
- ☆ **Phytostabilization:** Phytostabilization refers to the process in which plant roots immobilize contaminants in the soil by reducing their mobility, thereby preventing them from leaching into groundwater or being taken up by other organisms. Example: Acacia species, especially Acacia mangium, have been used to stabilize soils contaminated with heavy metals like aluminum and nickel, preventing further environmental degradation.
- ☆ **Phytovolatilization:** Some plants can absorb toxic substances, metabolize them, and release the products in a gaseous form through transpiration. This process is known as phytovolatilization. Plants involved in this mechanism can volatilize heavy metals or organic compounds into the atmosphere, reducing soil contamination. Example: Certain species of willows and poplars are known for their ability to volatilize contaminants like selenium, mercury, and volatile organic compounds (VOCs).
- ☆ **Mycorrhizal and Rhizobacterial Associations:** Many MPTS form symbiotic relationships with soil fungi (mycorrhiza) and bacteria (rhizobacteria), which enhance their ability to degrade pollutants, fix nitrogen, and improve soil structure. These associations help in the remediation of both organic and inorganic pollutants by increasing the soil's microbial diversity and activity. Example: Eucalyptus species and other MPTS, such as Acacia and Casuarina, often form beneficial associations with soil fungi like Arbuscular Mycorrhizal Fungi (AMF) that assist in nutrient uptake and pollutant degradation.

Types of Problematic Soils and MPTS for Remediation

Different types of soil problems require specific tree species for effective bioremediation. The following are some of the common soil problems and the MPTS used to address them:

Saline and Sodic Soils: Soils that are high in salts or sodium can be toxic to plants and degrade agricultural productivity. Certain MPTS are salt-tolerant and can improve soil quality by extracting excess salts, stabilizing the soil structure, and enhancing water infiltration. Example MPTS for saline soils:

- ☆ *Tamarix* spp. (Tamarisk) – Tolerant of high salinity, used for saline soil reclamation.
- ☆ *Atriplex* spp. (Saltbush) – Known for its salt-tolerant properties.
- ☆ *Casuarina equisetifolia* – Helps in the stabilization of saline coastal environments.

Acidic Soils: Acidic soils, typically caused by the accumulation of aluminum and iron oxides, can restrict plant growth and nutrient availability. Some MPTS can help neutralize soil acidity through their root systems, enriching the soil with organic matter and contributing to soil pH adjustments. Example MPTS for acidic soils:

- ☆ *Eucalyptus* spp. – Can tolerate low pH levels and enhance nutrient cycling.
- ☆ *Acacia mangium* – A nitrogen-fixing tree that contributes to soil fertility and pH neutralization.
- ☆ Pine species – Such as *Pinus radiata*, which can help in ameliorating acidic soils due to their ability to fix nitrogen.

Heavy Metal Contaminated Soils: Contaminated soils with heavy metals such as lead, arsenic, cadmium, and mercury pose a significant threat to environmental health. Certain MPTS can absorb, accumulate, and stabilize these toxic metals, reducing their bioavailability and movement into food chains. Example MPTS for heavy metal-contaminated soils:

- ☆ **Poplar trees** – Known for their ability to uptake and accumulate cadmium, lead, and zinc.
- ☆ **Willow species** – Have demonstrated the capacity to absorb and concentrate arsenic, cadmium, and lead.
- ☆ **Acacia species** – Known for their ability to remediate soils with high levels of aluminum and nickel.

Nutrient-deficient or Degraded Soils: Soils that have lost their organic matter content and become compacted or nutrient-deprived can benefit from MPTS that restore soil fertility and improve the soil's physical structure. Many MPTS are nitrogen-fixers, providing essential nutrients to surrounding plants. Example MPTS for degraded soils:

- ☆ ***Alnus* spp. (Alder)** – Fixes nitrogen, enriching poor soils and restoring fertility.
- ☆ ***Acacia* spp.** – Provides nitrogen fixation and organic matter to improve soil structure.

- ***Leucaena leucocephala*** – A fast-growing legume that helps restore nitrogen to poor soils.

Benefits of Using MPTS for Bioremediation

- **Cost-effectiveness:** Tree-based bioremediation is often more affordable than mechanical or chemical remediation techniques, such as excavation or soil washing.
- **Sustainability:** MPTS provide long-term environmental benefits by continuously improving soil health, preventing erosion, and supporting biodiversity.
- **Carbon Sequestration:** In addition to remediating soils, MPTS help mitigate climate change by sequestering carbon in biomass and soil organic matter.
- **Ecosystem Services:** Beyond soil remediation, MPTS can provide a range of ecosystem services, including habitat creation, water regulation, and improved air quality.

Challenges and Limitations

- **Slow Process:** Phytoremediation and bioremediation are relatively slow processes, which may take years or even decades to achieve significant results.
- **Limited Efficiency for Deep Contamination:** MPTS are most effective at remediating surface-level contamination; deep or widespread contamination may require additional remediation strategies.
- **Maintenance and Monitoring:** Bioremediation projects using MPTS require ongoing monitoring, care, and maintenance to ensure optimal conditions for plant growth and pollutant uptake.

Multipurpose tree species are versatile tools for bioremediation, offering an environmentally sustainable approach to restoring problematic soils. By selecting appropriate tree species for specific soil challenges, MPTS can help mitigate pollution, improve soil health, and contribute to ecosystem recovery. Integrating these species into land management and agroforestry systems can support long-term soil rehabilitation and enhance ecological resilience.

Advantages of Bioremediation

- Bioremediation is a natural process and is therefore perceived by the public.
- Bioremediation is useful for the complete destruction of a wide variety of contaminants.
- Instead of transferring contaminants from one environmental medium to another, for example, from land to water or air, the complete destruction of target pollutants is possible.

- Bioremediation can often be carried out on site, often without causing a major disruption of normal activities.
- Bioremediation can prove less expensive than other technologies that are used for clean-up of hazardous waste.

Disadvantages of Bioremediation

- Bioremediation is limited to those compounds that are biodegradable. Not all compounds are susceptible to rapid and complete degradation.
- There are some concerns that the products of biodegradation may be more persistent or toxic than the parent compound.
- Biological processes are often highly specific to the microbial populations, suitable environmental growth conditions, and appropriate levels of nutrients and contaminants.
- It is difficult to extrapolate (deduce) from bench and pilot-scale studies to full-scale field operations.
- Bioremediation often takes longer than other treatment options.

Chapter 6

Soil Taxonomy

Introduction

Soil is a natural entity consisting of solids (minerals and organic matter), liquids, and gases found on the terrestrial surface. Soil occupies a certain volume. Soil is defined by its horizons, or layers. These are differentiated from the original material due to the additions, losses, transfers, and transformations of energy and matter, or their capacity to sustain rooted plants in a natural environment. Soil is unevenly distributed. The characteristics of soil differ as a result of several soil-forming factors. Soil scientists have established a series of classification or taxonomy systems to identify, comprehend, and manage soils. Soil Taxonomy is a fundamental system for classifying soils to facilitate the creation and interpretation of soil surveys. This approach facilitates the classification of soils into several categories. The taxonomic classifications established in Soil Taxonomy categorize soils with analogous features that developed through comparable pedogenic processes.

The Soil Pedon

- ☆ The term "Pedon" refers to a unit of sampling. Several soil qualities can be assessed from the surface. To ascertain the characteristics of soil, one must examine its horizons or strata. This investigation necessitates pits or alternative methods for obtaining material samples from the surface to the soil's base.
- ☆ A pedon is the minimal volume necessary to characterize and sample soil, effectively representing the characteristics and organization of its horizons and the variety of qualities retained in samples. A pedon is analogous in several respects to the unit cell of a crystal.

The Epipedon

- The epipedon (from Greek, meaning 'over' or 'upon' and 'soil') is a soil horizon that develops at or near the surface, characterized by the significant alteration of the underlying rock structure. It is obscured by organic material or exhibits signs of eluviation, or both. The rock structure comprises fine stratification (less than 5 mm) in unconsolidated sediments (aeolian, alluvial, lacustrine, or marine) and saprolite originating from consolidated rocks, where unweathered minerals and pseudomorphs of weathered minerals maintain their relative positions.
- An epipedon differs from an A horizon. Soil Taxonomy identifies nine epipedons: folistic, histic, melanic, mollic, anthropic, umbric, ochric, plaggen, and grossarenic. The three predominant epipedons are ochric, mollic, and umbric. They are primarily differentiated by organic matter deposition, coloration, thickness, and level of base saturation.
- Instead of defining classes solely on theories of soil genesis, the classifications primarily rely on the presence of diagnostic horizons and traits. The diagnostic horizons and characteristics signify the significant pedogenic processes (i.e., additions, removals, transfers, and transformations) that are currently occurring or have transpired in the past, resulting in the soil profiles observed today.

The Endopedon

- The diagnostic subsurface layers are referred to as endopedons (from Greek endodermis, meaning sub-surface or deep-seated, and pedon, meaning soil). Nineteen endopedons are recognized: argillic, natric, agric, spodic, sombric, cambic, kandic, oxic, sulfuric, salic, placic, albic, glossic, calcic, gypsic, duripan, fragipan, petrocalcic, and petrogypsic. Among them, eight types—argillic, natric, cambic, kandic, oxic, salic, calcic, and gypsic—are frequently found in India.

Purpose of Soil Classification

- Soil categorization systems are developed to facilitate the prediction of soil behavior and to establish a standardized terminology for soil scientists. The National Cooperative Soil Survey and the USDA established the Soil Taxonomy classification system, utilized globally.
- Soils are categorized and classed into 12 orders based on the physical and chemical features of their horizons (layers). Soil taxonomy is utilized by farmers, builders, engineers, agricultural extension agents, homeowners, community planners, and governmental entities.

Farmers

- Understanding soil categorization aids in anticipating soil behavior. Soil behavior aids in forecasting soil performance for agricultural crop cultivation. County extension officers furnish soil survey maps detailing local soil classes to assist farmers in determining suitable crops for specific regions.

- Vertisol soil is characterized by a high clay content that undergoes significant shrinkage and swelling. The minimal organic-matter content of vertisol renders it inadequate for agriculture, unless substantially altered.

Fertile Soil

- Mollisol is the most fertile soil for agricultural purposes. Mollisol soil and its eight suborders constitute 21.5 percent of the land area in the United States. The fertile agricultural region of the Great Plains consists of mollisol soil.
- The topsoil layer is dense in organic matter and exhibits a black hue. The elevated fertility is attributed to the prolonged incorporation of organic matter from the native grassy plant roots of the area.

Engineers and Community Planners

- Engineers employ soil categorization studies to ascertain the potential behavior and constraints of soil. Pipelines, bridges, buildings, recreational spaces, and landfills must be constructed on soil that meets the engineering requirements of each project.
- Community planners utilize soil classification expertise to design community gardens, green belts, recreational areas, septic absorption fields, and optimize the use of local natural resources.

Maintaining Soil Fertility

- Inadequate land-use practices, including deforestation, overgrazing, and soil-degrading construction activities, exemplify issues mitigated by an understanding of soil kinds and classification. Erosion of previously productive soil strata impacts 65 percent of the Earth's surface area.
- Degraded soil is agriculturally unproductive. The soil categorization system elucidates the biological factors influencing soil formation, productivity, and erosion prevention.

Evolution of Soil Classification System

There are four categories:

Category I: Early Systems of Soil Classification

Initial soil classification methods were rudimentary and pragmatic.

1. Economic Classification

One of the initial soil classification systems, implemented by the Revenue Department, categorized soils based on their productivity for revenue purposes.

The parameters employed were soil color, soil texture, and irrigation potential. This system was insignificant and became outdated with the alteration in land use.

2. Physical Classification

It is among the earliest systems and was founded on soil texture, a characteristic linked to soil productivity and management. This soil classification scheme was relevant to the zonal soil category.

3. Chemical Classification

The classification of soils according to their chemical composition has not been extensively utilized for practical applications. Soils were categorized as calcareous, acidic, and sodic, among others. However, these features do not allow for the classification of all soil types seen in nature.

4. Geological Classification

This soil classification system was founded on the inferred characteristics of the underlying parent rock or substance.

This method identified two major categories of soils.:

- Residual or sedentary soils are formed in situ from the parent rocks.
- Soils formed from unconsolidated sediments, such as alluvium, colluvium, or aeolian materials.

The soil classification method, however, is not applicable to soil formation processes governed by active elements such as climate and vegetation, which overshadow the influence of parent material and consequently disrupt geological soil classification.

5. Physiographic Classification

This method took into account the landscape characteristics, leading to the introduction of several geomorphic words for soil classification, including levee soils, basin soils, terrace soils, mountain soils, hilly soils, upland soils, and lowland soils. A classification solely based on physiography may possess limited utility, as multiple soil groups with distinct features could be included under a single group.

6. Other Systems

Various soil classification schemes based on differing criteria have been utilized intermittently.

Based on Organic Matter as:

- Inorganic or mineral soils, and
- Organic soils

Based on Soil structure as:

- Single-grained soils, and
- Aggregated soils.

Additionally, soil classification has been conducted depending on characteristics such as humidity, vegetation, and temperature.

Category II: Recent Systems of Soil Classification

- In the latter half of the 19th century and the early 20th century, Dokuchaiev (1900), Joffey (1912), and Marbut (1935) classified soil based on zonality, intrinsic properties, and morphology, underscoring the necessity for the examination of actual soils to ascertain characteristics such as soil color, texture, structure, consistency, and drainage conditions. Marbut also underscored the notion of pedalfers and pedocals.
- The principal shortcomings of the Marbut Morphogenetic soil classification system were from its reliance, in part, on assumptions regarding soil genesis. Numerous soil series identified in the United States are absent from the Marbut classification system.

Category III: Baldwin and Associate's Genetic Approach

Marbut's morphogenetic soil classification system was refined and expanded by Baldwin, Kellogg, and Thorp in 1938.

The system's salient traits were:

- A reversion to the zonality principle of the Russian educational system. The pedocal-pedalfer idea was not emphasized.
- Increased attention was placed on soil as a three-dimensional entity and its properties.
- A novel group, termed Soil Family, was established between Great Soil Group and Soil Series; nevertheless, neither the Family nor the superior category was delineated about soil attributes.
- Significant issues arose when certain soil series did not conform to any established Great Soil Groups, while others could be classified equally into two distinct Great Soil Groups.

Recognizing this issue, Thorp and Smith (1949) referenced the soil classification of Baldwin et al. (1938), which is displayed in Table 6.1.

Orders of Genetic Approach

Post-modification, soils can be categorized into three classifications: zonal, intra-zonal, and azonal, based on the zonality principle formulated by Russian soil scientist Dokuchaiev.

Zonal Soils

Zonal soils are classifications of soils that have formed under analogous climatic conditions and are located within a specific climatic zone. Soils in this category have well-defined profiles indicative of climatic and vegetative influences. The sole distinctions arising from the parent material are overshadowed by prevailing climatic factors. Examples include laterite soil, podzol soil, and chernozem soil.

Table 6.1: Genetic Soil Classification System Indicating Corresponding Categories in Soil Taxonomy (Revised by Thorp and Smith, 1949)

Order	*Sub-order*	*Great Soil Groups*	*Some general Corresponding Classes in Soil Taxonomy*
Zonal Soils	1. Soils of cold zone	Tundra soils	Cryaquepts
	2. Light coloured soils of arid regions	Desert soils	Orchids, Argids
		Red desert soils	Argids, Orchids
		Sierozem	Argids, Orchids
		Brown soils	Argids, Some Ustoll
		Reddish brown soils	Argids, Some Ustoll
	3. Dark-coloured soils of semi-arid subhumid and humid grasslands	Chestnut soils	Ustolls Borolls
		Reddish chestnut soils	Ustolls Borolls
		Chemozem soils	Ustolls
		Prairie soils	Udolls
		Reddish Prairie soils	Udolls
	4. Soils of the forest grassland transition	Degraded cternozem Non-calcic Brown or Shantung Brown soils	Boralfs, Borolls Xeralfs
	5. Light-coloured podzolized Soil of the timbered regions	Podzol soils	Onhods Boralfs
		Gray wooded or Gray podzolic soils	Scone ochrepts
		Grey-brown podzolic soils	Udults
		Red-yellow podzolic soils	Udults
	6. Lateritic soils of forested warm-temperature and tropical regions	Reddish-Brown Lateritic soils	Udults
		Yellowish-Brous Lateritic soils	Udults
		Laterite soils	Some oxisols
Intrazonal soils	1. Halomorphic (saline and alkali) soils of imperfectly drained arid regions and littoral deposits	Solonchak or Wine soils	Sal orchids
		Salonetz soils	Natr great groups
		Soloth soils	Natr great groups

Order	Sub-order	Great Soil Groups	Some general Corresponding Classes in Soil Taxonomy
	2. Hydromorphic soils of marshes. seep areas and flats	Humic-gley soils	Aqu sub orders
		Alpine Meadow soils	Aqu sub orders
		Bog soils	Aqu sub orders
		Half Bog soils	Aqu sub orders
		Low-Humic grey soils	Aqu sub orders
		Planosols	Aqu sub orders
		Groundwater podzol soils	Aquods
		Groundwater laterite soils	Plinthaquex. Plinth-aqults
Azonal	3. Calcimorphic soils	Brown forest soils	Eutrochrepts
		Rendzina soils	Rendolls
		Lithosols	Lithic orthents Ochrepts
		Regosols	Orthents (Psamments)
		Alluvial soils	Fluv classes (Fluvents)

Intra-Zonal Soils

The soils exist within a certain zone, yet exhibit the impact of certain local factors, including topography and/or parent material. Under these conditions, the characteristics of soils imparted by the local conditions and dominant viz. excess of water, salt, calcium carbonate etc. Calcimorphic and Hydromorphic Soils, for instance.

Azonal Soils

The soils have underdeveloped profiles due to timing constraints. The group comprises immature soils lacking horizon differentiation, preventing their classification within either of the two orders (zonal and intra-zonal), thereby categorizing them under a distinct order termed azonal soils. For instance, alluvial soils.

The three orders were subsequently categorized into nine suborders based on distinct meteorological and vegetative zones. Each sub-order was subsequently categorized into major soil groupings that reflect more detailed circumstances. The major soil groups were then categorized into various soil families, series, and types.

Limitations in Genetic Systems of Soil Classification

The principal deficiencies in the genetic system of soil classification are as follows:

- The two highest categories are delineated in genetic terms rather than based on the inherent qualities of the soils.

- The terminology and notions pertaining to the highest category (order) concerning soil properties are ambiguous.
- The classifications and definitions of significant soil groupings have been established only on environmental considerations, rather than on soil qualities.
- The classification of arable soils within this approach is rendered confusing due to variations between virgin and developed soil conditions.
- Opportunities exist to define units in the lower categories for a limited number of soil attributes deemed significant for a certain interpretation. The results obtained in this context were inapplicable for other interpretations.
- Greater emphasis has been placed on soil color or vegetation rather than on soil attributes for the nomenclature of the highest levels within this system.
- In this system, naming intergrades proved challenging due to the nomenclature's evolution from a blend of multiple languages incorporating nouns and adjectives.

The soil classification system must rely on combinations of soil properties that are crucial to formation and behavior, and the classification should be explicable in terms of genesis and behavior, with genesis and behavior being secondary to the classification itself.

Category IV: New Comprehensive System of Soil Classification (Soil Taxonomy)

- A new comprehensive approach has been designed to address many abnormalities in the previous soil categorization system. Commencing in 1951, numerous approximations were developed following critical recommendations from pedologists across various nations.
- Since 1953, an approximation has been generated and evaluated annually. The 7th Approximation was released in 1960, with addenda in 1964 and 1967. Soil Taxonomy: A Basic System of Soil Classification for Making and Interpreting Soil Surveys was published in 1975, with its second edition released in 1999.

1. Advantages of New Comprehensive System of Soil Classification over the Earlier System

The new comprehensive system offers various advantages, which are as follows:

- The new comprehensive approach, in contrast to genetic systems, relies on quantifiable soil qualities. An effort has been made to categorize all classes based on soil attributes.

- ☆ The new comprehensive approach takes into account factors influencing soil genesis, which serves as the foundation of this system.PA class in a taxonomic system is defined as a type or orthotype.
- ☆ The majority of nomenclature terms originate from Greek and Latin, which are highly logical, facilitating the understanding of a taxon's position within this system and aiding in its interpretation.
- ☆ A new comprehensive system has incorporated a sub-group (new category) for the expression and recognition of soils that exhibit a continuum with gradual changes in various attributes.
- ☆ In contrast to the genetic system, it is a systematic framework devoid of biases, yet it promotes effortless memorization of items.

Taxonomy is the aspect of classification focused on relationships, involving the systematic differentiation and orderly nomenclature of type groups within a specific domain. Therefore, soil classification encompasses both soil taxonomy and other information.

The names of soil orders in the new comprehensive soil classification (7th approximation), together with their derivation, are shown in Table 6.2.

Table 6.2: Soil Orders in the 7th Approximation and their Derivation

Name of Soil Order	*Formative Elements*	*Derivation of Formative Elements*	*Approximation Equivalents in the Genetic System*
Entisol	Ent	Nonsense symbol	Azonal, some Low-Humic Gley soils
Vertisol	Ert	L. Verto, turn	Grumusols
Inceptisol	Ept	L. inception, beginning	Some Brown Forest, Low-Humic Gley and Humic Gley soils
Aridisol	Id	L. aridus, dry	Desert, Reddish Desert, Sierozem Solonchak, Solonetz
Mollisol	oll	L. Mollis, soft	Chestnut, Chernozem, Some Brown Forest, Solonetz and Humic Gley soils.
Spodosol	Od	Gk. Spodas, wood ash	Podzols, Brown podzolic soil, and Ground water podzols
Alfisol	Alf	Nonsense Symbol	Grey-Brown podzolic. Grey wooded and Non-caleic brown soils, Degraded Chemozems. Planosols and Half-Beg soils
Ultisol	Ult	L. Ultimus, last	Red-yellow podzolic soils, Reddish-brown Lateritic soils, Planosols and Half-Bog soils
Oxisol	Ox	F. oxide	Laterite Soils. latosols Bog
Histosol	Ist	Gk. histos, tissue	Soils
Andisol	and	Modified from Ando	Volcanic ash soils
Gelisol	el	Frost Churning or Mixing	Frozen Tundra soils exhibiting Cryoturbation

Note that all orders end in sol (Latin solum, soil).

Prevalent Soil Classification Systems

- Geological Classification
- Classification by Structure
- Classification based on Grain-size
- Unified Soil Classification System
- Preliminary Classification by soil types.

Standard Class systems: Australian Soil Classification./Canadian system of soil classification./French soil classification./FAO soil classification (1974-1998)/ International Committee on Anthropogenic Soils (ICOMANTH)/Unified Soil Classification System. USDA soil taxonomy/Indian Standard Classification (IS: 1498-1970).

Geological Classification

Soil types can be categorized according to their geological origin. The origin of soil can pertain to its components or the processes that have shaped its current condition. Soil may be classed as either inorganic or organic based on its contents. Soils may be categorized into the following types based on the agencies responsible for their current condition: Residual Soils. Transported Soils: Alluvial or sedimentary soils, Aeolian soils, Glacial soils; Deposited Soils: Lacustrine soils, Marine soils.

Classification by Structure

Soils may be classified into many categories based on their structure, contingent upon the average grain size and the conditions of their formation and deposition in their natural state. Soils with a single-grained structure/Soils with a honeycomb structure/Soils with a flocculent structure.

Classification Based on Grain-Size

Soils are classified based on grain-size or particle-size in the grain-size classification. Terms like gravel, sand, silt, and clay denote certain ranges of grain sizes. Given that natural soils have a blend of various particle sizes, it is preferred to refer to these fractions as sand-sized, silt-sized, etc. The spectrum of particle sizes found in soils is extensive, ranging from boulders exceeding 300 mm to clay particles measuring less than 0.002 mm. Certain clays comprise particles smaller than 0.001 mm, which function as colloids, meaning they do not precipitate in water.

In the Indian Standard Soil Classification System (ISSCS), soils are categorized into groups based on particle size, which are subsequently subdivided into coarse, medium, and fine sub-groups. The grain-size range serves as the criterion for categorizing soil particles into boulder, cobble, gravel, sand, silt, or clay.

Very coarse soils	Boulder size		> 300 mm
	Cobble size		80 - 300 mm
Coarse soils	Gravel size (G)	Coarse	20 - 80 mm
		Fine	4.75 - 20 mm
	Sand size (S)	Coarse	2 - 4.75 mm
		Medium	0.425 - 2 mm
		Fine	0.075 - 0.425 mm
Fine soils	Silt size (M)		0.002 - 0.075 mm
	Clay size (C)		< 0.002 mm

Gravel, sand, silt, and clay are denoted by the symbols G, S, M, and C, respectively.

Unified Soil Classification System

The unified soil classification system was initially devised by Casagrande in 1948 and was referred to as the airfield classification system. It was accepted with certain modifications by the U.S. Bureau of Reclamation and the U.S. Army Corps of Engineers. This technique relies on the grain size and plasticity properties of soil.

Coarse-grained soils are defined as those containing above 50 percent of particles larger than 0.075 mm in size. Coarse-grained soils are categorized into gravels (G) and sands (S). The gravels and sands are classified into four types based on gradation, silt, or clay concentration.

Fine-grained soils are defined as those in which over 50 percent of the particles are less than 0.075 mm in sieve size. They are categorized into three subdivisions: silt (M), clay (C), and organic salts and clays (O). Due of their plasticity characteristics, they are designated with the symbols L, M, and H to signify low plasticity, medium plasticity, and high plasticity, respectively.

These are the Examples with their notations:

GIN: well graded gravel	GP: poorly graded gravel
GM: silty gravel	SP: poorly graded sand
SW: well graded sand	SC: clayey sand
SM: silty sand	OH: organic silt and clays of high plastic.
CL: clay of low plastic	CI: clay of medium plastic
CH: clay of higher plastic	ML: silt of medium plastic
MI: silt of medium plastic	MH: silt of higher plastic
OL: organic silt and clays of low plastic I	OI: organic silt and clays of medium plastic

Fine-grained soils have been categorized into three subgroups based on compressibility: low, medium, and high, rather than the two categories of the original Unified Soil Classification System.

Preliminary Classification by Soil Types

An comprehension of soil behavior necessitates familiarity with prevalent soil types. This method categorizes soils by designations including Boulders, Gravel, Sand, Silt, Clay, Rock flour, Peat, China Clay, Fill, Bentonite, Black Cotton soil, Boulder Clay, Caliche, Hardpan, Laterite, Loam, Loess, Marl, Moorum, Topsoil, and Varved Clay.

Soil Taxonomy has Six Categories

- These classifications, in descending order, are: order, suborder, great group, subgroup, family, and series. There are twelve classes at the order level. The criteria employed to distinguish orders are broadly generalized and mostly rely on the types and intensities of soil-forming processes. Primarily, these criteria encompass characteristics that indicate significant disparities in soil formation.
- A suborder is a subset of an order that highlights genetic uniformity. The soil features employed to differentiate suborders within an order differ across various orders. The large group category is a subdivision of a suborder. They are differentiated by the type and arrangement of soil horizons. All soils classified inside a suborder will possess many horizons.
- Soils with these supplementary horizons are categorized into distinct large groups. Major group categories are classified into three types of subgroups: typic, intergrade, and extragrade. A typical subgroup embodies the fundamental notion of the larger group from which it originates. An intergrade subgroup comprises soils from one great group while exhibiting some characteristics of soils from another great group or class. The qualities are insufficiently developed or articulated to encompass the soils within the corresponding great group to which they transition. Extragrade subgroup soils possess anomalous characteristics that do not transition into any recognized soil type.
- A soil family category comprises a collection of soils within a subgroup that exhibit analogous physical and chemical characteristics influencing their management and manipulation responses. The primary attributes utilized to distinguish soil groups are texture, mineralogy, and temperature. Soil texture classes often categorize soils as clayey, loamy, or sandy. For certain soils, the requirements additionally delineate the quantities of silt, sand, and coarse pieces, including gravel, cobbles, and rocks.The soil series represents the most specific classification within soil taxonomy.

Soil Orders

- Soil scientists have established a classification or taxonomy system to identify, comprehend, and manage soils. Similar to the categorization systems for flora and fauna, the soil classification system encompasses multiple tiers of information, ranging from the most broad to the most specific. The highest level of classification in the United States soil system is the soil order.

- The U.S. Department of Agriculture classifies soils into 12 distinct orders. Each order is characterized by one or two predominant physical, chemical, or biological features that distinctly identify it from other orders. Each order is determined by one or two predominant physical, chemical, or biological features that distinctly identify it from the other orders.
- All 12 soil orders conclude with "sol," originating from the Latin term "solum," signifying soil or ground. Many of the orders possess characteristics that provide insights on the specific soil. For instance, "mollisol" derives from the Latin "mollis," signifying soft.

1. Gelisols

- Gelisols (derived from the Latin "gelare" meaning "to freeze") are soils that are permanently frozen, containing "permafrost," or exhibit indications of permafrost near the soil surface. Gelisols are located in the Arctic and Antarctic regions, as well as at very high altitudes.
- Diagnostic horizons may be absent or present, as the processes of thawing and freezing significantly influence their development.
- Permafrost affects land use by impacting the vertical infiltration of water, freeze-thaw cycles, and plant root depth. Gelisols, characterized by cryoturbation, exemplify a significant process involving frost heaves. Permafrost can also impede root development, constituting around 9 percent of the Earth's glacier-free land surface.

2. Histosols

- Histosols, derived from the Greek term histos meaning tissue, are mostly constituted of organic material in their upper layer. The Histosol order mostly comprises soils referred to as bogs, moors, peatlands, muskegs, fens, or peats and mucks. These soils develop when organic matter, including leaves, mosses, or grasses, decomposes at a slower rate than it accumulates, attributable to a reduction in microbial decay rates. The requisite minimum organic matter is 20 percent, provided the clay content is absent. This phenomenon typically transpires in very saturated environments or aquatic settings; hence, the majority of these soils remain perpetually saturated.
- Drained Histosols can serve as very productive agricultural land; nevertheless, they may degrade swiftly and have significant subsidence. They are also unsuitable for foundations or roadways and may exhibit significant acidity. Histosols constitute around 1 percent of the Earth's glacier-free terrestrial area.

3. Spodosols

- ☆ Spodosols, derived from the Greek term spodos meaning wood ash, are considered among of the most appealing soils. Their surface is often dark, resting on an ashy gray layer, which is further supported by a subsurface horizon that is reddish, rusty, coffee-colored, or black.
- ☆ These soils develop when precipitation interacts with acidic plant litter, such as pine needles, resulting in the formation of organic acids. These acids solubilize iron, aluminum, and organic substances in the topsoil and ashy gray (eluvial) layers. The dissolved minerals subsequently migrate (illuviate) to the vibrant subsoil horizons.
- ☆ Spodosols typically form in coarsely grained siliceous parent material (sands and loamy sands) beneath coniferous plants in cool, humid locations globally. They are often acidic, exhibiting low fertility and minimal clay content. Spodosols comprise around 4 percent of the Earth's glacier-free terrestrial area.

4. Andisols

- ☆ Andisols, derived from the Japanese term "ando" meaning black soil, generally originate from the weathering of volcanic elements like ash, leading to the presence of minerals in the soil characterized by suboptimal crystal structure. These minerals possess an exceptional ability to retain both nutrients and water, rendering these soils highly productive and fertile. These soils are generally black in color, possess low bulk density, lack an albic horizon, and must have andic features. The low bulk density and high fertility of these soils facilitate cultivation.
- ☆ Andisols comprise both weakly weathered soils rich in volcanic glass and more intensely weathered soils. They generally manifest in regions characterized by moderate to high precipitation and chilly temperatures. They are also prone to significant erosion on inclines. These soils constitute around 1 percent of the glacier-free terrestrial surface.

5. Oxisols

- ☆ Oxisols, derived from the French term for oxide, are soils found in tropical and subtropical locations characterized by a predominance of iron oxides, quartz, and extensively worn clay minerals like kaolinite. These soils are generally located on ancient, gently sloping land surfaces that have remained stable for an extended period. Oxisols predominantly exhibit a brick-red hue; however, some may also seem yellow or grey.
- ☆ Primarily, they are largely uniform soils lacking distinctly defined strata or horizons. Due to extensive weathering, they exhibit low natural fertility; nevertheless, their productivity can be enhanced with judicious

application of fertilizers and lime. Oxisols occupy around 8 percent of the glacier-free terrestrial surface.

6. Vertisols

- ☆ Vertisols, derived from the Latin term verto meaning "to turn," are clay-dominant soils comprising over 30 percent clay, characterized by a specific type of expansive clay that undergoes significant shrinkage and swelling. Consequently, these soils contract during desiccation and expand with saturation. Upon drying, vertisols develop extensive fissures that can exceed one meter in depth and several centimeters in width, often linked to gilgai microrelief or intersecting slickensides.
- ☆ The displacement of these soils can fracture building foundations and distort roadways. Vertisols possess significant fertility attributed to their elevated clay content; nonetheless, water tends to accumulate on their surfaces when saturated. Vertisols are found in regions where the underlying parent materials facilitate the development of expanding clay minerals. They constitute around 2 percent of the glacier-free terrestrial surface.

7. Aridisols

- ☆ Aridisols, derived from the Latin term aridus meaning dry, are soils found in areas that are excessively arid for mesophytic plants, which are accustomed to neither excessively moist nor excessively dry conditions, to thrive. The climate in which Aridisols are found also limits soil weathering processes.
- ☆ Aridisols frequently exhibit accumulations of salt, gypsum, or carbonates and are located in both hot and cold deserts globally. The soils stay arid for the majority of the year, leading to the accumulation of salts at the surface and/or within the solum, which results in the formation of salic, gypsic, or calcic horizons. They encompass around 12 percent of the Earth's glacier-free terrestrial expanse, including several arid valleys in Antarctica.

8. Ultisols

- ☆ Ultisols, derived from the Latin term ultimus meaning "last," are soils that develop in humid regions and undergo significant weathering. They generally possess a subsurface horizon with a significant quantity of translocated clay and exhibit relative acidity.
- ☆ These soils are base-deficient, forested, mineral soils of humid climates formed under substantial precipitation and arboreal vegetation. Distinguished by a clay-rich underlying argillic or kandic layer.PUltisol soils predominantly retain nutrients in the upper cm and are typically characterized by low fertility; however, their productivity can be enhanced

with the use of fertilizers and lime. Ultisols constitute around 8 percent of the glacier-free terrestrial surface.

9. Mollisols

- Mollisols, derived from the Latin term mollis meaning soft, are prairie or grassland soils characterized by a dark-colored surface horizon, enriched with bases such as calcium and magnesium, possessing a well-structured profile, and abundant in organic matter, rendering them particularly fruitful in subhumid to humid environments. The black surface horizon results from the annual accumulation of organic materials in the soil due to the roots of prairie vegetation.
- Mollisols are typically located in regions characterized by distinct dry seasons. They constitute around 7 percent of the glacier-free terrestrial surface.

10. Alfisols

- Alfisols, derived from the soil science term Pedalfer (indicating aluminum and iron), resemble Ultisols but exhibit less intense weathering and lower acidity, with a base saturation exceeding 35 percent. They tend to be more inherently fertile than Ultisols and are located in similar climatic regions, typically under forest vegetation.
- They are more prevalent than Ultisols, covering around 10 percent of the glacier-free terrestrial surface.

11. Inceptisols

- Inceptisols, derived from the Latin term inceptum meaning "beginning," demonstrate a moderate level of soil development, characterized by an absence of substantial clay accumulation in the subsurface. They manifest across diverse source materials and climatic situations, resulting in a broad spectrum of features.
- Typically, they are not arid and possess one or more diagnostic horizons (cambic, umbric, or mollic with poor base status).
- They are vast, covering over 17 percent of the Earth's glacier-free area.

12. Entisols

- Entisols, the most recent order in soil classification, display minimal soil development aside from the existence of a recognizable topsoil horizon (ochric or anthropic epipedon). These soils are found in regions of newly deposited sediments, typically in locations where deposition occurs at a rate exceeding soil growth.
- Entisols are typically found in dynamic floodplains, dunes, landslide regions, and areas behind retreating glaciers. They are prevalent in all habitats. Entisols constitute the second biggest category of soils following Inceptisols, covering around 16 percent of the Earth's surface.

Classification of Soils in India

ICAR (Indian Council of Agricultural Research) classifies soils into eight groups. These include alluvial soils, black cotton soils, red soils, laterite soils, mountainous or forest soils, arid or desert soils, saline and alkaline soils, and peaty and marshy soils.

Alluvial Soils

- ☆ Alluvial soils constitute 24 percent of India's land surface and are located in the Indo-Ganga-Brahmaputra plain, coastal plains, and river valleys. These are also abundantly located in foothill locations. It comprises sand, loam, and clay in varying proportions.

  There are two types of alluvial soils:
- ☆ KP Khadar – referring to more recent alluvial deposits. It is the alluvial soil deposited annually by rivers and is more fertile than Bangar soil.
- ☆ Bangar refers to ancient alluvial deposits. It comprises pebbles and calcium carbonates.
- ☆ Alluvial soils may exhibit hues of grey, light brown, or yellowish tones. These soils are abundant in chemicals such as potash, phosphoric acid, lime, and carbon compounds, although they are low in nitrogen and humus.

Black Cotton Soils

- ☆ The regional designation for cotton soil in Maharashtra is regur. It is also referred to as cotton soil.
- ☆ It is created from the weathering and erosion of volcanic lavas, primarily located in the Deccan lava plateau. This soil is optimal for dry farming.
- ☆ Their intense black coloration is attributed to the presence of 'Titaniferrous Magnetite,' a combination of iron and aluminum. Regur soils exhibit significant swelling at wetting and considerable shrinkage upon drying, attributed to the presence of Montmorillonite mineral, formed from basaltic material in semi-arid to sub-humid climatic conditions.
- ☆ Churning induces vertical mixing in deep soils, resulting in the formation of deep (>50 cm) and wide (>1 cm) fissures, gilgai micro-relief, and/or closely crossing slickensides.
- ☆ Regur soils provide a significant moisture retention capacity, making them good for cotton, coarse grains, sunflower, and citrus fruits; nevertheless, they are not appropriate for extensive irrigation.These soils lack nitrogen, humus, and phosphoric acid, however are abundant in lime, potash, iron, aluminum, calcium, and magnesium.

Red Soils

- ☆ Red soils typically form from the weathering of crystalline and metamorphic rocks in regions with substantial precipitation. Consequently, these soils are abundant in ferro-magnesium minerals, and their red hue is attributed to the excessive abundance of sesquioxides (iron and aluminum oxides).

- ☆ They are established on well-drained, stable, and elevated landforms in hot, semi-arid to humid subtropical climatic conditions.
- ☆ These soils are permeable, coarse-grained, and possess low moisture retention capability. The land is comparatively less fertile and requires substantial and frequent irrigation for agricultural purposes.
- ☆ In low-lying regions, it is present as clay, while in elevated elevations, it exists as unconsolidated soil. In elevated regions, bajra, groundnut, and potato are cultivated, while low-lying areas are utilized for rice, ragi, and vegetable cultivation.
- ☆ These soils are abundant in iron, magnesium, and aluminum compounds, however deficient in lime, humus, nitrogen, and phosphoric acid.

Laterite Soils

- ☆ Laterite soils are formed in regions of substantial rainfall (200 cm or more) due to the leaching of lime and silica. These soils are predominantly located on elevated terrains such as the Deccan Plateau, Western Ghats, Eastern Ghats, Assam, Mysore, Malabar, and Rajmahal hills in humid climates.
- ☆ These soils, abundant in sesquioxides and lacking bases and primary silicate minerals, are hard or can harden like bricks upon drying after precipitation.
- ☆ These are residual in nature, developed under normal monsoonal conditions. The process of formation is referred to as leaching. In the leaching process, siliceous soil compounds are eroded by intense rainfall in hot climates and sloped terrains, resulting in a deficiency of these compounds in the soil.
- ☆ These soils exhibit low fertility, rendering them suited solely for shrubs and pastures. This soil can be rendered appropriate for the cultivation of plantation crops such as coffee, coconut, cashew nut, and ragi with the application of fertilizers.
- ☆ These soils are abundant in iron and aluminium oxides but lacking in nitrogen, phosphoric acid, lime, and potash.

Mountainous or Forest Soils

- ☆ Variations in these soils arise from the climatic and biological situations in which they exist. These soils contain kankars, stones, and pebbles, varying by location according to climate. These soils are still undergoing the formation process. These are found in the Himalayas, Vindhyas, Satpura, Nilgiri, and other mountain ranges of India.

☆ These exhibit acidity because to the abundant presence of inadequately degraded humus. These soils necessitate fertilizers for agricultural applications. Regions with substantial rainfall has a higher concentration of humus. Consequently, these places are conducive to the cultivation of many fruits, wheat, maize, and barley.

☆ These soils are abundant in organic matter and humus but deficient in potash, phosphoric acid, and lime.

Table 6.3: Diverse Soil Varieties in India: Comprehend the Distinctions

Types of Soils	*States where found*	*Rich in*	*Lacks in*	*Crops Grown*
Alluvial	Mainly found in the plains of Gujarat, Punjab, Haryana, UP, Bihar, Jharkhand *etc.*	Potash and ime	Nitrogen and Phosphorous	Large variety of rabi and kharif crops such as wheat, rice, sugarcane, cotton, jute *etc.*
Black (Regur soil)	Deccan plateau-Maharashtra, Madhya Pradesh, Gujarat, Andhra Pradesh, Tamil Nadu, Valleys of Krishna and Godavari.	Lime, Iron Magnesia and Alumina Potash	Phosphorous, Nitragen and organic matter	Cotton, sugarcane, jowar, tobacco, wheat, rice *etc.*
Red	Eastern and Southern part of Eastern and Southern part of the deccan plateau, Orissa Chattisgrah and Southern parts of the middle Ganga plain.	Iron and Potash	Nitrogen, Phosphorous and humus	Wheat, rice, cotton, sugarcane and pulses
laterite	Karnatakan, Kerala Tamil Nadu, Madhya Pradesh Assam and Orissa hills	Iron oxide and potash	Organic matter, Nitrogen, phosphate and calcium	Cashewnuts, tea, coffee, rubber
Arid and Desert	Western Rajasthan, north Gujarat and southern Punjab	Soluble salts, phosphate	Humus, Nitrogen	Only drought resistant and salt tolerant crops such as barley, rape, cotton, millets maize and pulses
Saline and Alkaline	Western Gujarat, deltas of eastern Saline and coast, Sunderban areas of West Bengal, Punjab and Haryana	Sodium, Potassium, Magnesium	Nitrogen and calcium	Unfit for agriculture

Source: https://www.clearias.com/different-soil-types-india-understand-differences/

Arid and Desert Soils

- Arid and desert soils are predominantly found in the arid and semi-arid regions of Rajasthan, characterized by either a cold or hot temperature regime and an aridic moisture regime. It consists of substantial amounts of sand due to aeolian processes and is conducive to the production of crops such as jowar and bajra. However, in regions where irrigation facilities are accessible, such as Sri Ganganagar in Rajasthan, both wheat and cotton are cultivated.
- These soils exhibit a light hue (pale brown to yellowish brown), possess a weak subangular blocky structure, and contain elevated levels of soluble salts and phosphorus, yet are deficient in carbon compounds and nitrogen. These are typically deficient in nutrition and moisture retention capabilities.

Saline and Alkaline Soils

- These soils are located in the arid and semi-arid regions of Rajasthan, Punjab, Haryana, Uttar Pradesh, Bihar, Maharashtra, and Tamil Nadu. These soils occur in spots and are locally referred to as Reh, Kallar, and Usar. These soils are unproductive and unsuitable for agricultural use.
- These soils exhibit elevated sodium levels in the exchange complex, inadequate physical conditions, particularly regarding soil structure and drainage, limited nutrient and water availability, and deficiencies in micronutrients.
- These soils can be ameliorated through irrigation combined with the use of lime and gypsum, as well as by cultivating salt-resistant crops such as rice, sugarcane, cotton, wheat, and tobacco.
- Excessive levels of sodium and magnesium result in soil salinity. Therefore, these are inappropriate for agricultural use. These exhibit nitrogen deficiency and possess inadequate water retention capacity.

Peaty and Marshy Soils

- These soils are created by the deposition of substantial amounts of organic matter under prolonged water-logged conditions. These are predominantly located in coastal regions such as Orissa, West Bengal, Tamil Nadu, and Kerala in diverse segments. In Kerala, such soils are referred to as Kari.
- These soils exhibit significant acidity and are low in phosphate and potash. Consequently, unsuitable for agricultural use. The hue of marshy soils is dark grey or black owing to the presence of organic materials.

Chapter 7

Soil as a Source of Essential Plant Nutrients

Introduction

Of the one hundred and nine elements classified in the periodic table, the majority are found in the Earth's mantle, crust, and soil. The size and occurrences of these elements fluctuate both geographically and temporally. An element is deemed important to plants when it participates in metabolic operations, and its deficiency hinders the completion of the plant life cycle (i.e., growth and reproduction). Seventeen elements have been identified as important nutrition for plants. The elements are carbon (C), hydrogen (H), oxygen (O), nitrogen (N), potassium (K), calcium (Ca), magnesium (Mg), phosphorus (P), sulfur (S), chloride (Cl), iron (Fe), boron (B), manganese (Mn), zinc (Zn), copper (Cu), molybdenum (Mo), and nickel (Ni). Additional elements recognized as advantageous, while not universally needed for all plants, are cobalt (Co), selenium (Se), silicon (Si), sodium (Na), vanadium (V), and aluminum (Al). N, P, and K are classified as primary macronutrients, whereas the others are categorized as micronutrients; nevertheless, this does not imply that they are of lesser significance. Plants require all essential nutrients in balanced amounts for optimal growth and reproduction. To enhance understanding, many significant terminology utilized in this chapter are presented.

Nutrient

A nutrient element is essential for the completion of an organism's life cycle, and its relative lack results in distinct deficiency symptoms.

Plant Nutrient

A plant nutrient is a substance made of specific chemical ingredients, commonly known as 'plant nutrients' or plant food elements, which are crucial for the growth and development of plants.

Nutrition

The provision and assimilation of chemical substances essential for the regular growth and metabolism of an organism.

- ✰ Crops rely on both extrinsic and intrinsic variables for growth, as well as their surroundings to supply essential requirements for photosynthesis.
- ✰ Essential variables for plant growth include light, heat, air, water, nutrients, and physical support.
- ✰ If any single component, or a combination of variables, is in short supply, plant development will be negatively impacted. The significance of each plant development component and the appropriate combination of these elements for optimal plant growth is exemplified by the notion of limiting factors. This principle asserts that crop production is constrained by the most limiting fundamental plant development component. The notion of limiting factors might be likened to a barrel composed of staves of varying lengths, with each stave symbolizing a factor influencing plant growth.
- ✰ Crop yield and quality are contingent upon critical growth factors and the numerous connected variables of soil, plant, environment, and agronomy. In this system, certain aspects are uncontrollable, while others are controllable and managed.
- ✰ Soil is a critical determinant of plant growth, as illustrated in the figure. The primary functions of soil are to supply plants with nutrients, water, and oxygen.

Seven Crucial Soil Attributes (SQs) Influence Crop Development (FAO)

	Soil Properties	*Soil Characteristics*
SQ1	Nutrient availability	Soil texture, soil organic carbon content, soil pH, total exchangeable cations
SQ2	Nutrient retention capacity	Soil organic carbon, Soil texture, base saturation, cation exchange capacity of soil and of clay fraction
SQ3	Rooting conditions	Soil textures, bulk density, coarse fragments, vertic soil characteristics, and soil phases influencing root penetration, soil depth, and soil volume.

Soil Properties		*Soil Characteristics*
SQ4	Oxygen availability to roots	Soil drainage and the phases influencing soil drainage
SQ5	Excess salts	Soil salinity, soil sodicity, and soil phases affecting salinity conditions
SQ6	Toxicity	Calcium carbonate and gypsum
SQ7	Workability (Constraining field management)	Soil texture, effective soil depth/volume, and soil phases that limit soil management (including soil depth, rock outcrop, stoniness, gravel/concretions, and hardpans)

- **Soil fertility:** Soil fertility Soil fertility refers to the soil's capacity to provide nutrients essential for plant growth under optimal environmental conditions, including light, temperature, and soil physical characteristics. i.e., "the soil's ability to provide accessible nutrients to plants.
- **Soil productivity:** Soil productivity refers to the soil's capacity to yield a specific crop or sequence of crops under a defined management system, essentially indicating the crop-producing potential of the soil.

History: Soil Fertility Development

- Francis Bacon (1591-1624) claimed that the primary sustenance of plants was water, and the principal function of soil was to support plants' upright growth and shield them from temperature extremes.
- Jan Baptiste Van Helmont (1577 - 1644) asserted that water was the exclusive nourishment for plants.
- Robert Boyle (1627 – 1691), an English scientist, validated Van Helmont's discoveries and demonstrated that plants synthesize salts, spirits, and oils from H_2O.
- Anthur Young (1741 - 1820), an English agriculturist, conducted pot experiments utilizing barley as a test crop under sand culture conditions. He used charcoal, train oil, poultry manure, wine spirits, oyster shells, and various other substances, concluding that some of these components promoted enhanced plant development.
- Priestly (1800) demonstrated the necessity of O_2 for plant growth.
- J.B. Boussingault (1802-1882), a French chemist, conducted field experiments and maintained a balance sheet. He was the first scientist to undertake a field experiment. He is regarded as the progenitor of field experiments.
- Justus Von Liebig (1835) proposed that the majority of carbon in plants originates from atmospheric CO2. Hydrogen and oxygen are derived

from water, while alkaline metals are required for the neutralization of acids produced by plants during their metabolic processes. Phosphorus is needed for seed production; plants absorb all substances from the soil but excrete non-essential ones through their roots.

- The field may have certain nutrients in excess, some at optimal levels, and others in minimal amounts; yet, the limiting element for growth is the nutrient that is least available. The law of minimum, articulated by Liebig in 1862, serves as a straightforward yet rational framework for anticipating crop reactions to fertilization. This law asserts that "the extent of plant production cannot exceed that permitted by the most restrictive of the essential plant growth factors." Liebig's contributions to agricultural development were significant, earning him recognition as the founder of agricultural chemistry.
- J.B. Lawes and J.H. Gilbert (1843) initiated a permanent manurial experiment at the Rothamsted Agricultural Experiment Station in England. They performed field studies over a twelve-year period, concluding that crops necessitate both phosphorus and potassium; nevertheless, the composition of plant ash does not accurately reflect the quantities of these elements needed by the plant. No leguminous crop necessitates nitrogen. In the absence of this element, growth will not occur, irrespective of the amounts of P and K available. The atmospheric contribution of ammonium is inadequate for the crop's requirements. Soil fertility can be sustained for several years by chemical fertilizers, but the advantageous impact of fallow is attributed to the enhancement of accessible nitrogen molecules in the soil.
- Robert Warrington's proposition England demonstrated that nitrification could be facilitated by carbon disulfide and chloroform, but would be inhibited by the addition of a little quantity of unsterilized soil. He established that the reaction was a two-step event. Initially, NH_3 is transformed to nitrites, followed by the conversion of nitrites to nitrates.

Essential Elements

A mineral element is deemed important for plant growth and development if it participates in metabolic functions and is indispensable for the plant to complete its life cycle.

There are seventeen essential elements required for plant growth *viz.*, C, H, O, N, P, K, Ca, Mg, S, Fe, Mn, Zn, Cu, B, Mo, Cl, Ni.

The Essentiality Criteria Outlined by Arnon and Stout (1939)

- A plant must be incapable of completing its life cycle without the mineral element.

- ✰ The element's function must be irreplaceable by another mineral element.
- ✰ The element must be directly implicated in plant metabolism.
- ✰ D. J. Nicholas originated the term functional or metabolic nutrient. Any mineral element that plays a role in plant metabolism, regardless of whether its function is particular. (Chlorine, Silicon, Sodium, Vanadium, Cobalt, Selenium).

Essentiality of the Elements Established by

Elements	*Scientists*
Carbon	Priestly (1800)
Nitrogen	Theodore De saussure (1804)
Ca, Mg, K, P	Carl sprengel (1839)
Sulphur (S)	Von Liebig (1844) -Sachs and knop
Iron (Fe)	E. Greiss (1844)
Manganese (Mn)	J. S. McHargue (1922)
Zinc(Zn)	Sommer and Lipman (1926)
Copper (Cu)	Sommer, Lipman and McKinney (1931)
Molybdenum (Mo)	Arnon and Stout (1939)
Sodium (Na)	Brownell and wood (1957)
Cobalt(Co)	Ahamed and Evans (1959)
Boron(B)	Warington (1923)
Chlorine (Cl)	Broyer (1954)
Nickel (Ni)	Brown et.al. (1987)

Classification of Essential Elements of Soil

Essential nutrients are categorized into two primary classes based on their relative usage or absorption by plants, as well as their biochemical behavior and physiological roles.

I. According to the proportionate usage or absorption by the plants

A. Macro or Major Nutrients

B. Micronutrients

Macronutrients are categorized into two categories.

- ✰ ***Primary Nutrient***: Nitrogen, phosphorus, and potassium. These three elements are also referred to as fertilizer elements.

- ***Secondary Nutrients:*** Calcium, Magnesium and Sulphur.

II. Classification According to Biochemical Behavior and Physiological Functions

Group	*Nutrient Elements*	*Biochemical Functions*
Group I	C,H,O, Ca	**They are basic structural elements.** They are primary components of plants (carbohydrates, proteins, and lipids) and organic matter. These elements are also crucial to enzymatic processes. They supply energy for growth and development through oxidative degradation.
Group-II	N,P & S	Auxiliary structural components of the more dynamic and essential biological tissues. Crucial element of metabolically active substances such as amino acids, proteins, enzymes, and non-protein molecules. They participate in energy storage (ATP and ADP) and transfer (phosphate esters).
Group-III	K, Ca, Mg	Regulators and transporters mostly govern plant metabolism. They participate in the production and transport of carbohydrates, regulate ionic charge balance, and facilitate enzyme activation.
Group-IV	Fe, Mn, Zn, Cu, B, Mo, Cl.	**Catalysts and activators**. These elements participate in oxidation-reduction reactions, chlorophyll synthesis, and also exist in organic compounds.

http://ecoursesonline.iasri.res.in/mod/page/view.php?id=7777

Carbon, Hydrogen and Oxygen

- Carbon, hydrogen, and oxygen constitute around 95 percent of the dry mass of plants, sourced from CO_2 and H_2O.
- They are transformed into simple carbohydrates by photosynthesis and subsequently synthesized into complex amino acids, proteins, and protoplasm. The primary constituents of carbohydrates, proteins, and fats.

Functions of Carbon, Hydrogen and Oxygen

- They exert a pivotal influence in the processes of photosynthesis and respiration in plants.
- They participate in the synthesis of both simple and complex organic substances, such as carbohydrates, starch, and proteins.
- Preserving the architecture of plant cells. They supply the energy necessary for plant growth and development through the oxidative breakdown of carbohydrates, proteins, and lipids during cellular respiration.

Nitrogen

- Nitrogen is essential for plant nourishment. It is a primary nutritional element that promotes growth. Green plants are more significantly affected by nitrogen deprivation than by any other nutrient.
- Most plants absorb it in the ionic form of NO_3^-. Certain plants necessitate the NH_4^+ form, such as rice. When utilized as foliar nutrition, NH_2 (derived from amide) is also assimilated. It plays a highly acknowledged role in plant metabolism by performing the following essential tasks.

Functions in Plants

- Nitrogen plays a crucial role in the photosynthesis of plants, as it is a fundamental component of chlorophyll, the green pigment vital for this process.
- It is a fundamental component of plant life, as it constitutes an important part of proteins, nucleotides, phosphatides, alkaloids, enzymes, hormones, vitamins, and more.
- It enhances vegetative growth, and a sufficient nitrogen supply fosters quick early development, resulting in a dark green hue in plants, while also improving the quality and succulence of leafy vegetables and fodder crops.
- It promotes the development of fruit buds, enhances fruit set, and improves fruit quality. It regulates the enhanced consumption of potassium, phosphorus, and other elements.

Nutrient Deficiency Symptoms in Plants

Nutrient shortages arise when inadequate amounts are present to satisfy the needs of growing plants, typically evident through visual symptoms in various plant parts. These visual indications may manifest as stunting, chlorosis, discolouration, or necrosis. Nutrient mobility inside the plant determines their presence in either the older or younger regions of the plants.

Nitrogen

Nitrogen is a highly mobile element in plants, resulting in insufficiency manifesting in the older or lower leaves.

The striking deficiency symptoms are:

- Yellowing of mature leaves resulting from the suppression of chloroplast and chlorophyll production. Severe nitrogen deficiency results in leaf chlorosis.
- Plants exhibit dwarfism or limited development.
- Accelerates the timing of flower bud development and diminishes yield.

- Fruits exhibit hardness, diminutive size, and reduced bearing capacity of trees. Decreases fertilization, causes early fruit drop, and may result in seedless fruits.
- Severe shortage results in necrosis of plant leaves (total leaf death). Excessive nitrogen in plants results in increased vegetative growth. Leaves become increasingly succulent and more vulnerable to pest and disease infestations. Plant lodging may occur.
- Decreases the sugar content in plants, enhances the storage and shelf life of fruits or leaves, extends the growth period, and postpones the reproductive phase and crop maturity.

Symptoms of Nitrogen (N) Deficiency.

Phosphorus

- Phosphorus is a component of vital cellular constituents like phytins, phosphoproteins, phospholipids, nucleic acids (DNA, RNA), coenzymes (NAD and NADP), ATP, and other high-energy molecules.
- It serves as a structural element of the cell membrane, chloroplasts, mitochondria, and meristematic tissues. Plants assimilate phosphorus in the ionic forms of $H_2PO_4^-$ and HPO_4^{2-}. Phosphate molecules function as the "energy currency" in plants. It has high mobility in plants but remains stationary in soils.

Functions in Plants

- Engaged in energy storage and transmission. Additionally, facilitate numerous metabolic processes in plants.
- Participates in cell division and the production of meristematic tissue, hence enhancing vegetative growth in plants.
- Crucial for root formation and promotes root growth.

- Phosphorus aids in initial development, flowering, seed formation, fruit ripening, seed germination, and the early maturation of crops.
- It is crucial for the production of starch, proteins, nucleic acids, photosynthesis, nitrogen metabolism, carbohydrate metabolism, glycolysis, respiration, and fatty acid synthesis.

Symptoms of Nutrient Deficiency in Plants

- Stunted and sluggish plant growth resulting from its impact on cell division and the development of meristematic tissue.Leaves are diminutive, with defoliation commencing from the older foliage and resulting in early leaf abscission.
- Purplish foliage discolouration resulting from anthocyanin pigment. Plants exhibit necrotic lesions on the leaves, petioles, or fruits.
- Characteristics symptoms include a slender, woody stem and underdeveloped roots.
- Postponement of flowering and fruit maturation, substandard quality, abscission of blossoms, diminutive inflorescence, and early fruit drop.
- Inhibit sugar synthesis or too elevated sugar levels in plants.

Symptoms of Phosphorus (P) Deficiency.

Potassium

- Potassium is essential for plant nutrition and must be provided in substantial amounts to fruit and field crops. Plants assimilate potassium from the soil in the form of K^+ ions, which exhibit mobility within the plant system. Potassium is not a component of any elements of plant cells, including proteins, chlorophyll, lipids, and carbohydrates.
- It predominantly exists as soluble inorganic salts and, on occasion, as salts of organic acids. It is a prevalent cation in the cytoplasm, meristematic areas, and cell sap. It is regarded as a quality factor for numerous crops.

Functions

- Potassium is essential for osmoregulation and regulates cell turgor pressure.
- It plays a crucial role in pH stabilization, enzyme activation, protein synthesis, stomatal movement (opening and closing), cell elongation, and photosynthesis.
- Confer drought, heat, and frost resistance to plants via regulating transpiration and water conditions within plant cells. It enhances water utilization efficiency.
- Confer pest and disease resistance to plants
- Essential for ATP generation and enhanced nitrogen utilization efficiency via promoting protein synthesis.
- Plants develop strength and rigidity, hence minimizing lodging.
- Crucial in the synthesis and transference of starch and sugars, particularly in potato, sweet potato, turnip, banana, and tapioca.

Deficiency Symptoms

- Degradation of stems, crop lodging, and increased vulnerability to pests and diseases.
- Desiccation of foliage and charred appearance of leaf edges and apices. Suboptimal preservation of fruits. The quality of fruits and vegetables has diminished.
- Marginal necrosis and desiccation of leaf apices.
- Impeded growth, reduction of internodal length.
- It disrupts the water economy of plants, resulting in increased water loss per unit of dry matter.
- Inadequate germination of vines.
- Severe Botrytis cinerea infestation in grapes resulting from potassium deficit.

Symptoms of Potassium (K) Deficiency.

Secondary Nutrients

- Calcium, magnesium, and sulfur are vital minerals for plants. These nutrients are termed "secondary" because plants necessitate them in lesser amounts compared to nitrogen, phosphorus, and potassium.
- Conversely, plants necessitate these nutrients in greater amounts than the "micronutrients" like boron and molybdenum.
- Calcium, magnesium, and sulfur are typically sufficient in the majority of Mississippi soils with optimal pH and organic matter content. They influence pH upon application to the soil.
- Calcium and magnesium elevate soil pH, but sulfur from some sources diminishes soil pH. Compounds that contain one or more of these nutrients are frequently utilized as soil amendments rather than solely as sources of plant nourishment.

Calcium

- Calcium's principal role in plant growth is to furnish structural support to cell walls. Calcium functions as a secondary messenger during instances of physical or metabolic stress in plants.
- Calcium deficits infrequently manifest in Mississippi soils. Soils with optimal pH values typically do not exhibit calcium deficiency. Acidic soils containing 500 pounds of calcium per acre or less are inadequate for legumes, particularly peanuts, alfalfa, clovers, and soybeans. At this stage, crops with limited root systems, such as tomatoes, peppers, and cucurbits, would require supplementary calcium.
- Soluble calcium exists as the Ca^{2+} ion and is essential for peanuts during pegging, as well as for peppers and tomatoes to avert blossom end rot.

Symptoms of Calcium (Ca) Deficiency.

- Available calcium may be depleted from the soil through (a) dissolution and subsequent removal in drainage water, (b) uptake by plants, (c) absorption by soil organisms, (d) leaching via rainfall, or (e) adsorption by clay particles. Deficiency symptoms encompass mortality at the apical meristem, excessively dark green leaves, compromised stems, flower abscission, and any amalgamation of these indicators.
- Limestone serves as the principal source of calcium. Additional prevalent sources comprise basic slag, gypsum, hydrated lime, and calcined lime. These sources are advised for peanuts, peppers, and tomatoes. In peanuts, they inhibit popping and promote pegging. In tomatoes and peppers, they inhibit fruit splitting and blossom end rot. Hydrated lime and burned lime possess a higher availability of calcium compared to basic slag and gypsum. Gypsum does not influence soil pH despite its calcium content.

Magnesium

- Magnesium is assimilated as the Mg^{2+} ion and exhibits mobility in plants, translocating from older to younger leaves. It leaches from the soil similarly to calcium and potassium. Magnesium serves as the center atom surrounded by four nitrogen atoms in the chlorophyll molecule, thereby playing a role in photosynthesis.
- It functions as an activator for numerous enzymes essential for plant growth processes and stabilizes nucleic acids.
- Interveinal chlorosis is a sign of deficiency observed in crops including beans, corn, sorghum, cotton, and some green vegetables. Interveinal chlorosis refers to the yellowing of tissue between the veins, although the veins themselves stay green. The leaves may exhibit a color range from pink to light crimson and may curl upwards at the edges.

Symptoms of Magnesium (Mg) Deficiency.

- To rectify magnesium deficiency in soil, apply dolomitic lime when lime is required; utilize soluble magnesium sources when lime is unnecessary. Magnesium supplements may be necessary for cotton cultivation.
- Cattle frequently experience grass tetany when the magnesium concentration in feed is insufficient. Additional considerations encompass nitrogen, calcium, and potassium concentrations, development stage (often in spring), lactation status of cattle, and seasonal conditions.
- Dolomitic limestone is advised as a liming agent in areas where grass tetany has occurred. Administer supplementary magnesium and calcium to grazing animals in cases of grass tetany.
- The predominant soluble magnesium sources utilized as fertilizers include magnesium sulfate (comprising 10 percent Mg and 14 percent S, commonly referred to as Epsom salt), sulphate of potash magnesia (containing 11.2 percent Mg, 22 percent S, and 22 percent K_2O, marketed as K-Mag), and magnesium oxide (consisting of 55 percent Mg, also known as magnesia).

Sulfur

- Sulfur is required in substantial amounts by the majority of crops. It is a fundamental component in chlorophyll formation and protein synthesis. Sulfur is essential for rhizobia bacteria in legumes for nitrogen fixing.
- Crops typically extract approximately equal amounts of sulfur and phosphorus. Grasses extract sulfur more effectively than legumes, and clovers frequently vanish from pasture mixtures in conditions of low sulfur.
- The sulfate ion, SO_4, is the predominant type absorbed by plants. Sulfate is soluble and readily removed from soils by leaching. Sulfate leaches into the soil and accumulates in denser subsoils with higher clay concentration. Consequently, assessing sulfur levels in topsoil is an inconsistent indicator of sulfur availability throughout an extended growth season.
- Numerous coarse-textured, sandy soils and low organic matter, silty soils across Mississippi exhibit sulfur deficiency for agricultural output. Numerous acidic soils harbor metallic sulfides that liberate sulfur during the weathering process.
- Symptoms of sulfur insufficiency initially manifest on young leaves. The leaves exhibit a pale green to yellow coloration. The plants exhibit slender, diminutive forms with stunted development and postponed fruiting. To swiftly rectify a shortage, choose one of the easily accessible sulfate sources.
- Sulfur may be advised for significant crops in Mississippi at a rate of 8–10 pounds per acre annually under certain circumstances. Consult local MSU Extension Service offices or regional agronomists for more detailed information regarding crops and specific sites.

☆ Numerous sources of fertilizer sulfur exist. Organic matter serves as the primary source of organic sulfur compounds and constitutes the principal source of soil sulfur in the majority of soils. Additional sources of sulfur include precipitation and sulfur-containing fertilizers.

Sulfur (S)-Deficiency Symptoms.

☆ Accessible sources comprise ammonium sulfate (21% N and 24% S), potassium sulfate (50% K_2O and 17.6% S), gypsum (32.6% CaO and 16.8% S), and zinc sulfate (36.4% Zn and 17.8% S). There exist multiple sulfate sources in addition to less accessible sulfur sources in elemental or sulfide forms.

Table 7.1: Mean Percentage of Chemical Composition of Principal Sources of Calcium, Magnesium, and Sulfur

Material	*Calcium*	*Magnesium*	*Sulfur*	*Other*
Calcium				
Calcitic lime	31.7	3.4	0.1	0
Dolomitic lime	21.5	11.4	0.3	0
Gypsum	22.5	0.4	16.8	0
Basic slag	29.0	3.4	0.3	No longer a P source
Magnesium				
Magnesium sulfate	2.2	10.5	14.0	0
Sulfate of potash magnesium	0	11.2	22.7	22 K_2O
Magnesium oxide	0	55	0	0
Sulfur				
Ammonium sulfate	0.3	0	23.7	21 nitrogen
Potassium sulfate	0	1.2	17.6	50 K_2O
Gypsum	22.5	0.4	16.8	0

Material	Calcium	Magnesium	Sulfur	Other
Magnesium sulfate	2.2	10.5	14.0	0
Zinc sulfate	0	0	17.8	36.4 zinc
Prilled sulfur	0	0	33-39	0

- Elemental sulfur serves as an effective acidifying agent. The application of 500 pounds of sulfur per acre on sandy loam soil decreases the pH from 7.5 to 6.5. Approximately 3 pounds of lime are required to neutralize the acidity produced by 1 pound of sulfur.

Micronutrients

- Boron (B), chlorine (Cl), copper (Cu), iron (Fe), manganese (Mn), molybdenum (Mo), and zinc (Zn) are important micronutrients utilized by plants in minimal amounts. For the majority of micronutrients, crop absorption is below one pound per acre.
- Despite this minimal requirement, the absence of micronutrients restricts essential plant processes, leading to deformities, diminished development, and impaired reproduction. In such instances, costly, high-demand agricultural inputs like nitrogen and water may be squandered. Due to increased yields, elevated commodity prices, and rising crop input costs, cultivators are assessing all possible obstacles to optimal grain production, including vitamin deficits.
- The seven micronutrients are generally adequate in most soils to fulfill crop requirements. Certain sandy soils and other soils with limited organic matter are inherently deficient in micronutrients, whereas high pH soils may render some micronutrients less accessible, resulting in deficiencies.
- The availability of micronutrients is governed by the balance of the soil solution, soil organic matter, cation exchange sites, and insoluble micronutrient complexes due to intricate chemical interactions in the soil. Soil pH significantly influences the binding of micronutrients and their accessibility to plants.

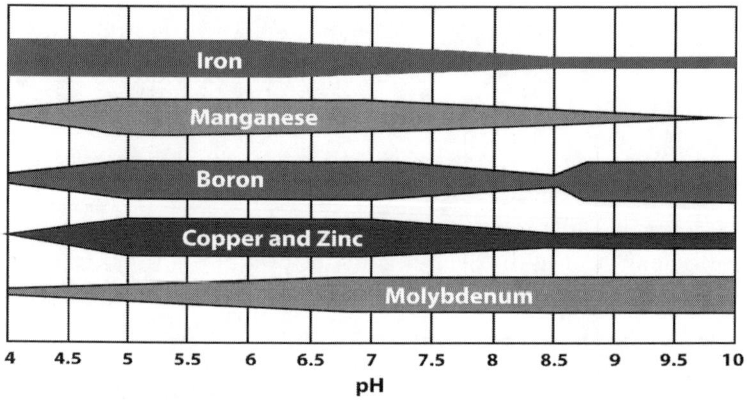

Figure 7.1: Relative Availability of Micronutrients by Soil pH.

- Micronutrients are predominantly accessible in acidic soils and frequently inaccessible at elevated pH levels (Figure 7.1).

Identifying Micronutrient Deficiencies

- Micronutrient deficits can be identified through visible signs on crops and by analyzing soils and plant tissues.
- Understanding visual symptoms necessitates knowledge of the function of each vitamin in plant growth and development.

Functions of Micronutrients

Micronutrients vary in their absorption forms, functions, movement within the plant, and specific deficiency or toxicity symptoms.

Table 7.2: Roles of Micronutrients in Plants

Element	*Function in Plant*
B	Sugar transport, cell division, amino acid production, Pollen germination
Cl	Turgor regulation, resisting disease and photosynthesis reactions
Cu	Component of enzymes, role in photosynthesis
Fe	Component of enzymes, Chlorophyll synthesis, photosynthesis
Mo	Nitrogen metabolism, nitrogen fixation in legumes
Mn	Chlorophyll production, cofactor of several reactions, activates enzymes
Zn	Component of many enzymes, essential for plant hormone balance and auxin activity

Symptoms of Micronutrient Deficiency

- With the exception of molybdenum, micronutrients are regarded as weakly mobile or immobile in plants. This indicates that deficiency symptoms manifest initially or most acutely on the youngest plant tissues, presuming that the plants began with a certain nutrient supply that was depleted as they grew.

☆ In molybdenum deficiency, symptoms initially manifest in the oldest plant tissues. Symptoms differ by crop, but generalized symptoms are presented in Table 7.3.

Table 7.3: Symptoms of General Micronutrient Deficiency

Element	*General Micronutrient Deficiency Symptom*
B	Subtle overall chlorosis, necrosis of the apical meristem, malformed leaves exhibiting regions of discolouration
Cl	Chlorosis and wilting of juvenile leaves, a deficit infrequently observed in field crops.
Cu	Moderate general chlorosis, necrosis of leaf tips with twisting, and diminished turgor in younger leaves.
Fe	Chlorosis or yellowing between the veins of young foliage
Mo	Analogous to typical nitrogen deficit, characterized by broad chlorosis (yellowing) of young plants and chlorosis of the oldest leaves.
Mn	Chlorosis, characterized by yellowing between the veins of young leaves, similar to iron insufficiency.
Zn	Impeded growth, reduced internode length, and smaller-than-normal juvenile leaves

Symptoms of Boron (B) Deficiency.

Symptoms of Chlorine (Cl) Deficiency.

Symptoms of Zinc (Zn) Deficiency.

Symptoms of Copper (Cu) Deficiency.

Symptoms of Iron (Fe) Deficiency.

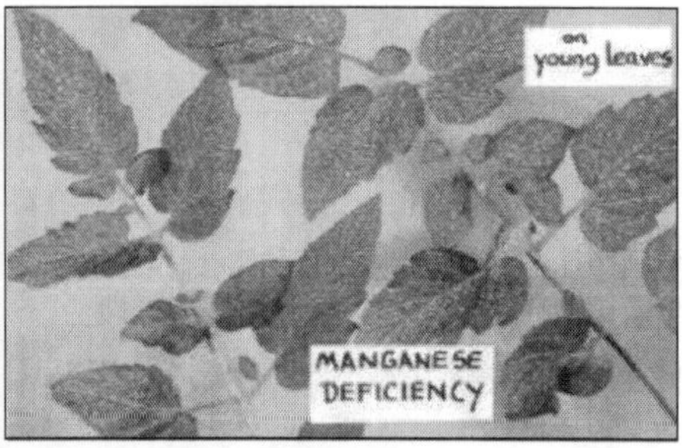

Symptoms of Manganese (Mn) Deficiency.

Symptoms of Molybdenum (Mo) Deficiency.

Common Micronutrient Deficiencies

The likelihood of micronutrient insufficiency is significantly heightened in particular soil types and specific crops, as illustrated in Table 7.4.

Table 7.4: Soil Conditions That May Result in Micronutrient Deficiencies for Diverse Crops

Micronutrient	*Soil Properties*	*Crops*
B	Sandy soils or extensively weathered soils deficient in organic matter	Alfalfa, Clover
Cl	Infrequently on sandy soils in regions with less precipitation	
Cu	Acidic muck peats with a pH of less than 5.3 or dark sands	Wheat, corn
Fe	Soils exhibiting elevated pH, soluble salts, and/or calcium carbonate concentrations	Corn, soybean
Mn	Peats and mucks with a pH more than 5.8, black sands, and lakebed or low-lying soils with a pH above 6.2	Soybean, wheat, sugar, eets, cron
Mo	Acid prairie soils	Soybean
Zn	Peats, mucks and mineral soils with pH >6.5	Corn, soybean

Rising Concern Regarding Nutrient Deficiencies

Identifying nutrient deficiencies in crop plants can be challenging, even for the most seasoned observers. Concurrent nutritional deficiencies may arise, with certain symptoms exhibiting similarities across various elements, hence increasing confusion. Nutrient shortages may be misinterpreted as manifestations of disease, drought, waterlogging, genetic variability, herbicide and pesticide exposure, and insect-pathogen occurrences. Each deficient symptom is associated with a specific function of the nutrient in the plant. Stunting in plants happens when a nutrient essential for functions such as stem elongation, photosynthesis, and protein synthesis is lacking. A comprehensive understanding of the fundamentals of nutrient deficit, coupled with adequate field trials, can significantly facilitate the assessment of nutritional deficiency.

The primary issue is the yield loss of agricultural crops due to various nutritional deficiencies. Immediate focus must be directed towards rectifying nutrient deficiencies through accurate diagnosis to prevent yield loss. Timely identification of nutritional deficiencies may provide prompt rectification without jeopardizing crop output. Occasionally, factors outside soil insufficiency may contribute to yield loss, necessitating interdisciplinary collaboration to address the underlying issues.

Chapter 8

Chemistry of Soil Acidity

- Soil pH, or soil response, quantifies the concentration of hydrogen ions in the soil solution. A lower soil pH indicates a higher level of acidity.
- The pH must be sustained above 5.5 in the topsoil (0-15 cm) and above 4.8 in the subsoil (> 15 cm).
- Properly maintained soil pH preserves soil resource value, optimizes crop and pasture selection, and prevents production losses associated with low pH levels.

Introduction

Soil acidity constitutes a significant environmental and economic issue. Acidic soils result in considerable reductions in agricultural yield, and when crop selection is confined to acid-tolerant species and cultivars, lucrative market prospects may be constrained. In pastures cultivated on acidic soils, output will fall, and certain legume species may not endure. Degradation of the soil resource is also of wider concern and off-site impacts must be considered. Off-site effects primarily stem from diminished plant growth. Elevated runoff and ensuing erosion adversely affect soil and water quality. Elevated nutrient leaching may pollute groundwater. In regions with elevated rainfall, increased soil acidity occurs naturally and can fluctuate based on geological landscape, clay mineralogy, soil texture, and buffering capacity. Soil acidity is a natural phenomenon, exacerbated by some agricultural management methods. Removal of plant material from the paddock results in the elimination of alkalinity. This elevates soil acidity. The extraction of grain, pasture, and animal products from a paddock results in increased soil acidity. The removal of hay is notably acidifying due to the substantial quantity of product extracted.

Soil acidity primarily results from nitrate leaching. Nitrogen is introduced to the soil by various methods, including:

- Nitrogen absorbed by leguminous plants;
- As nitrogen based fertilisers; from breakdown of organic matter; and dung and urine.
- Acidification may also occur in agricultural activities in soils due to the removal of plant and animal products, leaching of excess nitrate, application of certain nitrogen-based fertilizers, and accumulation of predominantly plant-derived organic matter.

Soil pH

- Soil pH quantifies the quantity of hydrogen ions in the soil solution and determines soil acidity. Soil acidity is attributed to the presence of ionizable hydrogen ions or protons.
- pH (Power of Hydrogen) is defined as the negative logarithm of hydrogen ion activity, while pOH is defined as that of the hydroxyl ion. The pH scale for water ranges from 1 to 14. In pure water, the sum of pH and pOH equals 14, with both pH and pOH measuring 7. Consequently, an acidic solution has a pH of less than 7, an alkaline solution has a pH greater than 7, and a solution with a pH of 7.0 is considered neutral. A one-unit change in pH corresponds to a tenfold increase in acidity.

Sources of Soil Acidity

The primary contributors to soil acidity are exchangeable H^+, Al_3^+, Al oxides, organic matter, and clay minerals.

The primary contributors to soil acidity are two adsorbed cations: hydrogen and aluminium. The majority of hydrogen ions are retained on the surfaces of clay minerals and organic substances.

Active acidity results from H^+ ions present in the soil solution. The potential acidity arises from H^+ ions included in diverse chemical compounds and those adsorbed on the surfaces of solid particles. The exchangeable hydrogen ions in soil offset the negative charge resulting from the isomorphous replacement of cations. The hydrogen ions result from the permanent charge on the mineral surfaces. The pH-dependent charge may originate from the structural OH^- groups located at the corners and edges of soil minerals, which dissolve into H^+ ions.

At low pH (<5.0), aluminum becomes soluble and is absorbed by silicate clays or is strongly bonded by organic materials. The Al_3^+ ions are subsequently hydrolyzed in the soil solution, producing H^+ ions that contribute to soil acidity.

$$Al^{3+} + 2H_2O = Al(OH)_2^+ + 2H^+$$

Consequently, in highly acidic soils, the activity of H^+ ions in the soil solution is significantly elevated. The impact of adsorbed hydrogen ions is diminished in very acidic soils.

In moderately acidic soils, the majority of aluminium is transformed into aluminium hydroxyl ions.

$$Al^{3+} + OH^- \longrightarrow Al(OH)_2^+$$

$$Al(OH)_2^+ + OH^- \longrightarrow Al(OH)_2^+$$

Certain aluminum hydroxyl ions are adsorbed and function as exchangeable ions. They maintain balance in the soil solution. They hydrolyze in the soil solution, generating hydrogen ions and so increasing soil acidity.

$$Al(OH) + H_2O \longrightarrow Al(OH)_2^+$$

$$Al(OH)_2^+ + H_2O \longrightarrow Al(OH)_2^+$$

At a pH below 5.0, iron exhibits behavior akin to that of aluminum; nevertheless, the acidity produced by iron is significantly lower than that of aluminum. In the acid sulphate soils of the Kuttanad region in Kerala and the Sundarbans in West Bengal, dissolved or free acidic compounds such as H_2SO_4, $Al_2(SO_4)_3$, and $Fe_2(SO_4)_3$ contribute to acidity. At elevated pH levels (>5.0), amorphous aluminum and iron hydroxyl complexes may experience hydrolysis, hence releasing hydrogen ions into the solution.

Fine-textured soils, containing high amounts of organic matter have more hydrogen ions than lighter soils with lower amounts of clay and organic matter. Moreover, different kinds of clay in soils hold variable amounts of hydrogen.

High aluminium ion concentration at the root surface may prevent the root from taking up phosphate, and the aluminium inside the living cell may interfere with sugar phosphorylation.

Soil organic matter or humans contains reactive carboxylic and phenolic groups that behave as weak acids. They dissociate, releasing H^+ions. In organic soils and in mineral soils with high organic matter content, organic acids contribute significantly to soil acidity.

Problems Associated with Soil Acidity

The principal issues associated with soil acidity are the toxicities of aluminum (Al) and manganese (Mn). Both toxicities, if unmitigated, may inflict significant harm on crop metabolism and output.

Aluminum Toxicity

- ☆ PAluminium toxicity may arise in soils with high concentrations of aluminium-bearing minerals. In certain soils, aluminum can leach into the soil solution when the pH falls below 5.4.
- ☆ Conversely, aluminium solubility significantly diminishes when soil pH exceeds 5.4. Consequently, effective management of soil pH helps avert issues related to aluminium toxicity.
- ☆ Excessive amounts of aluminum can inhibit root development and limit crop growth.

Soil acidification refers to the accumulation of hydrogen cations, resulting in a decrease in soil pH. This occurs chemically when a proton donor is introduced to the soil. The donor may be an acid, including nitric acid, sulfuric acid, or carbonic acid. It may also be a substance like aluminium sulfate, which interacts in the soil to liberate protons. Acidification also transpires when basic cations, including calcium, magnesium, potassium, and sodium, are leached from the soil. Soil acidity happens naturally as lichens and algae decompose rock surfaces. Acids persist in this dissolution throughout the soil formation process. Over time and via weathering, soils in natural environments grow increasingly acidic. Soil acidification rates can fluctuate and escalate due to causes such as acid rain, agricultural practices, and pollution.

Aluminum saturation denotes the relative concentration of aluminum in the soil. Aluminum saturation, akin to base saturation, represents the fraction of the cation exchange capacity (CEC) occupied by aluminum. As with all cations, aluminum retained by the cation exchange complex is in equilibrium with aluminum present in the soil solution. While tolerance to aluminum differs among plant species, the majority of plants cannot withstand metal saturation over 15 percent. Nevertheless, several crops cultivated in Hawaii, like sugarcane, pineapple, and corn, may endure relatively elevated levels of aluminum saturation.

Manganese Toxicity

- Manganese poisoning may arise in soils having manganese-rich minerals. Upon dissolution of these minerals, manganese ions are liberated into the soil solution. Manganese is a vital nutrient for plants; yet, high levels can adversely affect plant growth.
- Symptoms of manganese toxicity manifest as chlorosis in elder leaves, which then develop tiny, dark brown patches. While the tolerance of crops to manganese toxicity differs, the majority exhibit sensitivity to elevated manganese levels. Manganese toxicity induces the "sudden crash" phenomenon in watermelon, characterized by abrupt wilting and mortality of the plants.

Effects of Soil Acidity

- Soil pH levels between 5.5 and 8 promote plant growth and facilitate many soil processes, such as nutrient availability and microbial activity.
- Acidic soil, especially in the subsurface, will impede root access to water and nutrients.
- As soil pH decreases, aluminum becomes soluble. A minor decrease in pH can lead to a significant rise in soluble aluminum. In this manner, Al inhibits root development, limiting access to water and nutrients.

☆ Insufficient water and nutrients lead to suboptimal crop and pasture development, diminished yields, and reduced grain size. The impact of aluminum poisoning on crops is typically most apparent in seasons with a dry conclusion, since plants have limited access to stored subsoil moisture for grain development.

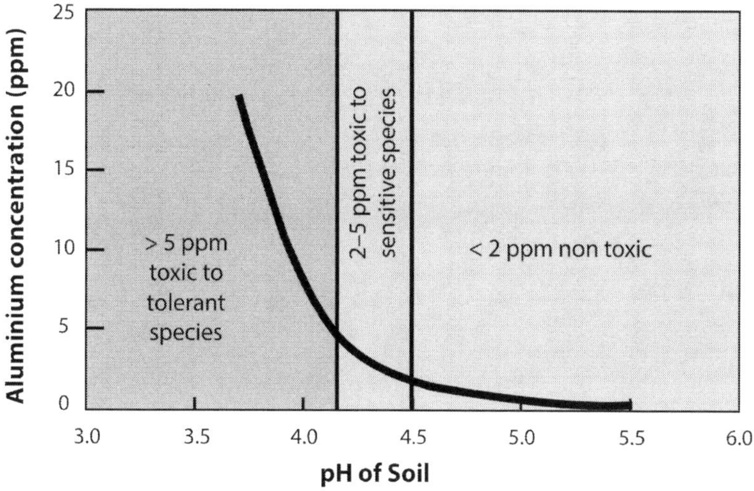

Figure 8.1: Aluminium (Al) and pH Graph with Rule of Thumb Al Toxicity.

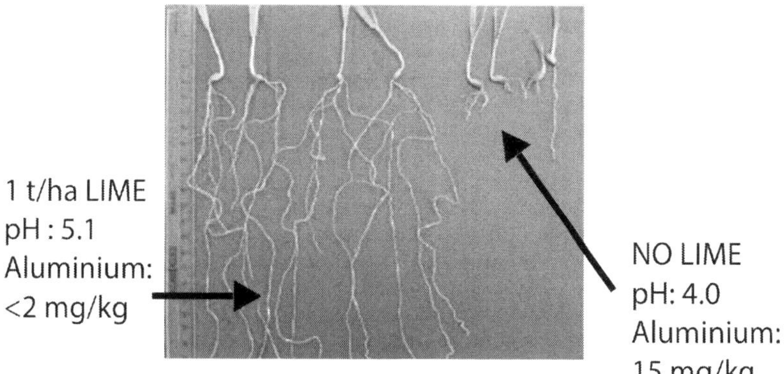

Figure 8.2: Roots of Barley Grown in Acidic Subsurface Soil are Shortened by Al Toxicity.

Nutrient Availability

☆ In very acid soils, all the major plant nutrients (nitrogen, phosphorus, potassium, sulphur, calcium, manganese and also the trace element molybdenum) may be unavailable, or only available in limited quantities.

- Plants can show deficiency symptoms despite adequate fertiliser application.

Microbial Activity

- Low pH in top surface soils (0-15 cm) may affect microbial activity, most notably decreasing legume nodulation. The resulting nitrogen deficiency may be indicated by reddening of stems and petioles on pasture legumes, or yellowing and death of oldest leaves on grain legumes.
- Rhizobia bacteria are greatly reduced in acid soils. Some pasture legumes may fail to persist due to the inability of reduced Rhizobia populations to successfully nodulate roots and form a functioning symbiosis.

Major Causes of Soil Acidity

Acid Rain

Rainfall is naturally acidic due to carbonic acid forming from carbon dioxide in the atmosphere. This compound causes rainfall pH to be around 5.0-5.5. When rainfall has a lower pH than natural levels, it can cause rapid acidification of soil. Sulfur dioxide and nitrogen oxides are precursors of stronger acids that can lead to acid rain production when they react with water in the atmosphere. These gases may be present in the atmosphere due to natural sources such as lightning and volcanic eruptions, or from anthropogenic emissions. Basic cations like calcium are leached from the soil as acidic rainfall flows, which allows aluminum and proton levels to increase.

Biological Weathering

Plant roots acidify soil by releasing protons and organic acids so as to chemically weather soil minerals. Decaying remains of dead plants on soil may also form organic acids which contribute to soil acidification. Acidification from leaf litter on soil is more pronounced under coniferous trees such as pine, spruce and fir, which return fewer base cations to the soil, rather than under deciduous trees.

Parent Materials

Certain parent materials also contribute to soil acidification. Granites and their allied igneous rocks are called "acidic" because they have a lot of free quartz, which produces silicic acid on weathering. Also, they have relatively low amounts of calcium and magnesium. Some sedimentary rocks such as shale and coal are rich in sulfides, which, when hydrated and oxidized, produce sulfuric acid which is much stronger than silicic acid. Many coal soils are too acidic to support vigorous plant growth, and coal gives off strong precursors to acid rain when it is burned. Marine clays are also sulfide-rich in many cases, and such clays become very acidic if they are drained to an oxidizing state.

Soil Amendments

Soil amendments such as fertilizers and manures can cause soil acidification. Sulfur based fertilizers can be highly acidifying, examples include elemental sulfur and iron sulfate while others like potassium sulfate have no significant effect on soil pH. While most nitrogen fertilizers have an acidifying effect, ammonium-based nitrogen fertilizers are more acidifying than other nitrogen sources. Ammonia-based nitrogen fertilizers include ammonium sulfate, di-ammonium phosphate, monoammonium phosphate, and ammonium nitrate. Organic nitrogen sources, such as urea and compost, are less acidifying. Nitrate sources which have little or no ammonium, such as calcium nitrate, magnesium nitrate, potassium nitrate, and sodium nitrate, are not acidifying.

Pollution

Acidification may also occur from nitrogen emissions into the air, as the nitrogen may end up deposited into the soil. Anthropogenic sources of sulfur dioxides and nitrogen oxides play a major role in increase of acid rain production. The use of fossil fuels and motor exhaust are the largest anthropogenic contributors to sulfuric gases and nitrogen oxides, respectively.

Other Causes

- In time, excessive rainfall leaches the soil profile's basic cations (calcium, magnesium, sodium, and potassium) results in decrease base saturation caused soil acidity. Soils that develop from weathered granite are likely to be more acidic than those developed from shale or limestone.
- Organic matter decay produces H^+ ions, which are responsible for acidity (an ion is a positively or negatively charged element). Like that from rainfall, acidic soil development from decaying organic matter is insignificant in the short term. Harvest of high-yielding crops plays the most significant role in increasing soil acidity.
- During growth, crops absorb basic elements such as calcium, magnesium, and potassium to satisfy their nutritional requirements. As crop yields increase, more of this lime like nutrients is removed from the field. Compared to the leaf and stem portions of the plant, grain contains minute amounts of these basic nutrients. Therefore, harvesting high-yielding forages and grasses leads to development of soil acidity.
- Nitrogen fertilizer has been blamed for the increase in soil acidity problems, when ammoniacal fertilizer materials are applied to the soil, acidity is enhanced, but the form of nitrogen removed by the crop is similar to that found in fertilizer. In reality, nitrogen fertilizer elevate soil acidity by increasing crop yields, thereby large extraction of basic cations by the crop plants.

Pools of Soil Acidity

There are three general pools, or sources, of acidity: active, exchangeable and residual.

- **Active acidity** is the quantity of hydrogen ions that are present in the soil water solution. The active pool of hydrogen ions is in equilibrium with the exchangeable hydrogen ions that are held on the soil's cation exchange complex. This pool most readily affects plant growth. Active acidity may be directly determined using a pH meter, such as an electron probe.
- The second pool, **exchangeable acidity**, refers to the amount of acid cations, aluminum and hydrogen, occupied on the CEC. When the CEC of a soil is high but has a low base saturation, the soil becomes more resistant to pH changes. As a result, it will require larger additions of lime to neutralize the acidity. The soil is then buffered against pH change.
- **Residual acidity** comprises of all bound aluminum and hydrogen in soil minerals. Out of all pools, residual acidity is least available.

Buffering Capacity

- The quantity of aluminum and hydrogen in each of the 3 pools of acidity is not permanently fixed. Instead, the relative amounts of aluminum and hydrogen can change, as aluminum and hydrogen moves from pool to pool. Thus, the soil is said to have a *buffering capacity*.
- Buffering capacity is the ability of the soil to resist change. In the case of acidity, it is the ability of the soil to resist change in pH. Thus, aluminum and hydrogen of one pool will replenish the aluminum and hydrogen of another pool as these acid cations are removed.
- For example, as aluminum and hydrogen are removed from soil solution, the acid cations of the CEC replenish the soil solution. Likewise, minerals containing aluminum and hydrogen dissolve and release these cations as they are removed from the exchangeable pool.

Outline of Buffering Reactions

- Exchangeable acidity will buffer changes in active acidity
- Residual acidity will buffer changes in exchangeable and active acidity
- Each soil has a unique buffering capacity. As a rule of thumb, finely-textured clay soils tend to have greater buffering capacities than coarse-textured soils.

Rule of Thumb

- Finely-textured clay soils tend to have greater buffering capacities than coarse-textured soils.

- Recall that 90 per cent of Hawaii's soils fall into this category. As a result, most Hawaii soils largely buffer soil acidity. This has great implications on **nutrient management** since buffering capacity determines the amount of resources, such as lime, that must be added to correct soil acidity.
- Soils that have high buffering capacities require larger amounts of liming resources to raise the pH to a target value than soils with low buffering capacities.

Impact of Soil Acidification

Management of Acidic Soils

Soil Testing

- Knowledge of how soil pH profiles and acidification rates vary across the farm will assist effective soil acidity management.
- Ideally, soil samples should be taken when soils are dry and have minimal soil biological activity. It is standard to measure pH using one part soil to five parts 0.01 M $CaCl_2$. Soils with low total salts show large seasonal variation in pH if it is measured in water. pH measured in water can read 0.6 – 1.2 pH units higher than in calcium chloride.
- Soil sampling should take paddock variability into consideration. For example, clays have greater capacity to resist pH change (buffering) than loams, which are better buffered than sands. Samples should be taken at the surface (0-15cm) and in the subsurface (15-30 or 45cm) to determine a soil pH profile. This will detect subsurface acidity, which may underlie top soils with an optimal pH.
- Samples need to be properly located (*e.g.* GPS) to allow monitoring. Sampling should be repeated every 3 – 4 years to detect changes and allow adjustment of management practices.

Interpreting pH Results

- Depending on soil pH test results, agricultural lime may need to be applied to maintain pH, or to recover pH to an appropriate level. If the topsoil pH is above 5.5 and the subsurface pH above 4.8, only maintenance levels of liming will be required to counter on-going acidification caused by productive agriculture.
- If the topsoil pH is below 5.5, recovery liming is recommended. Keeping the topsoil above 5.5 will treat the on-going acidification due to farming and ensure sufficient alkalinity can move down and treat subsurface acidity.
- Liming is necessary if the subsurface pH is below 4.8, whether or not the topsoil is acidic. If the 10 – 20 cm layer is below 4.8 but the 20 – 30 cm layer above 4.8, liming is still required. In this case the band of acidic soil will restrict root access to the more suitable soil below.

Liming

- Liming is the most economical method of ameliorating soil acidity. The amount of lime required will depend on the soil pH profile, lime quality, soil type, farming system and rainfall.
- Lime sand, from coastal dunes, crushed limestone and dolomitic limestone are the main sources of agricultural lime. Carbonate from calcium carbonate and magnesium carbonate is the component in all of these sources that neutralises acid in soil.
- The key factors in lime quality are neutralising value and particle size. The neutralising value of the lime is expressed as a percentage of pure calcium carbonate which is given a value of 100 per cent. With a higher neutralising value, less lime can be used, or more area treated, for the same pH change.
- Lime with a higher proportion of small particles will react quicker to neutralise acid in the soil, which is beneficial when liming to recover acidic soil.

Difference between Agricultural Lime and Other Lime

- Limes and, limestone and dolomitic lime (dolomite) are the most common agricultural limes. They are natural, mined products comprised mostly of calcium carbonate with some magnesium carbonate.
- Burnt lime or quicklime is produced by heating (burning) calcium carbonate materials, such as limestone, to above 825°C, causing chemical changes and forming calcium oxide. Hydrated or slaked lime is produced by adding water to calcium oxide, forming calcium hydroxide.
- Calcium oxide and calcium hydroxide are not recommended for use as agricultural lime because they are difficult to handle and store. They are caustic and can burn skin if contacted and cause respiratory problems if inhaled. Burnt lime is unstable and will revert to calcium carbonate if stored exposed to air.

Management of Soil Acidity

Land managers can manage soil acidity by raising the pH to a desired value through several methods:

- **Flooding**: In lowlands systems, flooding may be an effective technique in raising the pH of the soil. However, this effect is only good for the time for which the soil is flooded. Flooded or paddy mineral soils are 'self-liming'. When they are flooded and become anaerobic (lack of oxygen in the soil atmosphere) for a period of time, the pH rises toward neutrality even when the soil pH was originally acidic. If the soil is subsequently drained and becomes more aerobic (more oxygen in the soil atmosphere), the pH will return to an acidic state

- ❖ However, care must be taken if the soil contains manganese-oxide minerals, since flooding conditions may lead to manganese toxicity.
- ❖ Crop consideration is also required. Flooding conditions create oxygen deficiency within the soil. As a result, crops that do not tolerate high amounts of water and low oxygen levels are not be suited for flooded conditions. Taro and rice are examples of crops that grow well in flooded lowlands.

☆ **Additions of organic matter**: Additions of organic matter is a viable option to manage problems associated with soil acidity.

- ❖ Organic matter increases the cation exchange capacity of the soil. As the base saturation increases, the relative amount of "acidic cations" decreases.
- ❖ In addition, organic matter forms strong bonds, known as "chelates," with aluminium. Chelation reduces the solubility of aluminium and soil acidity. Again, if your soil is prone to manganese toxicity, it is not suggested to add organic matter.

☆ **Additions of wood ash**: Like organic matter, wood ash increases base saturation and forms chelates with aluminium.

☆ **Conventional Liming**: Various liming materials may be added to the soil that neutralize, or counteract, soil acidity. Liming materials are bases that react with hydrogen ions in the soil solution to form water.

- ❖ Examples of common liming materials are limestone (calcium carbonate), dolomite (calcium/magnesium carbonate), hydrated lime (calcium hydroxide), and quicklime (calcium oxide). Calcium and magnesium silicates are also used as liming agents.

Some Useful tips

☆ Soil pH is measured in either water or in calcium chloride. When measured in calcium chloride, the result is lower than pH measured in water.

☆ The pHW may be higher by 0.6 to 1.2 in low salinity soils and higher by 0.1 to 0.5 in high salinity soils. Research has shown a difference of 0.7 for a wide range of soils.

☆ Soil testing will tell you the current acidity status of your paddock. If your soil pHCa is above 5.5 then there is little immediate risk of acidity.

☆ Lime can restore productivity in acid soils and should be considered once the pH drops below pHCa 5.0 if sensitive species are to be grown successfully.

☆ You are unlikely to get responses to lime if other nutrients are lacking. This should show up in a soil test or plant tissue analysis and should be corrected. Conversely, you may not get a response to some nutrients if the soils are too acid. A holistic balanced approached is necessary.

- ☆ Lime responses are generally seen in the first and second year for cropping systems, but can take up to five years depending on soil type, rainfall and lime quality for permanent pasture systems.
- ☆ It is necessary to re-lime your paddock about every 10 years, depending on the rate of re-acidification.
- ☆ If paddocks with an acidity problem are not limed, the soil pH will continue to fall and settle at pHCa3.8 to 4.2.
- ☆ The amount of lime you need to apply varies according to soil type. Field experiments have shown that up to 5 tonnes a hectare on clay loams and 1.5 tonnes a hectare on sandy soils is needed to increase pH by one unit.
- ☆ Lime moves slowly (0.5 to 1cm per year) through the soil profile via the soil macropore structure. Incorporation into the soil profile, where possible, will assist effective treatment.

In permanent pasture situations, spreading the lime on the surface and allowing it to work its way into the soil is acceptable. Surface application is better than no application.

Chapter 9

Chemistry of Salt Affected Soils

Introduction

- Salt-affected soils are present on all continents and in nearly all climatic circumstances. Their distribution is, however, comparatively more extensive in dry and semi-arid regions than in wet regions.
- Salt-affected soils mostly manifest as saline soils, sodic soils, and saline-sodic soils.
- Saline and sodic soils impede the capacity of plant roots to absorb water and compromise soil structure by disaggregating and dispersing soil particles. This may adversely impact seed germination and root development. These disorders represent distinct soil concerns that result in various symptoms and complications. Nonetheless, the management of saline and sodic soils necessitates only marginally distinct techniques.
- Saline and sodic soils exhibit elevated pH levels, with issues generally arising at a pH of 7.8 or under highly alkaline circumstances. The fundamental distinction is that salty soils possess elevated salt levels, whereas sodic soils contain high sodium concentrations.
- Soils can simultaneously include elevated levels of both salt and sodium. In the context of saline soils, salt content pertains to the soluble salts present in the soil, including sodium, calcium, and magnesium chlorides, along with carbonate salts. These levels are sufficient to impede plant growth and induce plant mortality. When referring to sodic soils, the level of sodium in the soil dominates the other soluble salts.

☆ Soils with a higher concentration of soluble salts negatively impact plant growth. The salinity of saline soils surpasses a threshold of 4.0 dSm^{-1}. These soils predominantly contain chloride and sulfate salts. These salts are neutral, hence the pH of these soils may not exceed 8.5. Saline soils are generated through a pedogenic process known as salinization in semi-arid and dry regions. Salinization denotes the deposition of soluble salts in the upper layers of soil.

Salt Affected Soils: Definition and Classification

The phrase salt-affected soil is increasingly utilized to encompass saline, saline-sodic, and sodic soils. Table 9.1 presents the detailed classification of salt-affected soils based on pH, electrical conductivity (EC), and exchangeable sodium percentage (ESP).

Table 9.1

Soil Classification	*EC (dSm^{-1})*	*ESP*	*pH*
Saline	> 4.0	< 15	< 8.5
Sodic	< 4.0	> 15	> 8.5
Saline – sodic	> 4.0	>15	< 8.5 or > 8.5

Saline Soils

Saline soils are characterized by pH levels typically below 8.5, electrical conductivity (ECe) of 4 dS m^{-1}, and an exchangeable sodium percentage (ESP) of less than 15. A high electrical conductivity with a low exchangeable sodium percentage typically causes soil particles to flocculate into aggregates. The soils are typically identified by the presence of a white salt crust at certain times of the year. Permeability is either more than or equal to that of comparable 'normal' soils.

Saline-Sodic Soils

Saline-sodic soils possess adequate soluble salts (ECe > 4 dS m^{-1}) that hinder the growth of most crops and have a significant exchangeable sodium percentage (ESP ≥ 15) that negatively impacts soil attributes and plant growth, mostly through the deterioration of soil structure. The pHs may be either below or above 8.5.

Sodic Soils

Sodic soils display an Exchangeable Sodium Percentage (ESP) of 15 and have an Electrical Conductivity (ECe) of less than 4 dS m^{-1}. The pH typically ranges from 8.5 to 10, and may go as high as 11. Low electrical conductivity and high exchangeable sodium percentage tend to de-flocculate soil aggregates, hence reducing their permeability to water.

Genesis of Salt Affected Soils

- **Parent material**: Soils derived from rocks with a high concentration of bases become saline or sodic due to the release of salts into the soil environment as a result of weathering. For example, basalt, sandstone, etc.
- **Low rainfall***:* One significant explanation for the formation of saline-sodic soils is inadequate water to leach bases from the soil horizon, resulting in salt accumulation in the soil. This phenomenon is more prevalent in semi-arid and desert locations characterized by little rainfall.
- **High Evaporation***:* Water and salts ascend to the surface from the subsurface of the soil by capillary action, driven by elevated evaporation rates in arid and semi-arid climates. This leads to the accumulation of salt on the soil surface, whereas water alone evaporates into the atmosphere.
- **Poor drainage***:* Waterlogged salinity and sodicity are commonly observed in low-lying inland areas, particularly in heavy clay soils. Inadequate drainage results in salt accumulation at the surface horizon, facilitating sodium infiltration into the clay complex.
- **Poor quality irrigation waters**: The persistent utilization of low-quality saline/sodic water for agriculture results in the accumulation of salts and sodium in the soil.
- **High water table:** The elevated water table in alluvial plains and similar regions results in inadequate drainage, causing salt accumulation in the soil.
- **Sea water intrusion:** In coastal areas, seawater infiltrates the land, contaminating both the soil and the groundwater of the region.
- **Base forming fertilizers:** The ongoing use of base-forming fertilizers in agriculture contributes to soil salinity and sodicity. For example, sodium nitrate ($NaNO_3$).

Impact of Salinity on Agricultural Crops

- Alkaline conditions restrict the availability of nutrients like iron, phosphorus, and zinc, resulting in symptoms of saline soil in plant leaves, such as yellow stripes on the upper leaves due to zinc and iron deficiency, or dark green or purple discoloration of stems and lower leaves caused by phosphorus deficiency.
- Symptoms of drought, such wilting or yellowing of leaves, are prevalent in saline soils. A powdery substance may be observed on the soil, as the salts are distributed to the surface.

Impact of Sodicity on Agricultural Crops

- Plants cultivated in sodic soils may exhibit dehydration symptoms similar to those in saline soils; however, sodic conditions significantly diminish seedling viability and adversely impact germination.

- ☆ Sodic soils can produce a brownish-black crust due to the dispersion of organic materials. When this crust becomes apparent, the issue is typically acute, and plants will exhibit indications of distress.

Diagnosis of Saline or Sodic Soils

Although visual signs assist in identifying saline or sodic soils, a soil test is essential for precise diagnosis and remediation of the issue. Obtain a soil sample from the upper 6 to 12 inches and submit it to a laboratory for professional analysis and subsequent recommendations.

Table 9.2: Diagnosis of Elevated pH, Salinity, or Sodicity Issues

Problem	*Potential Symptoms*
High pH	Nutrient deficits present as stunted, yellowing plants. Plants exhibiting a color range from dark green to purplish hues.
Saline soil	White crust on the earth surface. Plants experiencing water stress. Burning of leaf tips.
Saline irrigation water	Foliar necrosis. Suboptimal growth. Hydric stress.
Sodic soil	Inadequate drainage. Black powdery deposit on the soil surface.
Saline-sodic soil	Typically, identical symptoms as saline soil.

- ☆ *Basic – High pH. Problems usually start at $<$ 7.8 (alkaline).*
- ☆ *Saline – High salt content.*
- ☆ *Sodic – High sodium content.*
- ☆ *Saline/sodic – High salt and high sodium content*

Impacts of Soil Salinity

- ☆ The defining characteristic of saline soil is the white encrustation on the soil surface, resulting from the evaporation of water into the atmosphere, which leaves salts deposited on the surface.
- ☆ The presence of salts alters the osmotic potential of the soil solution. Consequently, water absorption by plants is limited, resulting in a diminished uptake of nutrients.
- ☆ In these soils, elevated salt concentrations diminish microbial activity, resulting in sluggish breakdown and limited nutrient availability, particularly nitrogen and sulfur.
- ☆ Osmotic potential changes cause water to migrate from plant cells to the soil, resulting in dehydration of the plants. This leads to the desiccation of leaves and ultimately the demise of plants.

- In addition to the aforementioned impacts, specific ion effects on plants are also observed due to the toxicity of ions such as chloride and sulfate.
- All saline soils can be readily reclaimed if high-quality water is accessible. The soluble salts in these soils can be dissolved and leached from the field using clean water.
- In the absence of high-quality water, it is essential to manage salt soils to enhance plant growth.

Impacts of Sodicity

- Due to the elevated levels of CO_2^{-3}, HCO^-_3, and high pH, damage to plants caused by carbonate, bicarbonate, and hydroxyl ions are detected. The elevated salt content in clay contributes to its scattered characteristics in areas with high moisture.
- The dispersed characteristics of clay result in a slippery soil texture, water stagnation, inadequate infiltration and percolation, and insufficient aeration. Sodic soils solidify into a hard mass during arid conditions.These soils exhibit inadequate workability in both wet and dry conditions.
- Additional detrimental effects of sodium on plants are also observed. Sodium carbonate in aqueous solution dissociates into Na^+, HCO_3^-, and OH– ions, which are detrimental to plant growth.

  $2Na^+ + CO_3^{2-} + H_2O \rightarrow 2\ Na^+ + HCO_3^- + OH^-$
- Soils enriched with high sodium in the exchange complex exhibit reduced availability of calcium and magnesium. Elevated pH results in the reduced availability of Fe and Zn.
- Zn deficiency is prevalent in certain soils. The availability of phosphorus is diminished due to its conversion into insoluble calcium and magnesium phosphates.
- All the aforementioned effects on plants lead to localized desiccation within a field. No vegetation is observed in these soils under harsh conditions.

The management of salinity and sodicity becomes more complex when both are present in the same soil. A saline-sodic soil typically does not display pronounced sodicity symptoms. As soil salinity diminishes due to enhanced drainage or management, calcium and magnesium salts are preferentially leached, while sodium is leached more slowly and ultimately predominates in the soil. The management will closely resemble that of sodic soils. Similar to sodic soils, the remediation of saline-sodic soils necessitates an additional procedure involving the use of calcium (Ca^{++}) supplements, typically gypsum, succeeded by salinity remediation techniques aimed at enhancing soil drainage and reducing the water table level.

Management of Salt Affected Soils

☆ Management of saline soil entails the "leaching" of soluble salts from the soil profile. Leaching necessitates the application of irrigation water at levels beyond the fundamental water requirements of the crop. Leaching occurs through the percolation of soil solutes (soluble salts) in saturated soil conditions.

☆ Management of sodic soils typically necessitates a two-step procedure:

- ❖ Substituting exchangeable sodium with a more advantageous ion such as calcium (Ca) and/or magnesium (Mg)
- ❖ Eluting the soluble sodium that has been substituted on the soil colloid through the application of surplus irrigation water. Consequently, saline soils do not necessitate additions as such.

☆ Sodic soils may necessitate the application of soil amendments, such as gypsum ($CaSO_4$), as a source of Ca^{2+}. Effective management of either situation necessitates accurate diagnosis of soil characteristics, along with proficient management of the irrigation water utilized in the respective field.

☆ The quality of the irrigation water utilized frequently serves as a principal source of the soluble salts introduced to the land. Irrigation waters with electrical conductivity (ECw) values ranging from 0.7 to 3.0 dS/m may provide minor usage limits, whereas waters with ECw values beyond 3.0 dS/m are linked to significant limitations.

Table 9.3: Classification of Saline and Sodium-Impacted Soils

Criterion	*Normal*	*Saline*	*Sodic*	*Saline-Sodic*
EC_e (dS/m)	<4	>4	<4	>4
SAR_e	<13	<13	>13	>13

Table 9.4: Classification of Saline and Sodium-Affected Soils

Crop	*EC (ds/m, 25 °C) at which Yield will be Decreased by*		
	10 per cent	*25 per cent*	*50 per cent*
Lettuce (*Latica saliva* L.)	2	3	5
Broccoli (*Brassica oieracea* L.)	4	6	8
Cotton (*Gossypium hirsutum* L.)	10	12	16
Barley (*Hordeum vulgare* L.)	12	16	18

Adapted from L. Bernstein, Salt Tolerance of Plants, USDA Bull. 283, 1964 and E. V Maas, 1984.

☆ The management of salinity and sodicity becomes more complex when both are present in the same soil. A saline-sodic soil typically does not have pronounced sodicity characteristics. As soil salinity diminishes due to enhanced drainage or management, calcium and magnesium salts are preferentially leached, while sodium is leached more slowly and ultimately predominates in the soil.PThe management will closely resemble that of sodic soils. Similar to sodic soils, the remediation of saline-sodic soils necessitates an additional procedure involving the use of calcium (Ca^{++}) supplements, typically gypsum, succeeded by salinity remediation techniques aimed at enhancing soil drainage and reducing the water table level.

A. Administration of Saline Soils

i. Crop Management

Cultivating crops that exhibit tolerance to elevated soil salt levels, such as cotton, ragi, barley, sugar beet, beetroot, curry leaf, Bermuda grass, saline grass, and spinach. Crops that exhibit tolerance to medium levels of soil salt include paddy, wheat, onion, maize, sunflower, castor, grape, pomegranate, tomato, cabbage, and potato. Crops that exhibit tolerance to low levels of soil salinity include garden beans, radishes, and limes. Black gram and green gram exhibit sensitivity to soil salinity. Crops should be selected according to the soil salinity level. The comparative salinity tolerance of several crops is presented in Table 9.5.

Table 9.5: Comparative Salinity Tolerance of Crops

Plant Species	*Threshold Salinity ($dS\ m^{-1}$)*	*Plant Species*	*Threshold Salinity ($dS\ m^{-1}$)*
Field crops			
Cotton	7.7	Sugarbeet	7.0
Sorghum	6.8	Wheat	6.0
Soybean	5.0	Groundnut	3.2
Rice	3.0	Maize	1.7
Sugarcane	1.7		
Vegetables			
Tomato	2.5	Cabbage	1.8
Potato	1.7	Onion	1.2
Carrot	1.0		
Fruits			
Citrus	1.7		

ii. Soil/Cultural Management

Cultivating crops in elevated beds will diminish the buildup of salt in the root zone. Planting seedlings or spreading seeds on sloped ridges reduces salt collection in the root zone. Mulching soil inhibits evaporation, hence diminishing the accumulation of salts resulting from the capillary rise of water at the soil surface. Implementing drainage in waterlogged regions also mitigates salt accumulation.

iii. Fertilizer Management

The incorporation of an additional nitrogen dose, up to 20-25 percent of the optimum level, will mitigate the low nitrogen availability in these soils. The incorporation of organic fertilizers such as farmyard manure and compost mitigates the detrimental effects of salt by releasing organic acids during decomposition. Green manuring (such as sun hemp, Dhaincha, and Kolingi) and/or green leaf manuring mitigates the impacts of salinity.

iv. Irrigation Management

The proportional blending of high-quality water, when accessible, with saline water for irrigation mitigates the impact of salinity. Alternate furrow irrigation promotes plant development more effectively than flooding. Drip and sprinkler irrigation systems seek to minimize water use, which benefits plant growth by decreasing salt accumulation associated with lower water usage.

The aforementioned four management approaches are effectively combined to mitigate soil salinity, hence promoting enhanced plant growth and ultimately improving yields. Management of saline soils is vital and unavoidable, especially in regions where both the soil and irrigation water are saline.

B. Management of Sodic Soils

i. Use of Chemical Amendments

- ✰ Gypsum, calcium chloride, calcium carbonates, and phospho-gypsum are utilized for reclamation, as they directly deliver calcium according to their constitution. Gypsum is the most often utilized among them. Calcium carbonate ($CaCO_3$) is insoluble and therefore ineffective in sodic soils that already contain precipitated $CaCO_3$. However, it can be utilized in degraded sodic soils that lack precipitated $CaCO_3$, as the low pH of these soils promotes the solubilization of $CaCO_3$.
- ✰ Certain indirect přoviders of calcium, such as elemental sulfur, sulfuric acid, and iron sulfate, are also utilized for sodic soils. These substances facilitate the solubilization of precipitated $CaCO_3$ in sodic soils, hence releasing Ca for reclamation through exchange reactions. The absence of precipitated $CaCO_3$ in degraded sodic soils renders the utilization of these materials unadvantageous. Lime sulfur supplies calcium both directly and indirectly.

Reactions (Direct Sources)

$$^{Na+}\text{Micelle}^{Na+} + CaSO_4 \longrightarrow \text{Micelle}^{Ca++} + Na_2SO_4 \text{ (leachable)}$$

$$^{Na+}\text{Micelle}^{Na+} + CaCl_2 \longrightarrow \text{Micelle}^{Ca++} + 2NaCl \text{ (leachable)}$$

Indirect Sources

Sulfur and iron sulfate are transformed into sulfuric acid, which solubilizes native $CaCO_3$. Consequently, Ca is rendered available for exchange reactions.

Sulphur – microbial oxidation

$$2\,S + 3\,O_2 \longrightarrow 2\,SO_3$$

$$SO_3 + H_2O \longrightarrow H_2SO_4$$

$$H_2SO_4 + 2\,CaCO_3 \longrightarrow CaSO_4 + Ca\,(HCO_3)_2$$

$$2Na - X + CaSO_4 \longrightarrow Ca\text{-}X + Na_2SO_4$$

ii. Cultivation of Sodicity-Tolerant Crops

The selection of appropriate crops for the reclamation of sodic soils is crucial. Developing crops that are tolerant to elevated exchangeable sodium can secure satisfactory yields during the early stages of reclamation or when irrigated with water that poses a sodicity risk. The sodicity tolerance thresholds for several crops are presented in Tables 9.6 and 9.7.

Table 9.6: Comparative Tolerance of Crops to Sodicity

ESP (Range)*	*Crop*
2-10	Deciduous fruits, nuts, citrus, avocado
10-15	Safflower, black gram, peas, lentil, pigeon pea
16-20	Chickpea, soybean
20-25	Clover, groundnut, cowpea, pearl millet
25-30	Linseed, garlic, cluster bean
30-50	Oats, mustard, cotton, wheat, tomatoes
50-60	Beets, barley, sesbania
60-70	Rice

*Relative yields constitute merely 50 percent of the potential within the different sodicity bands.

Table 9.7: Comparative Tolerance of Fruit Trees to Sodicity

Tolerance to Sodicity	*ESP*	*Trees*
High	40-50	Ber, tamarind, sapota, wood apple, date palm
Medium	30-40	Pomegranate
Low	20-30	Guava, lemon, grape
Sensitive	20	Mango, jack fruit, banana

- ☆ Grasses appropriate for sodic soils include Cynodon dactylon, Brachiaria mutica, and Pai korai.
- ☆ Fodder species appropriate for sodic soils include *Medicago sativa*, *Trifolium aleandrum*, *Avena sativa*, and *Medicago denticulate*.

- Trees appropriate for sodic soils include *Azadirachta indica, Prosopis juliflora, Tamarind, Albizzia, Zizyphus* species, *Acacia* species, and *Cassia* species, among others.

iii. Use of Organic Amendments

Organic amendments utilized in soil reclamation originate from several sources, including agriculture, forestry, and urban environments. Among agricultural byproducts, livestock manure—whether fresh, composted, or derived from solid fractions of anaerobic digesters—originating from diverse species such as cattle, hogs, and poultry, is the most predominant. Additional amendments originating from agriculture encompass agricultural leftovers (such as straw and legumes) and spent mushroom compost. Forestry generates organic amendments like deinking sludges, wood chips, and shavings. Organic additions influence soil qualities in many and multifaceted manners. Effects may be direct, stemming from the inherent characteristics of the organic amendments, or indirect, by altering the physical, biological, and chemical aspects of the soil. Soil salinity inhibits plant growth due to elevated salt concentrations, resulting in osmotic stress in the root zone. The utilization of green biomass enhances the release of salts into the soil solution due to mineral dissolution, driven by elevated partial pressure of carbon dioxide and organic acids, resulting in the leaching of salts beneath the root zone and fostering a conducive environment for crop plant roots. The synthesis of organic acids (amino acids, glycine, cysteine, and humic acid) during the mineralization of organic matter by heterotrophs and nitrification by autotrophs results in a reduction in soil pH. The utilized green manures generate CO_2 during decomposition, which dissolves in water to form carbonic acid. This acid enhances the solubility of calcium carbonate minerals by decreasing the pH, dissolving calcium carbonate, and generating various complex calcium ion pairs, thereby elevating the concentration of Ca^{2+} in the soil solution, which displaces Na+ on the exchange complex and consequently reduces the exchangeable sodium percentage (ESP). The utilization of green manures diminishes the concentration of specific cations and anions, including Cl^-, SO_4^{2-}, HCO_3^-, and CO_3^{2-}.

References

1. Agarwal, R.R., J.S.P. Yadav and R.N. Gupta. 1982. Saline and alkali soils of India. Indan Council of Agricultural Research, New Delhi.
2. Bandopadhyay, A.K. 1988. Coastal Saline Soils of India and their management. Bull. Cent. Soil Salinity Res. Inst. 13, CSSRI Regional Research Station, West Bengal and Karnal, Haryana.
3. Bhargava, G.P. 1989. Salt affected soils of India-A source Book, Oxford and IBH Publishing Co., New Delhi.
4. T.D. Biswas and S.K. Mukherjee. 1995. Text book of Soil Science. Tata McGraw-Hill Publishing Company Limited, New Delhi.

Chapter 10

Chemistry of Soil Organic Matter

Introduction

- Soils are classified as mineral or organic based on their organic matter concentration. Mineral soils constitute the majority of the world's arable land and may comprise between a trace and 30 percent organic stuff. Organic soils are inherently abundant in organic materials mostly due to climatic factors. Despite containing above 30 percent organic matter, this characteristic renders them non-essential as farming soils.
- Soil organic matter refers to any material generated by living organisms (flora or fauna) that is reintroduced to the soil and undergoes breakdown (Figure 10.1). At each moment, it comprises a variety of substances, from the intact original tissues of flora and fauna to the significantly disintegrated amalgamation of components referred to as humus (Figure 10.1).antially decomposed mixture of materials known as humus (Figure 10.1).

The term "Soil organic matter" (SOM) has been used in different ways to describe the organic constituents of soil. In this report, SOM will be used as defined by Baldock and Skjemstad (1999) as "all organic materials found in soils irrespective of origin or state of decomposition". Since SOM consists of C, H, O, N, P and S, it is difficult to actually measure the SOM content and most of the analytical methods determine the soil organic carbon (SOC) content and estimate SOM through a conversion factor.

Figure 10.1: Crop residues incorporated into the soil undergo decomposition by soil macrofauna and microorganisms, hence enhancing the organic matter content of the soil.

- Soil organic matter (SOM) is a component of soil comprising plant and animal wastes at different stages of decomposition. Organic matter comprises organic carbon and nitrogen. Carbon serves as an energy source, while nitrogen provides protein for soil microbes. Certain microbes are pathogens that induce plant diseases; nevertheless, in healthy soil, the majority of these organisms are advantageous and assist in preventing dangerous entities, such as plant infections, from becoming predominant. The SOM comprises three distinct components. Living organic matter comprises around 15 percent and is predominantly composed of decomposers:

Microbial Group	*Population per gram of soil*
Bacteria	300,000 – 200,000,000
Actinomycetes	100,000 – 100,000,000
Fungi	20,000 - 1,000,000
Protozoa	10,000 - 100,000
Algae	100 - 50,000

Additional living soil organic matter comprises nematodes, insects, earthworms, plant roots, and small mammals.

- **Dead organic matter (about 15 per cent)**: function as sustenance for live creatures and encompasses deceased microorganisms, aged plant roots, agricultural wastes, and the remains of larger insects and animals.

Soil health is fundamentally dependent on organic matter. Soil organic matter enhances various physical, chemical, and biological properties of the soil, such as water retention capacity, cation exchange capacity, pH buffering capacity, and the chelation of micronutrients. Moreover, well-decomposed SOM promotes soil structure by promoting aggregation, boosts biological activity in the soil, gradually releases nutrients, and mitigates certain diseases. The depletion of soil organic matter (SOM) can result in soil erosion, diminished fertility, compaction, and overall land degradation.

- **Highly decomposed organic matter**, comprising around 70 percent, is referred to as humus, characterized by its dark coloration. Humus persists in decomposing, but at an exceedingly gradual pace.

Factors that Influences Amount of SOM

- The typical soil organic matter (SOM) content in Massachusetts soils varies from 1 to 5 percent, with a minimum of 4 percent being preferable. The preservation and improvement of soil organic matter are essential for soil health and the sustainability of agricultural systems.
- The buildup of soil organic matter (SOM) is a balance between the input of plant and animal leftovers and their subsequent loss through microbial decomposition and soil mismanagement. Any factor influencing soil microbial activity also impacts the decomposition of soil organic matter (SOM).
    - **Soil temperature** significantly affects microbial activity. The ideal soil temperature for the majority of bacterial activity ranges from 70 to 100 °F, however some activity may occur as low as 40 °F, albeit at significantly diminished rates.
    - **Soil bacteria** want oxygen and water for respiration; when soil is compacted or waterlogged, respiration diminishes, thereby impeding the decomposition of soil organic matter (SOM).
    - **Soil pH**: Under acidic conditions, bacterial activity, which is primarily responsible for the breakdown of organic matter, is significantly diminished.
    - **Soil fungi** that facilitate the decomposition of soil organic matter are typically less impacted by low pH levels.

Soil Organic Matter: Components and their Function

- The majority of soil organic matter derives from plant tissues. Plant residues include a moisture content ranging from 60 to 90 percent. The residual dry matter comprises carbon (C), oxygen (O), hydrogen (H), and trace quantities of sulfur (S), nitrogen (N), phosphorus (P), potassium (K), calcium (Ca), and magnesium (Mg). Despite their minimal presence, these minerals are crucial for effective soil fertility control.

- Soil organic matter comprises several components. This comprises, in differing proportions and numerous intermediary stages, an active organic fraction containing microorganisms (10-40 percent) and resistant or stable organic matter (40-60 percent), commonly known as humus.

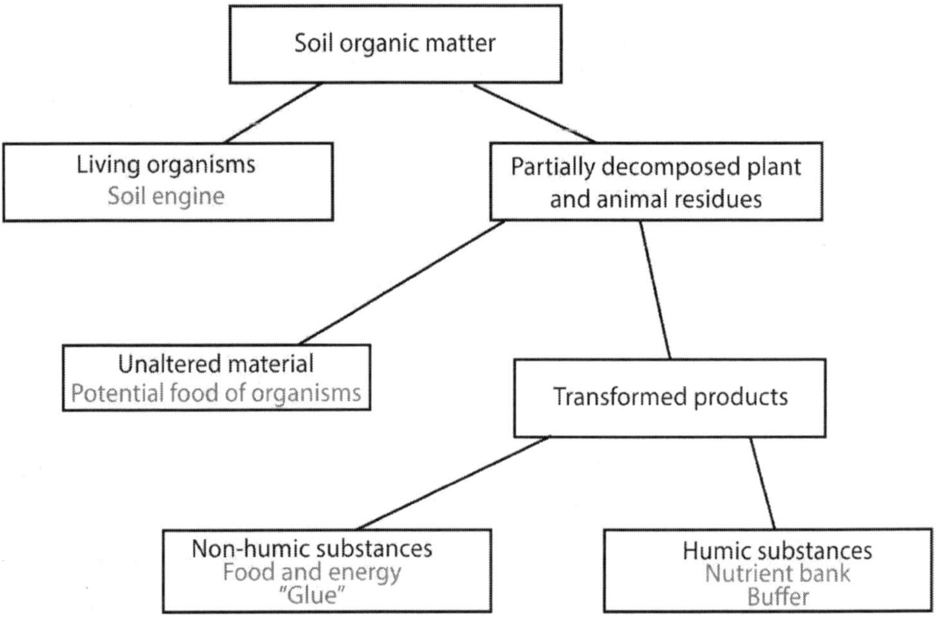

Figure 10.2: Components of Soil Organic Matter and their Functions.

- The forms and classification of soil organic matter have been delineated by Tate (1987) and Theng (1987).
- For practical purposes, organic matter can be categorized into aboveground and belowground portions. Aboveground organic matter includes plant and animal residues, while belowground organic matter consists of living soil fauna and microflora, partially decomposed plant and animal residues, and humic compounds.
- The carbon-to-nitrogen (C:N) ratio serves as an indicator of material type and decomposition rate; hard, woody materials with a high C:N ratio exhibit more resilience compared to soft, leafy materials with a low C:N ratio.While soil organic matter can be efficiently divided into several fractions, these do not constitute static end products. The quantities observed indicate a dynamic equilibrium.
- The overall quantity and distribution of organic matter in the soil are affected by soil characteristics and the volume of annual contributions from plant and animal leftovers to the ecosystem.

- The decomposition rate and buildup of soil organic matter in a certain soil ecosystem are influenced by factors such as texture, pH, temperature, moisture, aeration, clay mineralogy, and biological activities within the soil. A difficulty is that soil organic matter also affects numerous soil characteristics.
- Raw plant remains on the soil surface serve to shield the soil from the impacts of rainfall, wind, and sunlight. The removal, integration, or incineration of leftovers subjects the soil to adverse climatic effects, while removal or incineration deprives soil organisms of their essential energy supply.
- Soil organic matter performs multiple functions. From a pragmatic agricultural perspective, it is significant for two primary reasons: (i) as a "revolving nutrient fund"; and (ii) as a means to enhance soil structure, preserve tilth, and reduce erosion.
- Organic matter operates primarily as a cyclical nutrient reservoir.
    - Soil organic matter, primarily originating from plant remains, encompasses all required plant nutrients. Consequently, accumulating organic matter serves as a reservoir of plant nutrients.
    - The stable organic portion (humus) adsorbs and retains nutrients in a form accessible to plants.
- Organic matter liberates nutrients in a form accessible to plants during decomposition. To sustain this nutrient cycle system, the rate of organic matter input from crop residues, manure, and other sources must match the rate of decomposition, while considering plant uptake and losses due to leaching and erosion.
- When the rate of addition is inferior to the rate of decomposition, soil organic matter diminishes. Conversely, when the rate of accumulation exceeds the rate of decomposition, soil organic matter proliferates. The term steady state refers to a circumstance in which the pace of accumulation equals the rate of decay.
- The enhancement of soil structure involves active and certain resistant organic components, together with microorganisms, particularly fungus, which contribute to the aggregation of soil particles into bigger clusters. Aggregation is essential for optimal soil structure, aeration, water infiltration, and resistance to erosion and crusting.
- Historically, soil aggregation has been associated with either total carbon or organic carbon concentrations. Recent advancements have emerged in approaches to fractionate carbon based on lability (easy of oxidation), acknowledging that these carbon sub-pools may significantly influence soil physical stability and serve as more sensitive indicators of carbon dynamics in agricultural systems than total carbon values. The labile

carbon fraction serves as an indicator of essential soil chemical and physical qualities. This proportion is identified as the principal factor influencing aggregate disintegration in Ferrosols (non-cracking red clays), as determined by the percentage of aggregates smaller than 0.125 mm in the surface crust following simulated rainfall in a laboratory setting.

- ☆ The resistant or stable component of soil organic matter primarily contributes to nutrient retention capacity (cation exchange capacity [CEC]) and soil pigmentation. This portion of organic materials decomposes at a sluggish rate. Consequently, it exerts a diminished impact on soil fertility compared to the active organic component.

Function of SOM in Soils

- ☆ The functions of soil organic matter (SOM) can be categorized into three primary groups: biological, physical, and chemical. These groups are not fixed entities; dynamic interactions transpire among these three principal components.

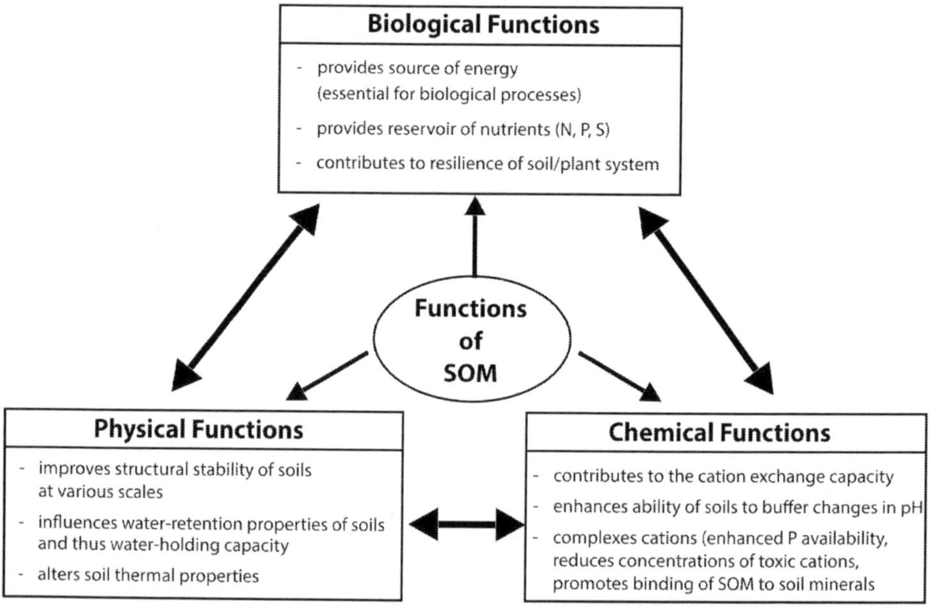

Figure 10.3: Functions Ascribed to SOM (Note that interactions occur between the different soil functions modified from Baldock and Skjemstad, 1999).

- ☆ The interconnections among soil functions, the distinct requirements for optimal soil organic matter (SOM) levels for each function, and the unique mineralogical features of soil prevent the establishment of a universal optimal SOM level.
- ☆ SOC requirements are expected to vary according on function and soil type. For instance, in sandy soils, the significance of soil organic carbon

(SOC) is greater than in clayey soils regarding cation exchange capacity (CEC). SOC is needed in greater quantities in sandy soils, as clayey soils can supply a significant share of cation exchange capacity through charges from clay minerals. Soil organic carbon (SOC) is essential for biological processes and nutrition provision, as well as for thermal characteristics, regardless of clay concentration.

- Baldock and Skjemstad (1999) and Skjemstad (2002) indicated that total soil organic carbon (SOC) may not serve as an effective metric for evaluating the performance of specific soil functions, primarily due to the significant variability in the physical and chemical properties of the numerous pools constituting the bulk SOC. The humic fraction is regarded as the primary contributor to the soil's cation exchange capacity (CEC), while both the humic and particulate organic carbon (POC) fractions are responsible for providing and sustaining soil structure. In sandy soils, the POC component significantly contributes to the physical aggregation of particles.
- Soils with elevated clay concentration necessitate both humic materials and particulate organic carbon (POC) to achieve optimal structural support, both chemical and physical binding are essential. In contrast, POC is crucial for supplying energy to biological processes, while humus serves as a significant source of vital soil nutrients. The thermal qualities of soil, namely its capacity to rapidly warm in cold climes, are influenced by color, with the inert carbon pool, comprising highly fragrant materials like charcoal, being the most significant factor.
- Fertile and healthy soil is fundamental for the well-being of plants, animals, and humans. Soil organic matter is the fundamental basis for healthy and productive soils. Comprehending the significance of organic matter in sustaining healthy soil is crucial for formulating ecologically responsible agriculture operations. How can organic matter, comprising only a minor fraction of most soils, be significant enough to warrant three chapters in this part for discussion? Organic matter favorably affects or alters the qualities of nearly all soil characteristics. This underscores the need of comprehending soil health and optimizing soil management practices. Organic matter is fundamentally the core of the narrative, while it is not the sole component. Soil organic matter plays a vital role in numerous essential functions that enhance soil processes and crop development, and it is integral to various global and regional cycles.
- It is indeed possible to cultivate plants in soils with minimal organic matter. In reality, the presence of soil is entirely unnecessary. While gravel and sand hydroponic systems devoid of soil can produce exceptional crops, large-scale implementations of such systems are typically neither economically viable nor ecologically sustainable. Additionally, it is accurate to assert that there are other significant factors beyond

organic matter that influence soil quality. As soil organic matter diminishes, cultivating plants becomes progressively challenging due to the heightened prevalence of issues related to fertility, water availability, compaction, erosion, pests, diseases, and insects.

- Increasing quantities of inputs—fertilizers, irrigation water, herbicides, and machinery—are necessary to sustain yields amid the decrease of organic matter. However, with diligent management of organic matter, the soil can sustain a robust crop without necessitating costly interventions.
- The organic matter level of agricultural topsoil typically ranges from 1 to 6 percent. A study of Michigan soils indicated potential crop yield enhancements of approximately 12 percent for each 1 percent increase in organic matter. In a Maryland study, researchers observed an increase of almost 80 bushels of maize per acre when organic matter rose from 0.8 to 2 percent. The substantial impact of organic matter on many soil properties—biological, chemical, and physical—renders it essential for soil health. The influence can be attributed to the diminutive particle size of the well-decomposed organic materials, known as humus.
- The extensive surface area-to-volume ratio indicates that humus interacts with a significant percentage of the soil. The close interaction of humus with the surrounding soil facilitates the swift occurrence of several reactions, including the release of accessible nutrients into the soil water. Nonetheless, the functions of living creatures render soil life a crucial component of the organic matter narrative.

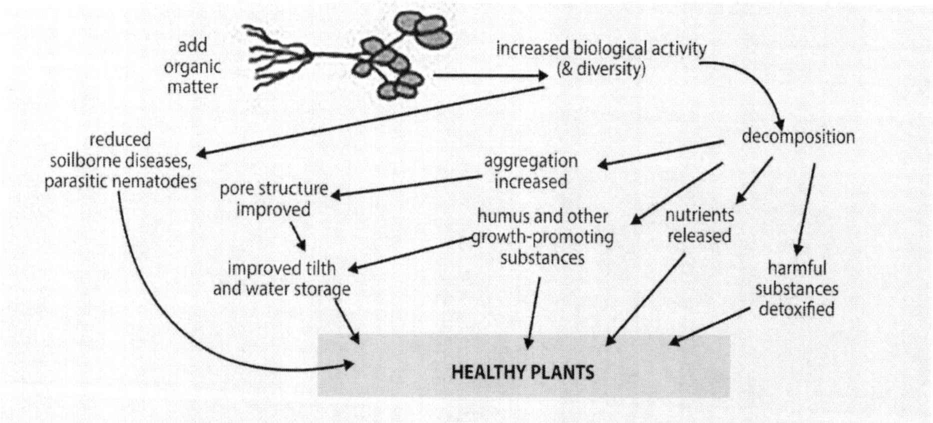

Figure 10.4: Adding Organic Matter Results in Several Alterations (Modified from Oshins and Drinkwater, 1999).

Plant Nutrition

- Plants require eighteen chemical elements for their growth: carbon (C), hydrogen (H), oxygen (O), nitrogen (N), phosphorus (P), potassium

(K), sulfur (S), calcium (Ca), magnesium (Mg), iron (Fe), manganese (Mn), boron (B), zinc (Zn), molybdenum (Mo), nickel (Ni), copper (Cu), cobalt (Co), and chlorine (Cl). Plants acquire carbon in the form of carbon dioxide (CO2) and oxygen mostly as oxygen gas (02) from the atmosphere. The remaining vital components are primarily sourced from the earth. The presence of organic matter directly or indirectly influences the availability of these nutrients. The elements required in substantial quantities—carbon, hydrogen, oxygen, nitrogen, phosphorus, potassium, calcium, magnesium, sulfur—are termed macronutrients. The remaining components, referred to as micronutrients, are vital substances required in minimal quantities. Sodium (Na) aids in the growth of some plants; nonetheless, it is not deemed essential for plant growth and reproduction.

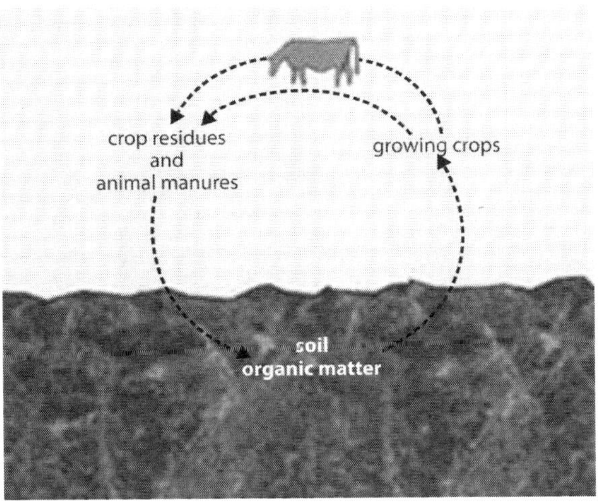

Figure 10.5: The Nutrient Cycle of Plants.

Nutrients from Decomposing Organic Matter

- The majority of nutrients in soil organic matter are unavailable to plants while they remain integrated within big organic molecules. Soil organisms degrade organic or mineral substances into forms readily accessible to plants. This process, known as mineralization, supplies a significant portion of the nitrogen required by plants by transforming it from organic forms.
- Proteins are transformed into ammonium (NH_4^+) and subsequently into nitrate (NO_3^-). The majority of plants absorbed most of their nitrogen from soils as nitrate.
- The mineralization of organic matter is a crucial process for providing plants with nutrients such as phosphorus, sulfur, and various micronutrients. The release of nutrients from organic matter through mineralization constitutes a component of a broader agricultural nutrient cycle.

Addition of Nitrogen

- ☆ Bacteria residing in the nodules of legume roots transform atmospheric nitrogen gas (N_2) into forms that the plant may utilize directly. Several free-living microorganisms also perform nitrogen fixation.

Storage of Nutrients in Soil Organic Matter

- ☆ Decomposing organic materials can directly nourish plants, while also indirectly enhancing their nutritional value. Numerous critical nutrients exist in soils as positively charged ions known as cations.

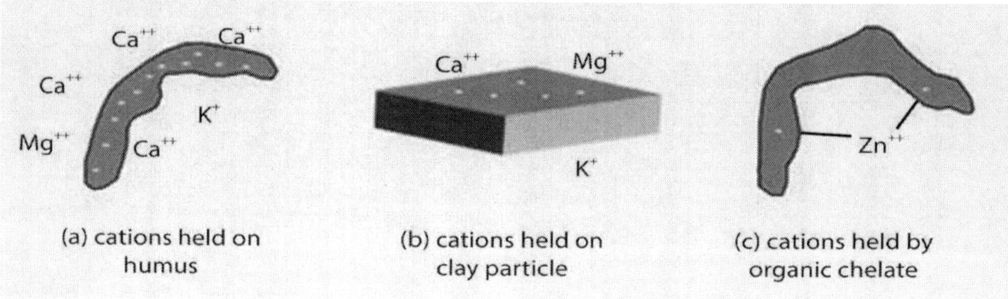

Figure 10.6: Cations are retained by negatively charged organic matter and clay.

- ☆ Cation exchange capacity (CEC) refers to the capacity of organic matter to retain cations, so making them accessible to plants. Humus possesses numerous negative charges. Due of the attraction between opposite charges, humus can retain positively charged nutrients, including calcium (Ca^{++}), potassium (K^{+}), and magnesium (Mg^{++}). This prevents them from percolating deeply into the subsurface as water traverses the topsoil. Nutrients retained in this manner can be progressively released into the soil solution and made accessible to plants over the growth season. Nonetheless, it is important to note that not all plant nutrients exist as cations. The nitrate form of nitrogen (NO_3^-) possesses a negative charge and is hence rejected by the negatively charged cation exchange capacity (CEC). Nitrate readily leaches as water percolates through the soil and surpasses the root zone.
- ☆ Clay particles possess negative charges on their surfaces; nonetheless, organic matter may serve as the primary source of negative charges in coarse and medium-textured soils. Certain clay varieties, particularly those located in the southeastern United States and tropical regions, exhibit minimal negative charge. In the presence of such clays, organic matter may serve as the primary source of negative charges that retain nutrients, even in fine-textured (high-clay-content) soils.

Protection of Nutrients by Chelation

- ☆ Organic molecules in the soil may retain and safeguard specific nutrients. These particles, termed "chelates" (pronounced key-lates), are by-products of the active breakdown of organic materials and are smaller than the constituents of humus. Generally, elements are more firmly retained by chelates than by the attraction of positive and negative charges. Chelates are effective due to their ability to bind the nutrient at several sites on the organic molecule.

- In certain soils, trace metals like iron, zinc, and manganese might be rendered inaccessible if not chelated. Low-organic-matter soils or exposed subsoils weak in certain micronutrients are frequently encountered.

Alternative Methods for Preserving Nutrient Availability

- Evidence suggests that organic matter in the soil may impede the transformation of available phosphorus into forms that are inaccessible to plants. One explanation is that organic matter envelops the surfaces of minerals that can adhere strongly to phosphorus. Once these surfaces are coated, the accessible forms of phosphorus are less prone to react with them.
- Furthermore, humic compounds can chelate aluminum and iron, both of which may interact with phosphorus in the soil solution. When these metals are chelated, they cannot form an insoluble mineral with phosphorus.

Positive Impacts of Soil Organisms

- Soil organisms are vital for ensuring that plants receive adequate nutrients, as they decompose organic debris. These organisms liberate nutrients from organic compounds.
- Certain microorganisms convert atmospheric nitrogen gas into a form accessible to plants. Other organisms solubilize minerals, hence enhancing the availability of phosphorus. In the absence of active soil organisms, an increased quantity of fertilizers will be required to provide essential plant nutrients.
- A diverse collection of organisms provides optimal protection against significant pest infestations and soil fertility issues. Soil abundant in organic matter and consistently enriched with various fresh residues supports a more wide array of species than soil without organic matter. The increased diversity of organisms ensures that fewer potentially hazardous species can establish substantial populations that diminish crop production.

Soil Tilth

- Soil is said to possess good tilth when it exhibits optimal physical conditions for plant growth. This type of soil is porous, facilitating the easy infiltration of water rather than surface runoff. A greater volume of water is retained in the soil for plant utilization between precipitation events, resulting in less erosion.

- Optimal tilth indicates that the soil is adequately aerated. Roots can efficiently acquire oxygen and eliminate carbon dioxide. A permeable soil facilitates root growth and exploration. Poor tilth in soil leads to structural degradation and disintegration of soil aggregates, resulting in heightened compaction and diminished aeration and water retention. A soil layer can become too compacted, inhibiting root growth. A soil exhibiting superior physical qualities will possess multiple channels and pores of varying dimensions.
- Research on both undisturbed and farmed soils indicates that an increase in organic matter correlates with reduced soil compaction and enhanced capacity for air infiltration and water retention. Adhesive compounds are generated during the breakdown of plant remnants. In conjunction with plant roots and fungal hyphae, they consolidate mineral particles into aggregates. Moreover, the adhesive exudates of mycorrhizal fungi—beneficial fungi that infiltrate roots and assist plants in acquiring additional water and nutrients—serve as crucial binding agents in soils. The configuration and accumulation of minerals as aggregates, along with the extent of soil compaction, significantly influence plant growth.

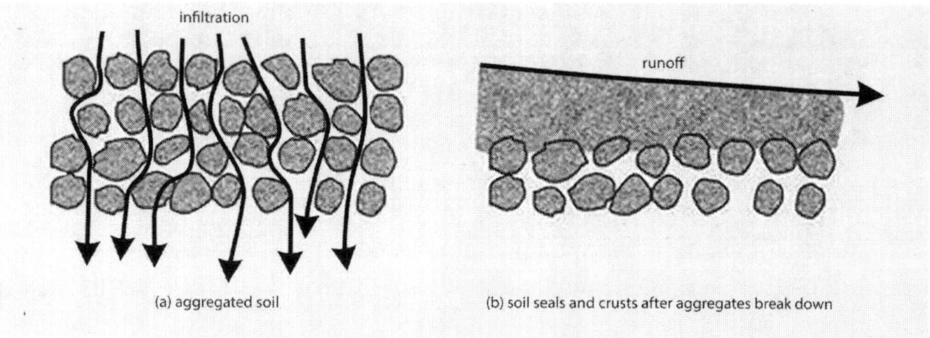

Figure 10.7: Alterations in Soil Surface and Hydrological Flow Patterns Upon the Formation of Seals and Crusts.

- The formation of aggregates is advantageous in all soil types as it enhances drainage, aeration, and water retention. The one exception is to wetland crops, such as rice, which require a compact, saturated soil to maintain flooding.
- Organic matter, present as surface residue or as a binding agent for near-surface aggregates, significantly contributes to the reduction of soil erosion. Surface residues capture raindrops and reduce their ability to dislodge soil particles. These surface residues impede water flow over the field, enhancing its potential for soil infiltration. Aggregates and extensive channels significantly improve the soil's capacity to transmit water from the surface to the subsurface.

- Most farmers may discern the superiority of one soil type over another by visual inspection, evaluating its texture during tillage, or by tactile assessment when stepped over or touched. What they are observing or perceiving is indeed excellent soil structure.

Refer to the photograph on the back cover of this book for an example. It demonstrates that variations in soil can be induced by diverse management practices. Farmers and gardeners would undoubtedly prefer to cultivate their crops on the more permeable soil illustrated in the photograph on the right.

- Erosion typically depletes the most productive soil, resulting in a substantial decline in crop output. In certain soils, the depletion of merely a few inches of topsoil can lead to a 50 percent decrease in output. The surface of certain soils deficient in organic matter may form a seal or crust as precipitation disintegrates aggregates and fills surface pores with particles. When this occurs, water that cannot permeate the soil flows off the field, transporting essential topsoil.
- Large soil pores, or channels, are very important because of their ability to allow a lot of water to flow rapidly into the soil. Larger pores are created through many mechanisms. Decomposed root channels may persist for an extended duration. Substantial soil creatures, including insects and earthworms, generate channels when traversing the soil. The mucus secreted by earthworms prevents their skin from desiccation and also maintains the integrity of their burrows over extended periods.

Protection of the Soil against Variations in Acidity

- Acids and bases are generated as minerals dissolve and organisms perform their routine activities of digesting organic matter or fixing nitrogen. Roots of plants excrete acids or bases, while acids are generated in the soil due to the application of nitrogen fertilizers. Optimal plant health is achieved when soil acidity, denoted as pH, remains stable throughout the season. The pH scale quantifies the concentration of free hydrogen ions ($H+$) in soil water.
- Lower numerical values signify more acidic environments, characterized by increased hydrogen concentrations. A soil with a pH of 4 is highly acidic. Its solution is tenfold more acidic than soil with a pH of 5. A soil with a pH of 7 is neutral, indicating an equal concentration of base and acid in the water. The majority of crops thrive in slightly acidic soil with a pH ranging from 6 to 7. Essential nutrients are more accessible to plants within this pH range compared to soils that are either more acidic or more alkaline. Soil organic matter can mitigate fluctuations in pH by sequestering free hydrogen ions when acids are generated or by releasing hydrogen ions when bases are formed.

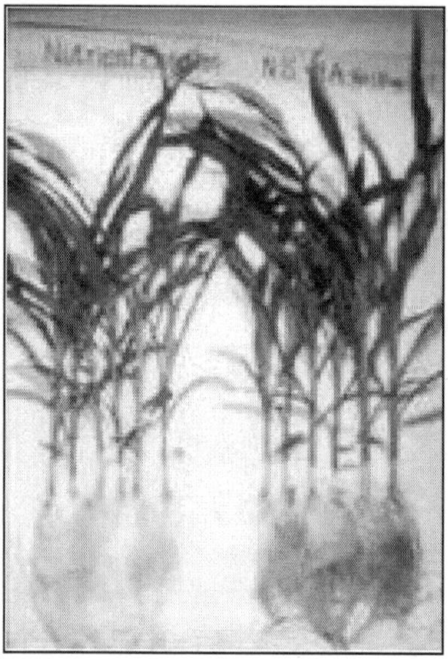

Figure 10.8: Corn Cultivated in Nutrient Solution with (Right) and without (Left) Humic Acids. Photograph by R. Bartlett. Rich Bartlett's experiment demonstrated that the incorporation of humic acids into a nutritional solution enhanced the growth of tomatoes and corn, as well as the quantity and branching of roots.

Stimulation of Root Development

- Soil microorganisms generate several compounds that promote plant growth. Humus exerts a directly advantageous influence on plants. This stimulation primarily enhances the availability of micronutrients to plants, leading to longer and more branched roots, which results in larger and healthier plants. Furthermore, numerous soil microorganisms synthesize various root-stimulating compounds that function as plant hormones.

Darkening of the Soil

- Organic stuff typically darkens soils. It is readily observable in coarse-textured sandy soils comprising light-hued minerals. In well-drained conditions, a darker soil surface facilitates a more rapid warming of the soil in spring.
- This offers a marginal benefit for seed germination and the initial phases of seedling growth, which is frequently advantageous in frigid areas.

Protection against Harmful Chemicals

- Certain naturally occurring soil pollutants can be detrimental to plants. Aluminum is a significant component of numerous soil minerals and, hence, does not endanger plants. As soil acidity increases, particularly at pH levels beneath 5.5, aluminum solubility rises. Certain soluble aluminum compounds, when present in the soil solution, are detrimental to plant roots. In such circumstances, the substantial presence of soil organic matter strongly binds aluminum, mitigating its potential harm.

- Organic matter is the paramount soil characteristic that mitigates pesticide leaching. It retains a variety of pesticides firmly. This mitigates or diminishes the leaching of toxic compounds into groundwater and provides time for microbial detoxification. Microorganisms can alter the chemical composition of certain pesticides, industrial oils, certain petroleum products (fuels and oils), and other potentially hazardous substances, therefore neutralizing their toxicity.

Benefits of Organic Matter

Organic matter is an essential element of soil due to its involvement in chemical, physical, and biological processes.

Chemical Processes

- **Nutrient Supply:** Organic matter serves as a reservoir of nutrients, such as nitrogen and phosphorus, which can be released into the soil via the process of mineralization. Every percent of organic matter in the soil corresponds to around 1000 kg of organic nitrogen, together with other nutrients. The progressive mineralization of this nitrogen offers a crucial source of nutrients for crop needs.
- **pH Buffering**: Organic matter effectively mitigates pH changes.

Physical Processes

- **Water Holding Capacity:** Organic matter functions akin to a sponge, has the capacity to absorb and retain substantial amounts of water. An significant benefit of the water-retention capacity of organic matter is its ability to release the majority of the absorbed water to cultivating plants. Conversely, clay can retain substantial amounts of water; yet, a significant portion remains inaccessible to plants.
- **Soil Structure Aggregation:** Organic matter induces soil aggregation, enhancing soil structure. Enhanced soil structure increases permeability, hence augmenting the soil's capacity to absorb and retain water.
- **Erosion Prevention:** Elevated organic matter levels generally diminish soil erosion due to enhanced water infiltration facilitated by a more stable soil structure. An optimal soil structure mitigates the likelihood of wind erosion.
- **Compaction:** Elevated organic matter concentrations generally diminish the likelihood of soil compaction.
- **Reduces surface crusting**: Elevated organic matter concentrations generally diminish the likelihood of soil capping, especially in fine-textured soils, by enhancing soil structure.

- **Reduces Erosion:** The majority of soil organic matter is located in the topsoil. Soil erosion caused by water movement results in the removal of organic materials. Proper tillage methods, especially on inclined terrains, can mitigate erosion and the loss of organic matter, such as contour plowing.

Biological Processes

- **Biodiversity:** Soil organic matter is essential for maintaining soil biodiversity by serving as a source of energy and nutrients for soil microbes.

Natural Determinants Affecting Organic Matter Quantity

- The transformation and movement of materials within soil organic matter pools is a dynamic process influenced by climate, soil type, plants, and soil organisms. All these components function within a hierarchical geographical framework. Soil organisms facilitate the decomposition and cycling of macronutrients and micronutrients, influencing the soil's structure, tilth, and productivity.
- In undisturbed natural humid and sub-humid forest ecosystems, the biotic and abiotic components maintain a dynamic equilibrium. The accumulation of litter on the soil surface beneath various canopy layers, along with elevated biomass output, typically leads to increased biological activity in the soil and its surface. Mollison and Slay (1991) identified five distinct mechanisms:
    - A persistent layer of living vegetation that, in conjunction with soil structure, enhances the collection and absorption of precipitation while safeguarding the soil.
    - A layer of decomposing leaves or waste serving as a continual energy source for macro- and microorganisms.
    - The roots of various plants, located at varied soil depths, facilitate efficient nutrient absorption and promote active interaction with microbes.
    - The primary interval of nutrient liberation by microorganisms aligns with the principal phase of nutrient requisition by plants.
    - Nutrients are recycled by deep-rooted plants and soil macrofauna and microfauna.
- This equilibrium facilitates nearly closed-cycle nutrient transfers between soil and vegetation suited to the specific site conditions, yielding optimal physical and hydric conditions for plant growth, such as a cool microclimate, enhanced evapotranspiration, favorable rooting conditions with adequate porosity, and sufficient soil moisture. This promotes water absorption and mitigates erosion and runoff. Consequently, this

leads to purified water in the streams originating from the region, a reasonably stable fluctuation in stream flow throughout the year, and the replenishment of groundwater.

- The environmental and edaphic parameters governing soil biota activity, and therefore the equilibrium between organic matter buildup and decomposition in the soil, are delineated below:

Temperature

- Numerous field investigations have demonstrated that temperature is a crucial determinant of the breakdown rate of plant leftovers. Decomposition typically transpires more swiftly in tropical regions than in temperate zones.
- Ladd and Amato (1985) indicated that, notwithstanding variations in plant material and climatic conditions, the decomposition of leguminous materials in southern Australia exhibited a similar pattern to that of ryegrass in Nigeria and the United Kingdom (Jenkinson and Ayanaba, 1977), albeit with differing time scales. Reaction rates doubled with each increment of 8-9 °C in the mean annual air temperature.
- The accelerated decomposition rate caused by persistent temperature in the tropics suggests that attaining elevated equilibrium levels of organic matter is challenging in tropical agro ecosystems. Consequently, substantial annual quantities of organic inputs are required to sustain a sufficient labile soil organic matter reservoir in cultivated soils. Soils in cooler temperatures typically contain higher organic matter due to reduced mineralization (decomposition) rates.

Soil Moisture and Water Saturation

- Soil organic matter levels often rise with an increase in mean annual precipitation. Elevated soil moisture conditions lead to increased biomass production, resulting in greater leftovers and thus more potential food for soil biota.
- Soil biological activity necessitates the presence of air and moisture. Optimal microbial activity transpires at about "field capacity," defined as 60 percent water-filled pore space (Linn and Doran, 1984).
- Conversely, intervals of water saturation result in inadequate aeration. The majority of soil organisms require oxygen; hence, a decrease in soil oxygen levels results in a diminished mineralization rate as these organisms become inactive or perish. Certain transformation processes become anaerobic, potentially resulting in root injury due to waste products or creating conducive circumstances for pathogenic organisms. Ongoing production and gradual decomposition can result in substantial organic matter accumulation in soils subjected to prolonged water saturation, such as peat soils and tea plantations in India.

- Aside from hyper-humid regions, the climates of extensive areas within the humid, sub-humid, and semi-arid tropics are marked by pronounced wet and dry seasons. on the wet-dry tropics, substantial quantities of nitrate frequently accumulate on the surface soil during the initial phase of the rainy season (Greenland, 1958; Mueller-Harvey, Juo and Wild, 1989). The rapid nitrogen mineralization, induced by a significant rise in microbial activity, results from the initial rains activating the labile soil organic matter.
- Farmers employing "slash and burn" agriculture frequently choose for early planting to capitalize on the surge of inorganic nitrogen before it is diminished by leaching and runoff. In these low-input systems, the concentration of nitrate in the soil at the onset of the rainy season is closely correlated with the organic matter content of the soil. The supply of N decreases in the latter portion of the rainy season.

Soil Texture

- Soil organic matter generally rises with an increase in clay concentration. This augmentation relies on two methods. The connections between clay particle surfaces and organic substances impede the breakdown process. Secondly, soils with elevated clay content enhance the likelihood of aggregate formation. Macroaggregates provide physical protection to organic matter molecules, preventing further mineralization due to microbial activity (Rice, 2002). For instance, when earthworm castings and the substantial soil particles they encompass are fragmented by the combined influence of several variables (temperature, vegetation, and other creatures), nutrients are liberated and rendered accessible to other soil microorganisms.
- In comparable climatic conditions, the organic matter content in fine-textured (clayey) soils is two to four times more than that in coarse-textured (sandy) soils (Prasad and Power, 1997).
- Kaolinite, the predominant clay mineral in several tropical upland soils, possesses a significantly lower specific surface area and nutrient exchange capability compared to most other clay minerals. Consequently, kaolinitic soils possess significantly fewer clay-humus complexes. Furthermore, the unprotected labile humic compounds are susceptible to breakdown at suitable soil moisture levels. Consequently, elevated concentrations of organic matter are challenging to sustain in cultivated kaolinitic soils within the wet-dry tropics, as climatic and soil conditions promote fast decomposition. Conversely, organic matter may endure as organo-oxide complexes in soils abundant in iron and aluminum oxides. These characteristics promote the development of soil microaggregates, which are characteristic of numerous fine-textured, oxide-rich, high base-status soils found in tropical regions (Uehara and Gilman, 1981). These soils are characterized by low bulk density, elevated microporosity, and significant

organic matter retention under natural vegetation, however they also have a high capacity for phosphate fixation on oxides when utilized for agricultural production. Current understanding indicates that while organic matter influences the dark coloration of Vertisols (Coulombe, Dixon and Wilding, 1996), it is deemed insignificant in affecting the growth, strength, or resilience of soil structure (McGarry, 1996). Organic matter concentrations in Vertisols are often minimal, reaching as low as 10 g/kg (Coulombe, Dixon and Wilding, 1996).

- Parent material affects organic matter accumulation not solely through its impact on soil texture. Soils derived from naturally fertile materials, such as basalt, exhibit greater fertility than those created from granitic materials, which include fewer mineral nutrients. Furthermore, the former exhibits greater organic matter accumulation due to prolific vegetative growth.

Topography

- Organic matter accumulation is frequently promoted at the base of hills. This buildup occurs for two reasons: the circumstances are more humid than at mid- or upper-slope locations, and organic matter is conveyed to the lowest landscape point by runoff and erosion.
- Soil organic matter concentrations are elevated on north-facing slopes in the Northern Hemisphere relative to south-facing slopes, and conversely in the Southern Hemisphere, due to lower temperatures.

Salinity and Acidity

- Salinity, toxicity, and high soil pH levels (either acidic or alkaline) lead to diminished biomass production, thus resulting in decreased organic matter contributions to the soil. pH influences humus formation through two mechanisms: breakdown and biomass generation.
- In very acidic or highly alkaline soils, the circumstances for microbial growth are suboptimal, leading to diminished biological oxidation of organic matter (Primavesi, 1984). Soil acidity affects the availability of plant nutrients, hence indirectly regulating biomass production and the food supply for soil biota. Fungi exhibit lower sensitivity to acidic soil conditions compared to bacteria.

Vegetation and Biomass Production

- The pace of soil organic matter accumulation is primarily influenced by the quantity and quality of organic matter input. In tropical settings, the use of easily degradable materials with low carbon-to-nitrogen ratios, such as green manure and leguminous cover crops, promotes decomposition and a temporary rise in the labile nitrogen pool during the growth season.

- Conversely, the utilization of plant materials characterized by high carbon-to-nitrogen ratios and lignin content, such as cereal straw and grasses, often promotes nutrient immobilization, organic matter accumulation, and humus formation, hence enhancing the potential for improved soil structure development.
- Plant components, including lignin and other polyphenols, inhibit decomposition. An experiment in southern Nigeria aimed at comparing management impacts on soil organic matter accumulation revealed that a three-year fallow with Guinea grass (*Panicum maximum*), characterized by high lignin content, sustained a carbon level similar to that of forest fallow. Nonetheless, the cultivation of leguminous species like pigeon pea (*Cajanus cajan*) resulted in a notable reduction in total soil carbon (Juo and Lal, 1977).
- Palm and Sanchez (1990) indicated that the breakdown rate and nitrogen release patterns of three tropical legumes (*Inga edulis, Cajanus cajan, and Erythrina* spp.) were associated with the concentration of polyphenolic chemicals, including lignin, in the leaves. The leaves of Erythrina had the lowest polyphenol contents and the highest breakdown rate among the three species examined.
- Root turnover significantly contributes to humus accumulation in the soil, hence playing a crucial role in carbon sequestration. In forests, the majority of organic matter accumulates as surface litter. In grassland ecosystems, as much as two-thirds of organic matter is contributed via root breakdown (Quideau, 2002).

Methods to Enhance Soil Organic Matter Concentrations

The accumulation of soil organic matter is a protracted process, however it confers numerous advantages to the soil. Available options to enhance organic matter content include the following:

- **Incorporate straw:** The chopping and reincorporation of straw enhances organic matter content.
- **Apply farmyard manures or mushroom compost:** Consistent applications of farmyard manure or mushroom compost will incrementally enhance organic matter content while supplying essential nutrients for plants. Nevertheless, under the nitrates limits, farmers must modify their fertilizer applications to consider the nutrients present in these organic fertilizers.
- **Apply organic fertilizers:** Poultry, swine, and bovine excrement contribute organic matter to soil. The nutrients in these fertilizers must be considered in accordance with the nitrates laws.
- **Cover Crops:** Increasing green cover in autumn with spring crops can enhance or sustain soil organic matter, especially when used as green

manure. Optimal outcomes can be attained when the cultivation of cover crops is integrated with additional remedial strategies.

- **Crop rotation:** Incorporating grass into crop rotation may restore organic matter levels, contingent upon specific circumstances.
- **Reduced Till options:** Min-till and no-till systems alone will not substantially elevate organic matter levels in soil but will mitigate the rate of loss. When integrated with additional practices such as straw incorporation and green manure, they will assist in sustaining or enhancing the organic matter content.
- **Reduce or Eliminate Tillage in the cropping rotation**: Tillage enhances soil aeration and stimulates microbial activity, accelerating the breakdown of organic materials. Tillage frequently exacerbates erosion, resulting in the depletion of organic materials.
- **Grass:** Grass-based farming typically elevates the organic matter content in soils. In areas with minimal organic matter, the most effective method to enhance levels is to revert the land to permanent grazing.

Problems with Incorporation

The integration of organic stuff may provide some challenges.

- It is challenging to integrate substantial amounts through cultivation.
- Green manure crops decompose rapidly and contribute minimally to soil organic matter levels.
- Ploughing accelerates the decomposition of humus and may negate the minor advantages provided by the crop itself.
- Incorporating organic matter into damp soil may lead to compaction, resulting in insufficient oxygen for microorganisms to breakdown the organic waste. This could impact agricultural productivity and nitrogen availability.
- The inclusion of straw may result in a transient deficiency of nitrogen for the cultivated crop, since microorganisms will utilize the finite nitrogen present in the decomposing straw.
- Chemicals emitted from organic debris may temporarily inhibit plant growth or exert harmful effects on young seedlings.

Chapter 11

Soil Testing and Indicators

- Soil testing is widely acknowledged as a reliable scientific method to evaluate the soil's intrinsic capacity to provide plant nutrients. A soil test involves the chemical extraction of a soil sample to assess nutrient availability.
- Soil tests extract a portion of the total nutrient content that correlates with, but does not equate to, the amount of plant-available nutrients. Consequently, a soil test serves merely as an indicator of nutrient availability.
- In contrast to plant analysis, soil testing assesses the relative nutrient status prior to planting. The assessment encompasses soil testing for various properties, including pH, electrical conductivity, cation exchange capacity, texture, structure, water retention capacity, and parameters for the remediation of chemically degraded soils, to recommend soil amendments such as gypsum for alkaline soils and lime for acidic soils.
- One purpose of soil testing is to distinguish nutrient-poor areas from those that are not deficient. This information is crucial for assessing if the soils can provide sufficient nutrients for optimal crop production.
- The principle of balanced crop nutrition dictates the application of plant nutrients in specific ratios necessary for the crops, achievable only with an understanding of the soil's nutrient availability.
- Soil testing aids in comprehending the intrinsic fertility status of the soils. Moreover, factors beyond inadequate soil fertility may contribute to suboptimal crop yields; however, the state of soil fertility remains paramount.

- Fertilizer recommendations derived from soil analyses must consider the soil test values obtained, the research on crop responses to fertilizer application in the specific region, and the management practices and proficiency of the farmer involved.
- The soil test conducted for the assessment of soil fertility, along with the subsequent fertilizer recommendations, serves as the crucial link between agronomic research and its practical implementation in farmers' fields.

Soil Sampling for Soil Testing

Soil testing comprises a three-step process: collecting a representative sample from each field or area, accurately analyzing that sample to ascertain the levels of available nutrients, and utilizing the results to establish optimal fertilizer rates. Maintaining records is essential to the soil-testing process, as they facilitate the assessment of whether soil test levels are rising, falling, or remaining stable over time.

When to Sample

Fields prepared for spring seeding ought to be sampled post-October. Fields designated for fall-seeded crops ought to be sampled one month prior to sowing. Forage fields designated for seed, pasture, or hay may be sampled post-September. Areas of problematic soil may be sampled at any time. Sampling of frozen and water-logged soils is inadvisable due to the challenges in acquiring a representative sample.

Types of Sampling

Fields can be divided into zones or grids when formulating a soil sampling plan. Soils inside those zones or grids may be collected randomly or sampled at or near the intersections. Soil test values derived from random and grid sampling are frequently utilized to yield a singular estimate for a whole field. This value can thereafter be utilized to determine fertilizer application rates.

1. Random Sampling

Uniform fields may be sampled randomly across the entire area. To observe long-term trends in soil nutrient data, these points must be geo-referenced using a global positioning system (GPS) receiver and collected at the same places in consecutive years.s.

2. Grid Sampling

Grid sampling is especially beneficial in situations with no prior understanding of within-field variability. It also mitigates sampling bias that may arise from the acquisition of an unrepresentative composite sample due to a significant proportion of subsamples obtained from the same region. Two prevalent forms of grid sampling are grid-cell and grid-point. Grid-cell soil sampling randomly gathers one or more sub-samples within the cell to create a composite sample. Grid-point soil sampling involves the collection of one or more subsamples surrounding a geo-referenced point within a grid or at a grid intersection.

3. Zone Sampling

The zone soil sampling technique posits that each field comprises distinct soils with specific qualities and crop characteristics, necessitating their division into separate management zones. Regions of fields with distinct crop histories, yields, fertilizer treatments, or significant variations in slope, texture, depth, and soil color should be sampled individually and designated as distinct zones. In contrast to grid sampling, the quantity, shape, and dimensions of zones will be contingent upon the extent of field variability. Moreover, zone sampling decreases the quantity of soil samples relative to grid or random sampling and facilitates variable rate fertilizer applications ("prescription" rates). Variable fertilizer application can enhance yields and decrease fertilizer expenses.

4. Soil Series

Soil series sampling delineates distinct zones within and between fields by utilizing soil survey and topographic maps. Each soil series exhibits distinct features and is likely to possess varying levels of accessible nutrients. Consequently, distinct soil samples for each soil series within a field are gathered. Soil test results may thereafter be area-weighted according to the acreage of each soil series.

5. Topographic/Geographic Unit Sampling

Fields exhibit diverse natural characteristics, including elevation, hilltops, slopes, and depressions. Topographic and geographic unit sampling presumes that these features exhibit variations in soil qualities, hence utilizing them to delineate distinct zones. Sampling of topographic/geographic units can be categorized into two primary types: area-based sampling and point-based sampling. Area-based soil sampling involves the collection and compositing of many soil samples from the vicinity of the center of each topographic zone, whereas point-based soil sampling entails the collection of a single sample from the center of each topographic zone.

6. Yield Sampling

Crop growth and yields fluctuate according on several soil factors, including texture, drainage, depth, and management practices, such as land shaping, spreader patterns, and prior land use. Yield sampling zones utilize crop yield maps produced from combine yield monitor data to identify locations for soil sampling. Yield data obtained from yield monitors can be integrated with GPS to create yield maps. Yield maps are most effectively utilized for zone delineation when the field is partitioned into arbitrary grids via a GIS application, with the yields inside each grid being averaged. Grids with yields exceeding the average are assigned a value of +1, those with yields below the average receive a value of -1, and grids with average yields are designated a value of 0. Repeating this technique for each year's yield data, irrespective of crop type, would generate a normalized yield frequency map upon the aggregation of multi-year normalized yield data in a spreadsheet and subsequent mapping. The resultant maps delineate areas that consistently produce high or low yields and those that do not. If a uniform factor governs yield variability in a field, then the distribution of this factor, and consequently the distribution of crop yield, might aid in identifying locations for soil sampling. If insufficient nutrient levels correlate with low yield regions, the application of that nutrient should enhance yield in those areas. If soil test results reveal sufficient or elevated nutrient levels in low-yielding regions, the soil should be assessed for compaction and other physical properties that may influence yield, especially those impacting water retention or drainage. Consequently, fertilizer application can be diminished in these areas.

7. Management Zones

The management zone strategy integrates various zone sampling techniques to delineate distinct management zones. Integrating prior experience, soil survey maps, yield maps, topography, electrical conductivity (EC) measurements from sensors like the Veris EC sensor or the EM-38 magnetic sensor, soil color, organic matter, soil nutrients, moisture, and remotely sensed vegetation indices is essential for creating distinct zones through multiple layers of information. These levels of information can be utilized independently (as previously explained) or in various combinations to delineate distinct zones.

Sampling Tools and Accessories

Depending on the intended purpose and desired accuracy, the following tools may be used for soil sampling:

- A soil auger can be a tube auger, post hole auger, screw type auger, or a spade for sampling purposes.
- A clean bucket or a tray or a clean cloth for mixing the soil and sub sampling.
- Cloth bags of specific size.
- Copying pencil for markings and tags for tying cloth bags.
- Soil sample information sheet.

Selection of a Sampling Unit

A visual assessment of the area should occur prior to the actual sample. Observe the differences in slope, color, texture, management, and cropping patterns while traversing the field. Divide the field into uniform sections, each of which must be sampled individually. If all these requirements are analogous, one field may be regarded as a singular sampling unit. The unit should not surpass 1 to 2 hectares, and it must be an area to which the farmer is prepared to devote distinct attention. The sample unit represents a balance between cost, labor, and time, contrasted with precision. Due to restricted soil testing facilities, it is proposed to implement an alternative method whereby a sample is obtained from an area of 20-50 hectares, designated as a composite area soil sample, and analyzed to formulate a unified suggestion for the entire region.

Sampling Procedure

Construct a map of the region designated for the survey, delineating the boundaries of various sampling units. A schematic detailing the quantity of samples and the method of composite sampling is inscribed on the map, with various fields identified by the letters A, B, C, etc. Each region is traversed individually. A segment of the plough layer is extracted at intervals of 15 to 20 paces or based on the region to be addressed. A composite sample typically requires the collection of 10 to 20 locations, contingent upon the field's size. Remove surface debris; get a consistently thick soil sample from the surface to the plough depth at each location. A V-shaped incision is executed with a spade to extract a 1 to 2 cm layer of earth. The sample can be gathered on the spade's blade and placed in a clean pail. Collect samples from all designated locations for a single sampling unit. In instances of compacted soil, samples are extracted using an auger from the plough depth and gathered in a bucket.

Transfer the soil from the bucket onto a clean sheet of paper or cloth and mix thoroughly. Distribute the soil uniformly and partition it into four quadrants. Eliminate two opposing quadrants and reconstitute the remaining soil. Continue the procedure until approximately half a kilogram of soil remains; then collect it and place it in a clean cloth bag. Each bag must be appropriately labeled to identify the sample. The sampling bag must consistently be clean and devoid of any contaminants. To reuse the bag, invert it and eliminate the soil particles. Document the specifics of the sample in the information sheet. Place a copy of this information leaflet in the bag. Securely fasten the opening of the bag.

Precautions to be taken during Soil Sampling

- Refrain from sampling atypical areas such as unevenly fertilized regions, swampy land, old pathways, ancient channels, dilapidated bunds, areas adjacent to trees, sites of former compost piles, and other non-representative locations.
- A tube auger or spade is deemed enough for soft and damp soil. For denser soil, a screw auger may be more advantageous.
- Collect samples from the center of the rows where crops are planted, avoiding areas where fertilizer has been applied in bands.
- Eliminate all forms of contamination at every level. Soil samples must not be stored alongside fertilizer products and detergents. Contamination is probable when soil samples are dried near stored fertilizers or on surfaces previously occupied by fertilizers.

- ☆ Soil samples must be assessed for cleanliness and integrity prior to being placed in bags.PThe information sheet must be clearly inscribed with a pencil.

Preparation of Soil Samples for Analysis

i. Handling in the Laboratory

Upon receipt of the samples at the soil testing laboratory, they must be verified against the accompanying information list. If the personnel at the soil testing laboratory gathered the samples independently, then comprehensive field notes may have been maintained. All unidentified samples must be disposed of. Details pertaining to samples must be recorded in a register, and each sample should be assigned a laboratory number in addition to its sample number, facilitating differentiation when many sources of samples are present.

ii. Drying of Samples

Samples received in the laboratory may exhibit moisture. These should be desiccated in wooden or enameled trays. It is imperative to preserve the identity of each sample during all phases of preparation. During the drying process, the trays may be numbered or a plastic tag may be affixed. The soils are permitted to desiccate in the atmosphere. The trays may also be positioned in racks within a hot air cabinet, maintaining a temperature not exceeding 350°C and a relative humidity between 30% and 60%.

iii. Post Drying Care

Upon drying, the samples are transferred to the preparation room, distinct from the primary laboratory. Air-dried samples are pulverized using a wooden pestle and mortar to smash the soil aggregates while preserving the integrity of the soil particles. Heavy clay soil samples may require grinding using an end runner mill equipped with a hardwood pestle and a rubber-lined mortar. Pebbles, concretions, and stones must remain intact during grinding. Subsequent to grinding, the soil is filtered through a 2 mm sieve. The method of allowing only a fraction of the ground sample to pass through the sieve while discarding the rest is flawed. This introduces a positive bias in the sample, as the rejected portion may contain soil components with varying fertility levels. The complete sample should be subjected to sieving, excluding concretions and stones over 2 mm in size. The coarse residue on the sieve should be reintroduced to the mortar for additional grinding. Continue sifting and grinding until all aggregate particles are sufficiently fine to pass through the sieve, leaving only stones, organic residues, and concretions behind. Containers composed of copper, zinc, and brass must be excluded during the grinding and handling of soil intended for trace element analysis. Sieves of various dimensions are available in stainless steel. Aluminum or plastic sieves serve as effective alternatives for most applications. Subsequent to the sample's passage through the sieve, it must be mixed well once more. Soil samples must be preserved in cardboard boxes within wooden drawers. The boxes must be numbered and organized in rows within the wooden drawers, which are installed in a cabinet located in the soil sample room.

Soil Testing

Soil testing is characterized as a systematic procedure for evaluating soil fertility through fast chemical analysis, specifically to determine the available nutrient status and pH of the soil. A soil test is a chemical procedure for assessing the nutrient supply capacity of soil. Soil testing offers a significant benefit over plant analysis by assessing the nutritional condition of the soil prior to crop planting. The outcome of a soil analysis is referred to as the soil test value. A soil test value quantifies a portion of the total nutrient availability in the soil and serves solely as an index of nutrient capacity. Soil tests do not quantify the precise amount of a nutrient that a crop may absorb. To forecast the nutrient requirements of crops, a soil test must be calibrated with the crop responses observed in nutrient rate studies conducted in greenhouses and fields. The interpretation and evaluation of soil test findings essentially underpin fertilizer recommendations. The significance of soil testing is in facilitating appropriate fertilization for profitable and sustainable crop production.

Error, Precision, Accuracy and Detection Limit

A vast array of definitions concerning detection limits and quantitation limits is employed in the literature and by governmental agencies. Regrettably, there are no commonly agreed methodologies for determining these boundaries. This can be exasperating and perplexing for both regulators and the regulated entities. The subsequent Numerical Criteria for selecting an analytical method and their definitions to address these concerns are as follows:

Accuracy

It is a synthesis of the bias and precision of an analytical method, indicating the proximity of a measured value to the true value. Standard Methods, 18th Edition. In the context of laboratory certification, accuracy refers to the proximity of a measured value to its widely recognized value or its value derived from an approved reference standard. NR 149.03(2). Bias quantifies systematic or determinative mistake in an analytical procedure. Bias is evaluated by measuring the percentage recovery of spiked samples. Historically, the term accuracy has been used synonymously with bias, but numerous authorities differentiate between the two (Standard Methods, 18th edition).

Enforcement Standard (ES)

It denotes a numerical number indicating the maximum concentration of a material in groundwater, as established under s. 160.07, Stats, and s. NR 140.10 or s. 160.09, Stats, and s. NR 140.12. These guidelines are derived from toxicological assessments to safeguard human health. Analytical values exceeding the ES initiate the procedure outlined in s. NR 140.26. s.NR 140.05(7) A False Positive, or Type I (alpha) error, refers to the erroneous conclusion that a chemical is present when it is, in fact, absent. A False Negative, or Type II (beta) error, occurs when one erroneously concludes that a chemical is absent when it is, in fact, present.

Instrument Detection Limit (IDL)

The concentration corresponding to a signal from the analyte of interest is the minimum signal that can be differentiated from background noise by a certain instrument. The IDL must consistently remain beneath the method detection limit and is not utilized for compliance data reporting; however, it may be employed for statistical data analysis and for comparing the characteristics of various instruments. The IDL resembles the "critical level" and "criterion of detection" as delineated in the literature. Standard Methods, 18th Edition.

Limit of Detection (LOD)

LOD, or limit of detection, is the minimal concentration level that can be statistically distinguished from a blank at a 99 percent confidence level. The limit of detection (LOD) is generally established in the range where the signal-to-noise ratio exceeds 5. Limits of detection are particular to the matrix, technique, and analyte. Sections NR 140.05(12) and 149.03(15).

Limit of Quantitation (LOQ)

The lower limit of quantitation (LOQ) is the threshold at which quantitative results can be achieved with a certain level of confidence. The Limit of Quantification (LOQ) is technically defined as ten times the standard deviation of data from a set of repetitions, establishing an acceptable detection limit. The limits of quantitation are particular to the matrix, technique, and analyte. Sections NR 140.05(13) and 49.03(16).

Linear Calibration Range (LCR)

The Range of Linearity (LCR) refers to the segment of a calibration curve where the relationship between the concentration of an analyte and its corresponding response is linear, with a correlation coefficient nearing 1 (often 0.995 for most analytes). The plot may be normal-normal, log-normal, or log-log, depending on the analytical method employed. At the upper and lower limits of quantitation, the analyte's signal response relative to concentration diverges from the linear relationship. The range between the limit of quantification (LOQ) to the limit of linearity (LOL) is referred to as the Dynamic Range.

Maximum Contaminant Level (MCL)

It is a quantitative measure indicating the highest allowable concentration of a pollutant in water supplied to any consumer of a public water system. Maximum contamination levels are specified in s. NR 809.09, Wisconsin Administrative Code (NR 809.04(34)).

Method Detection Limit (MDL)

The minimal concentration of a substance that can be quantified and reported with 99 percent confidence that the analyte concentration exceeds zero, derived through the analysis of a sample within a certain matrix containing the analyte.

Practical Quantitation Limit (PQL)

It is a quantitation limit that signifies a practicable and consistently attainable threshold with a high level of certainty (>99.9 percent confidence) in the outcomes.

Precision

It quantifies the random error linked to several measurements of the same parameter within a sample. Precision denotes the degree of agreement among multiple analyses of a certain sample and is occasionally termed reproducibility. Precision is assessed using absolute standard deviation, relative standard deviation, variance, coefficient of variation, relative percent difference, or the absolute range of a set of measurements. Refer to NR140.05 (16) and the 18th edition of Standard Methods. It can be computed mathematically as:

Terms	Definition
Absolute standard deviation, s	$s = \sqrt{\frac{\Sigma_i^N = (x_i - \bar{x})^2}{N-1}}$
Relative standard deviation (RSD)	$RSD = \frac{S}{x}$
Standard error of the mean, s_m	$S_m = \frac{s}{\sqrt{N}}$
Coefficient of variation (CV)	$CV = \frac{s}{x} \times 100\%$
Variance	s^2

$$Where, \bar{x} = \sqrt{\frac{\Sigma_{i=1}^N X_i}{N}}$$

Preventive Action Limit (PAL)

It is a numerical value indicating the maximum concentration of a substance in groundwater as established by s. 160.15, Stats., and s. NR 140.10, 140.12, or 140.20. Values exceeding the PAL activate the procedure outlined in s. NR 140.22. The permissible action level (PAL) is often established at one-tenth of the enforcement standard for substances that are carcinogenic, mutagenic, teratogenic, or exhibit a synergistic effect. The PAL constitutes 20 percent of the enforcement standard for other chemicals of public health significance. The Reporting Limit is a designated threshold beneath which data is not disclosed. The reporting limit may be scientifically established or may be an estimate derived from the analyst's experience and judgment. Analytical results below the reporting limit are denoted as "less than" the reporting limit. Reporting limitations are not permissible alternatives to detection limits unless explicitly authorized by the Department for a specific test.

Sample Matrix, or Matrix, delineates the overall physical-chemical composition of a certain sample. While the specific matrix of a sample differs between discharges and locations, the primary categories of matrices encompass: reagent (clean) water, wastewater, public drinking water, waste, surface water, groundwater, sediments, and soils. NR 149.03(28).

Sample Standard Deviation, or Standard Deviation (s)

It quantifies the extent of concordance or accuracy among repeated analyzes of a sample. Standard Methods, 18th Edition In this text, standard deviation refers to sample standard deviation (with n-1 degrees of freedom). The population standard deviation (n degrees of freedom) should be utilized exclusively when addressing a genuine representation of a population (e.g., beyond 25 data points). The standard deviation is defined as follows:

i) Sensitivity means the ability of a method or instrument to detect an analyte at a specified concentration. (NR 149.03(28m)).

ii) Signal to Noise Ratio (S/N) is a dimensionless measure of the relative strength of an analytical signal (S) to the average strength of the background instrumental noise (N) for a particular sample and is closely related to the detection level. The ratio is useful for determining the effect of the noise on the relative error of a measurement. The S/N ratio can be measured a variety of ways, but one convenient way to approximate the S/N ratio is to divide the arithmetic mean (average) of a series of replicates by the standard.

Soil Nutrient as an Index of Soil Fertility

- ☆ Soil testing laboratories often utilize organic carbon as an indicator of available nitrogen, employ Olsen's and Bray's methods for assessing available phosphorus, and apply the neutral normal ammonium acetate method for determining available potassium. In semi-arid tropics, nitrate nitrogen serves as an indicator of nitrogen availability in soil.
- ☆ The nutrient status in soils is typically categorized as low, medium, or high, a classification commonly adhered to at the national level:

Table 11.1 Available Soil Nutrient Status Ratings

Soil Nutrients	*Soil Fertility Ratings*		
	Low	*Medium*	*High*
Organic Carbon as a measure of available Nitrogen (per cent)	< 0.5	0.5-0.75	>0.75
Available N as per alkaline permanganate method (kg ha^{-1})	<280	280-560	>560
Available P by Olsen's method (kg ha^{-1})	<10	10-24.6	>24.6
Available K by Neutral N–ammonium acetate method	<108	108-280	>280

Micro and Secondary Nutrients: Soil Testing

- To guarantee the provision of inadequate micro and secondary nutrients to crops, a standardized testing procedure for evaluating available micronutrients has been implemented in soil testing laboratories.
- The quantity of nutrients that can be annually extracted by standard crop yields differs by element and crop type.
- The nutrient forms and their mobility in plants are presented in Table 11.2, while the essential limitations in soil, plant, and soil tests are outlined in Table 11.3.

Table 11.2: The Composition of Nutrients and their Relative Mobility in Plants

Element	*Form Absorbed*	*Mobility in Plants*
Sulphur	SO_4^{2-}, SO_2	Less immobile
Calcium	Ca^{2+}	Immobile
Magnesium	Mg^{2+}	Mobile
Boron	H_3BO_3	Immobile
Copper	Cu^{2+}	Less mobile
Iron	Fe^{2+}	Less immobile
Manganese	Mn^{2+}	Less immobile
Molybdenum	MoO_4^{2}	Less mobile
Zinc	Zn^{2+}	Moderately mobile
Chlorine	Cl^-	Less mobile

Plant can absorb SO_2 directly from the atmosphere.

Table 11.3: Soil Testing Techniques with Critical Thresholds in Soil and Plant

Nutrients	*Soil Test Method*	*Critical Level in Soil*	*Critical Level in Plant*
Sulphur	Hot water, $CaCl_2$	8-30 ppm	<0.15-0.2 per cent
Calcium	Ammonium Acetate	<25% of CEC or <1.5meq Ca/ 100g	<0.2%
Magnesium	Ammonium Acetate	<4 per cent of CEC or <1 meq Mg/100g	<0 1.0.2%
Boron	Hot water	0.5 ppm	<20 ppm
Copper	DTPA ammonium acetate	0.2 ppm	<4 ppm (3.10)
Iron	DTPA ammonium Acetate	2.5 - 4.5 ppm	<50 ppm(25- 80)
Manganese	DTPA	2 ppm	<20 ppm (10-30)
Molybdenum	Acid ammonium oxalate	0.2 ppm	<0.1 ppm
Zinc	DTPA	0.6 (0.4-1-.2) ppm	<15-20 ppm

Quality Control of Analytical Procedures

i. Independent Standards

Independent standards are included in the system to assess and confirm the accuracy of the analysis. The degree of divergence of the analytical value from a standard sample reflects the accuracy of the analysis. An independent standard can be generated in the laboratory using pure ingredients.

ii. Use of Blank

A blank determination is an analysis conducted without the analyte or attribute, essentially an analysis devoid of a sample, wherein all procedural steps are followed using only the reagents. Utilization of blank accounts to identify any contamination in the chemicals employed during the real analysis. The estimate of the blank is deducted from the sample estimates.

iii. Blind Sample

A sample with a known analyte concentration is introduced by the laboratory director in batches and at periods unknown to the analyst. Diverse categories of sample material can function as blind samples, including control samples or adequately sized residuals of test samples that have undergone multiple analyses. It is imperative that the analyst is cognizant of the potential existence of a blind sample while remaining unable to identify the material as such.

iv. Validation of Procedures of Analysis

Validation is the process of assessing the performance attributes of a method or approach. It is a prerequisite for assessing the appropriateness of generated analytical data for its intended application. This indicates that a method may be applicable in one context and inapplicable in another. If a method is highly precise and accurate yet costly to implement, it may be utilized solely when data of that level of precision is required. The data may be insufficient if the procedure is less precise than necessary. Two categories of validation are implemented.

v. Validation of Own Procedure

In-house validation of methods or procedures by specific user laboratories is a prevalent technique. Numerous laboratories employ their own adaptations of even widely recognized methods for reasons of efficiency, cost-effectiveness, and convenience. A modification in the liquid-to-solid ratio during extraction operations for accessible soil nutrients, as well as variations in shaking duration, leads to altered values, necessitating confirmation.

vi. Validation of the Standard Procedure

This pertains to the validation of novel or established methods and processes intended for utilization across several laboratories, including those recognized by national systems or ISO standards. This entails an inter-laboratory testing program conducted by representatives from selected prestigious laboratories, adhering to a methodology distributed to all participants.

vii. Inter-laboratory Sample and Data Exchange Programme

When a mistake is detected in the technique and doubt cannot be easily resolved, it is usual practice to have the sample analyzed in a different laboratory within the same system or organization. The findings of the alternative laboratory may be biased, hence uncertainty may remain. An additional analysis by a different authorized laboratory may be essential and beneficial to address the issue. An approved laboratory must engage in a minimum of one inter-laboratory exchange program. Such programs occur at local, regional, national, and worldwide levels.

Preparation of Reagent Solutions and their Standardization

Chemical reagents are produced and sold in various purity grades. Typically, the most refined reagents are sold as Analytical Reagent or "AR" grade. Additionally, the labeling may indicate "LR" for laboratory reagent or "CP" for chemically pure. The potency of substances is denoted as normalcy or molarity. Consequently, it is advantageous to possess knowledge regarding the potency of significant acids and alkalis frequently utilized in chemical laboratories. Acids and alkalis are fundamental compounds necessary in a laboratory setting.

Molarity

A one molar (M) solution comprises one mole or one gram equivalent of the substance's molecular weight per liter of solution. The molar technique of describing concentration is advantageous since equal volumes of equimolar solutions contain an identical number of molecules.

Normality

The normalcy of a solution is defined as the number of gram equivalents of solute per liter of solution. It is typically denoted by the letter 'N'. Semi-normal, penti-normal, desinormal, centi-normal, and milli-normal solutions are commonly denoted as 0.5N, 0.2N, 0.1N, 0.01N, and 0.001N, respectively.

Table 11.4

Sl.No.	*Reagent/ Chemical*	*Normality*
1	Nitric acid	16
2	Sulphuric acid	35
3	Hydrochloric acid	11.6
4	Phosphoric acid	45
5	Perchloric acid	10.5
6	Ammonium hydroxide	15

Table 11.5: Strength of Commonly Used Acids and Alkali

Molarity	*Formula by Weight*	*Per cent by Weight (Approx.)*	*Specific Gravity (Approx.)*	*Ml Required or 1N/Litre Solution (Approx.)*	*Ml Required or 1M/Litre Solution (Approx.)*
16	63	70	1.42	63.7	63.7
17.5	98	98	1.84	28	56
11.6	36.5	37	1.19	82.6	82.6
15	98	85	1.71	22.7	68.1
1.5	100.5	65	1.6	108.7	108.7
15	35	28	0.90	67.6	67.6

Equivalent Weight

The equivalent weight of a material is the mass in grams that corresponds to one gram atom of hydrogen, one gram atom of hydroxyl, half a gram atom of oxygen, or one gram atom of a univalent ion in its reaction. When one equivalent weight of a substance is dissolved in one liter, it produces a 1 N solution. The equivalent and molecular weights of several significant compounds are presented in Table 11.6.

Table 11.6: Equivalent and Molecular Weights of some Important Compounds

Compound	*Formulae*	*Mol Wt. (g)*	*Eq. Wt (g)*
Ammonium acetate	$NH_4C_2O_2$	77.08	77.08
Ammonium chloride	NH_4C	53.49	53.49
Ammonium fluoride	NH_4F	37.04	37.04
Ammonium nitrate	NH_4NO_3	80.04	80.04
Barium chloride	$BaCl_2 2H_2O$	244.28	122.14
Boric acid	H_3BO_3	61.83	20.61
Calcium acetate	$(CH_3 COO)_3Ca$	158.00	79.00
Calcium Carbonate	$Ca CO_3$	100.09	50.05
Calcium chloride (dehydrate)	$CaC_2.2H_2O$	147.02	73.51
Calcium hydroxide	$Ca(OH)_2$	74.00	37.00
Calcium nitrate	$Ca(NO_3)_2$	164.00	82.00
Calcium sulphate	$CaSO_4.2H_2O$	172.17	86.08
Ferrous ammonium sulphate	$(NH_4)_2 SO_4. FeSO_4.6H_2O$	392.13	392.13
Ferrous sulphate	$FeSO_4.7H_2O$	278.01	139.00
Magnesium chloride	$MgCl_2.6H_2O$	203.30	101.65
Magnesium nitrate	$Mg(NO_3)_2.6H_2O$	256.41	128.20
Potassium chloride	KCl	74.55	74.55
Potassium dichromate	$K_2Cr_2 O_7$	294.19	49.04

Compound	Formulae	Mol Wt. (g)	Eq. Wt (g)
Potassium hydroxide	KOH	158.03	158.03
Potassium permanganate	$KMnO_4$	158.03	31.60
Potassium nitrate	KNO_3	101.10	101.10
Potassium sulphate	K_2SO_4	174.27	87.13
Potassium hydrogen phthalate	COOH C_3H_3 COOK	204.22	204.22
Oxalic acid	$C_2H_2O_4.2H_2O$	126.00	63.00
Silver nitrate	$AgNO_3$	169.87	169.87
Sodium acetate(Anhydrous)	CH_3COONa	82.04	82.04
Sodium bicarbonate	$NaHCO_3$	84.01	84.01
Sodium carbonate	Na_2CO_3	106.00	53.00
Sodium chloride	NaC	58.45	58.45
Sodium hydroxide	NaOH	40.00	40.00
Sodium nitrate	$NaNO_3$	84.99	84.99
Sodium oxalate	$Na_2C_3O_4$	134.00	67.00
Sodium sulphate	$Na_2 SO_4$	142.04	71.02
Sodium thiosulphate	$Na_tS_2O_3.5H_2O$	248.18	248.18

Milli-equivalent Weight

Equivalent weight (Eq W), when articulated as milli-equivalent weight (meq W), signifies the equivalent weight in grams divided by 1000. It is frequently denoted as "meq". This figure is the most useful as it represents the weight of a substance contained in or equivalent to one milliliter of a -1 N solution. Consequently, it is a unit applicable to both volume and weight, facilitating the conversion of a solution's volume to its corresponding weight and a substance's weight to its comparable solution volume.

Number of meq = Volume x Normality

Buffer Solutions

Solutions comprising a weak acid and its corresponding salt or a weak base and its respective salt (e.g., $CH_3COOH + CH_3COONa$ and $NH_4OH + NH_4Cl$) have the distinctive ability to maintain stable pH levels upon the addition of an acid or base. These solutions are known as buffer solutions. The subsequent properties are significant for a buffer solution:

- It possesses a specific pH value.
- The pH value remains unchanged during extended storage periods.
- The pH value is minimally affected by the addition of a strong basic or strong acid.

Due to the aforementioned property, conveniently manufactured buffer solutions with known pH values are utilized to verify the accuracy of pH meters employed in the laboratory.

Titration

It is a procedure for ascertaining the volume of a chemical necessary to fully react with a certain quantity of another substance. The solution of known concentration utilized in titration is referred to as the titrant. The component to be analyzed in the solution is referred to as the titrate. The reaction's completeness is assessed using a suitable indication.

Indicator

An indication is a chemical that signifies the endpoint upon the completion of a reaction. The most often utilized markers in volumetric analysis are:

- Internal indicator
- External indicator
- Self indicator

i. Internal Indicator

Indicators such as methyl red, methyl orange, phenolphthalein, and diphenylamine, which are included into the reaction solution, are referred to as internal indicators. Upon the conclusion of the titrant's reaction with the titrate, a color shift occurs due to the indicator, which also signifies the completion of the titration. Examples of color change resulting from pH variation in solutions are presented in Table 11.7.

Table 11.7: ppH Transition Range of Various Indicators

pH Indicator	*pH Transition Interval*			
Name	*Colour*	*pH*	*pH*	*Colour*
Corsol red	Pink	0.2	1.8	Yellow
m-Cresol purple	Red	1.2	2.8	Yellow
Thymol blue	Red	1.2	2.8	Yellow
2.4-Dinitrophenol	Colourless	2.8	4.7	Yellow
Bromochlorophenol blue	Yellow	3.0	4.7	Purple
Bromophend blue	Yellow	3.0	4.6	Purple
Methyl orange	Red	3.1	4.4	Yellow-orange
Bromocresol green	Yellow	3.8	5.4	Blue
2.5-DIndrophenol	Colourless	4.0	5.8	Yellow
Methyl red	Red	4,4	6.2	Yellow-orange
Chlorophenol red	Yellow	4.8	6.4	Purple
Litmus extra pure	Red	5.0	8.0	Blue

pH Indicator	pH Transition Interval			
Name	Colour	pH	pH	Colour
Bromophenol red	Orange-yellow	5.2	6.8	Purple
Bromocresol purple	Yellow	5.2	6.8	Purple
4-Nitrophend	Colourless	5.4	7.5	Yellow
Bromoxylend blue	Yellow	5.7	7.4	Blue
Bromothymd blue	Yellow	6.0	7.6	Blue
Phenol red	Yellow	6.4	8.2	Rod
3-Nitrophenol	Colourless	6.6	8.6	Yellow-orang
Cresol red	Orange	7.0	8.8	Purple
1-Naphtholphthalein	Brownish	7.1	8.3	Purple
Thymol blue	Yellow	8.0	9.6	Blue
Phenolphthalein	Colourless	8.2	9.8	Red-violet
Thymolphthalein	Colourless	9.3	10.5	Blue

ii. External Indicator

Certain indicators are employed externally to the titration mixture. Potassium ferricyanide serves as an external indicator in the titration of potassium dichromate and ferrous sulfate in an acidic solution. During this titration, many drops of indicator are applied on a white porcelain tile. A glass rod, immersed in the titration solution, is removed and brought into contact with the indicator droplets on a white tile. The initial deep blue hue is observed, which transitions to a greenish tint with the end of the titration.

iii. Self Indicator

The titrant, upon completion of the reaction, retains its own color due to a tiny surplus in minute quantities. During the titration of $KMnO_4$ with ferrous sulfate, the introduction of $KMnO_4$ initiates a reaction with the colorless $FeSO_4$. Upon the conclusion of titration, a tiny excess of $KMnO_4$ imparts a pink hue to the solution, serving as a self-indicator that signifies the completion of the titration.

Standard Solution

A solution with a precisely known strength (or concentration) is referred to as a standard solution. It includes a certain number of gram equivalents or gram moles per liter of solution. If it contains 1 gram equivalent weight of a substance per liter, it is a 1N solution. All titrimetric procedures rely on standard solutions that contain precise quantities of reagents in a unit volume of the solution. A solution is formulated with an approximate required concentration. This solution is subsequently standardized through titration with another material that can be acquired in a highly pure state. Consequently, potassium permanganate solution can be standardized using sodium oxalate, which is readily obtainable in a high degree of purity due to its ease of drying and non-hygroscopic nature. A material with stable weight and purity is referred to as a 'Primary Standard.' A primary standard must possess the following attributes:

- ✰ It must be available in a pure form or in a condition of established purity. It must have a singular response throughout titration, with no ancillary reactions occurring.
- ✰ It must be non-hygroscopic. Salt hydrates are typically unsuitable as primary standards.
- ✰ Typically, it should possess a substantial equivalent weight to minimize weighing errors.
- ✰ An acid or base should ideally be strong, possessing a high dissociation constant to serve as standards.

A primary standard solution is one that can be directly prepared by weighing and used to titrate and standardize other solutions of approximate strength. The following are some primary standards:

Acids	*Bases*	*Oxidizing Agents*	*Reducing Agents*	*Others*
Potassium hydrogen phthalate	Sodium carbonate	Potassium dichromate	Sodium oxalate	Sodium chloride
Oxalic acid	Borax	Potassium bromate	Potassium Ferro cyanide	Potassium chloride

Secondary Standard Solutions are made by dissolving slightly more than the gram equivalent weight of a chemical per liter of solution, followed by precise standardization with a primary standard solution. Secondary standards for acids include sulfuric acid and hydrochloric acid, while sodium hydroxide serves as a secondary standard for bases. All reagents necessary for laboratory analysis must be prepared as standard solutions and made available prior to conducting any assessments. Nevertheless, their potency should be routinely assessed or new reagents generated before to analysis.

Chapter 12

Analysis of Soil Physical Properties

Introduction

Soil physical measurements typically encompass soil water content, infiltration, hydraulic conductivity, evapotranspiration, thermal properties, temperature, reflectivity, porosity, particle size, bulk density, aggregate stability, and particle size distribution.

Soil Water

- Water is the primary limiting element in agriculture; thus, determining soil moisture is of paramount importance.
- Soil moisture impacts crop growth by influencing plant physiological processes, serving as a solvent, transporting nutrients, and sustaining plant turgidity.

Apparatus

Electric oven with thermostat and Desiccator.

Procedure

- Weigh 10 g air-dry soil (< 2mm) into a previously dried (105 °C) and weighed metal can with lid.
- Dry in an oven with the lid unfitted at 105 °C overnight.

Calculations

$$\%\,Moisture\ in\ soil\,(Q) = \frac{Wet\ soil\,(g) - Dry\ soil\,(g)}{Dry\ soil\,(g)} \times 100$$

Particle Size Distribution

- Soil particles exhibit significant variability within any given soil type. Concurrently, as these particles are consolidated, a range of aggregate forms and dimensions emerges.
- For conventional particle size assessment, the soil fraction that traverses a 2 mm sieve is taken into account. Laboratory procedures typically quantify the proportions of sand (0.05 - 2.0 mm), silt (0.002 - 0.05 mm), and clay.

Apparatus

Soil dispersing stirrer: A high-speed electric stirrer with a cup receptacle. Hydrometer with Bouyoucos scale in g/ltr.

Reagents

- Dispersing agent: Dissolve 40 g sodium hexametaphosphate [$(NaPO_3)$ 6], and 10 g sodium carbonate (Na_2CO_3 in distilled water, and bring to 1 litre volume with distilled water. This solution deteriorates with time and should not be kept for more than 1 to 2 weeks.
- Amyl alcohol.

Procedure

- Weigh 40 g air-dry soil (2 mm) into a 600 ml beaker.
- Add 60 ml dispersing solution.
- Cover the beaker with a watch-glass, and leave overnight.
- Quantitatively transfer contents of the beaker to a soil-stirring cup, and fill the cup to about three-quarters with water.
- Stir suspension at high speed for 3 minutes using the special stirrer. Shake the suspension overnight if no stirrer is available.
- Rinse stirring paddle into a cup, and allow standing for 1 minute.
- Transfer the suspension quantitatively into a 1 litre calibrated cylinder (hydrometer jar), and bring to volume with water.

Determination of Blank

- Dilute 60 ml dispersing solution to 1 litre hydrometer jar with water.
  - Mix well, and insert hydrometer, and take hydrometer reading (Rb).
  - The blank reading must be re-determined for temperature changes of more than 2 °C from 20 °C.

Determination of Silt plus Clay

- Mix suspension in the hydrometer jar, using a special paddle carefully, withdraw the paddle, and immediately insert the hydrometer.

☆ Eliminate any foam, if necessary, by adding one drop of amyl alcohol, and obtain the hydrometer value 40 seconds after removing the paddle. The reading (RSC) and the proportion of [silt + clay] in soil can be computed as follows:

$$\%\,[Silt + Clay]\left(\frac{W}{W}\right) - (R_{sc} - R_b)x\frac{100}{Oven\ dry\ soil\ (g)}$$

Determination of clay

☆ Gently stir the suspension in the hydrometer jar with a paddle; carefully remove the paddle without disturbing the suspension.

☆ After 4 hours, insert the hydrometer and record the hydrometer reading (Rc); the percentage of clay in the soil can then be estimated as follows:

$$\%\,Clay\left(\frac{W}{W}\right) = R_c - R_b\ x\frac{100}{Oven\ dry\ soil\ (g)}$$

Thus, percentage silt in soil can be calculated as:

$$\%\,Silt\left(\frac{W}{W}\right)\left[\%\,Silt + Clay\left(\frac{W}{W}\right)\right] - [\%\,Clay\left(\frac{W}{W}\right)]$$

Determination of Sand

☆ After obtaining the necessary measurements for clay and silt, quantitatively discharge the suspension through a 50 mm screen. Rinse the sieve until the effluent is clean, then quantitatively transfer the sand from the sieve to a pre-weighed 50 ml beaker.

☆ Permit the sand in the beaker to settle, then decant the surplus water. Dry the beakers with sand overnight at 105°C, then cool them in a desiccator and reweigh the beakers with sand.

The percentage of sand in soil can be determined as follows:

$$\%\,Clay\left(\frac{W}{W}\right) = R_c - R_b\ x\frac{100}{Oven\ dry\ soil\ (g)}$$

Thus, percentage silt in soil can be calculated as:

$$\%\,Silt\left(\frac{W}{W}\right)\left[\%\,Silt + Clay\left(\frac{W}{W}\right)\right] - [\%\,Clay\left(\frac{W}{W}\right)]$$

Soil Texture

☆ Upon measuring the proportions of sand, silt, and clay, the soil can be categorized into a textural class with the USDA textural triangle.

☆ Within the textural triangle are various soil textures which depend on the relative proportions of the soil fractions.

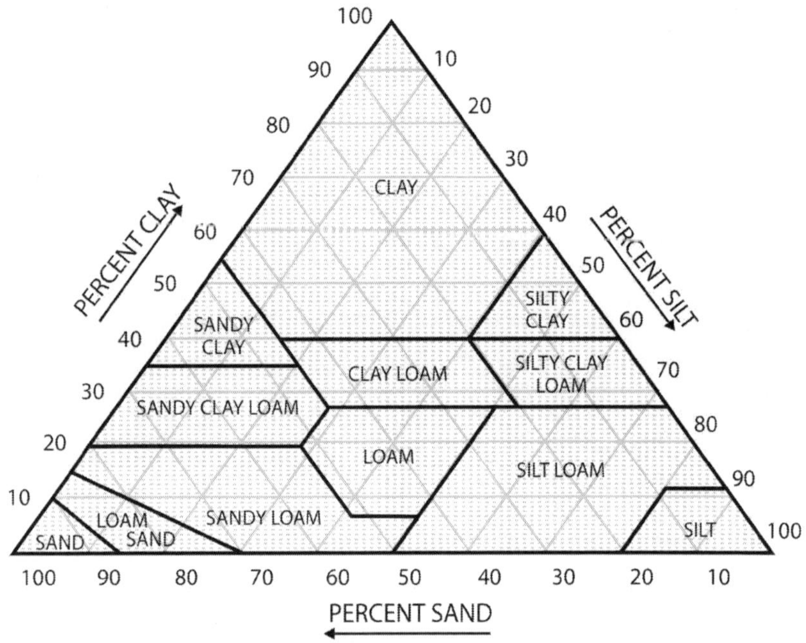

Figure 12.1: USDA Textural Triangle for Soil Textural Class Determination

Soil Structure

- ☆ Soil structure refers to the configuration of soil particles. Soil particles encompass not just sand, silt, and clay but also the aggregates or structural elements created by the aggregation of smaller mechanical fractions.
- ☆ Soil structure is classed according to size as follows:
  - ❖ Very coarse: >10 mm
  - ❖ Coarse: 5-10 mm
  - ❖ Medium: 2-5 mm
  - ❖ Fine: 1-2 mm
  - ❖ Very fine: <1 mm

Dry Aggregates Analysis (Gupta and Ghildyal, 1998)

The size distribution of dry clods is assessed using Dry Sieving Analysis conducted on an air-dry bulk soil sample, either manually or utilizing a rotary sieve shaker.

Apparatus

- ☆ Nest of sieves 20 cm in diameter and 5 cm in height, with screens having 25.0, 10.0, 5.0, 2.0, 1.0, 0.5 and 0.25 mm size round openings with a pan and a lid.

- Rotary sieve shaker and Aluminum cans.
- Balance, Spade and Brush
- Polyethylene bags and Labels.

Procedure

- A bulk soil sample is obtained from the tilled field using a ring with a diameter of 20 cm and a height of 10 cm. The ring is positioned on the cultivated soil and compressed until it is flush with the surface. The loose soil inside the ring is extracted and gathered in a polyethylene bag.
- One label indicating the depth and soil sample is put inside the bag and the other label is tied with the bag. The soil samples are subsequently transported to the laboratory and air-dried.
- Distribute the dirt onto a sheet of paper and prepare the subsamples by the method of quartering. The blended dirt is carefully coned near the middle of the mixing sheet to ensure symmetry between fine and coarse soil materials. The cone is compressed and bisected with a flat metal spatula or sheet, with one half being quantitatively displaced to the side. Each half is subsequently subdivided into halves, resulting in four quarters that are organized into distinct piles or 'quarters.'The sub-samples from two of these 'quarters' are weighed and utilized for clod size and aggregate distribution analysis as duplicates.
- The measured soil sample is placed in the uppermost sieve of a series of sieves including circular apertures of 5.0, 2.0, 1.0, 0.5, and 0.25 mm in diameter, with a pan positioned at the base. Enclose the upper sieve with the cover and position the stack of sieves on a rotary shaker.
- Activate the shaker for 10 minutes, thereafter detach the sieves, and gather the soil retained on each screen into pre-weighed aluminum cans using a small brush, then weigh the cans containing the soil. If the proportion of dry aggregates on the 5 mm sieve surpasses 25 percent, relocate these aggregates to a set of sieves comprising 25.0 mm, 10.0 mm, and 5.0 mm, accompanied by a pan.
- Enclose the upper sieve containing the aggregates with a lid and position the stack of sieves on the rotating sieve shaker. Activate the motor for 10 minutes and continue as previously described for the assessment of aggregate size distribution.
- Examine the duplicate sample using the identical methodology and compute the percentage distribution of dry aggregates retained on each sieve. A 100 g duplicate sample is subjected to drying in an oven at 105 °C for 24 hours to determine the oven-dry weight of the soil sample.

Calculation

- Weight of aggregates in each sieve group = (Wt. of Aggregates + Can) – Wt. of Can
- Percent distribution of aggregates in each size group =

$$\frac{\text{Weight of Aggregate in each size group} \times 100}{\text{Total weight of soil}}$$

$$\text{Oven-dry weight of Aggregate (\%)} = \frac{\text{Air-dry weight (\%)} \times 100}{100 + \text{Moisture\%}}$$

Wet Aggregate Analysis (Gupta and Ghildyal, 1998)

Apparatus

- A mechanical oscillator powered by a gear reduction motor having amplitude of oscillation 3.8 cm and frequency of oscillator 30-35 cycles per minute.
- Two sets of sieves, each having 20 cm diameter and 5 cm height with screen openings of 5.0, 2.0, 1.0, 0.5, 0.25 and 0.1 mm diameter.
- Two Buckner funnels 15 cm in diameter with rubber stoppers and two vacuum flasks of one litre capacity.
- Suction pump and rubber policeman.
- Twelve aluminium cans and perforated cans.
- Sand bath and filter papers.

Reagents

- 5 per cent Sodium hexametaphosphate
- 4 per cent Sodium hydroxide

Procedure

- The soil sample is collected while it is damp and crumbly. It is fragmented by the application of mild stress into smaller aggregates that can pass through an 8 mm screen.
- The sieved soil sample is placed on a watch glass for wetting using either the vacuum soaking or immersion method. The sample is positioned in a vacuum desiccator that contains de-aerated water at the base for vacuum wetting. The desiccator is emptied until the internal pressure decreases to around 3 mm, at which point water begins to boil.
- Water is thereafter let to enter via the upper section of the desiccator and pour into the watch glass containing the sample. Sufficient water is introduced to submerge the soil sample. The soil sample is thereafter removed from the desiccator.

- Prepare four soil samples, each weighing 25 g; place a duplicate set of samples in an oven to ascertain moisture content. A further collection of saturated duplicate soil samples is transferred to the uppermost sieve of the nested sieves (5.0, 2.0, 1.0, 0.5, 0.25, and 0.1 mm) and distributed using a glass rod and a gentle stream of water. The bottom pan is subsequently detached, and the assembly of sieves is affixed to the Yoder-type wet sieve shaker. Fill the drum containing the set of sieves with salt-free water at a temperature of 20-25 °C to a level just below the screen of the top sieve when the sieves are positioned at their highest point. Subsequently, immerse the nest of sieves to moisten the soil for 10 minutes. Return the nest of sieves to the original position and calibrate the water level to ensure that the screen in the upper sieve is submerged at its highest point. Activate the mechanical oscillator to oscillate the nest of sieves vertically at a frequency of 30-35 cycles per minute and a stroke length of 3.8 cm. Sieving is conducted for a duration of 10 minutes. Extract the cluster of screens from the water and permit it to drain for a period.
- Displace the soil on each screen using a stream of distilled water and a brush into a Buckner funnel with pre-weighed filter paper, which is attached to a suction pump. Transfer the soil and filter paper into an aluminum container and dry at 105 °C for 24 hours.
- Measure the soil with a precision of 0.01 g. Convey the oven-dried soil aggregates from each can of the set into the dispersion cup. Introduce a dispersion agent (10 ml of a 5 percent sodium hexametaphosphate solution for normal and Ca-saturated soils, or 10 ml of a 4 percent sodium hydroxide solution for acid soils) and sufficient distilled water to fill the cup to within 4 cm of the rim, then agitate the suspension for 10 minutes.
- Rinse the suspension using the same set of sieves as before with a stream of tap water and a brush, then transfer it to aluminum cans. The sand in each container is oven-dried and weighed using the same method as previously described. Determine the percentage distribution of soil particles (aggregates and sand) and the sand particles retained on each sieve.

Calculation

Size distribution of soil particles (aggregate + sand):

Soil particles in each size group (%) $\dfrac{W_{oven\ dry}(ag+s)i \times 100}{W_{oven\ dry}}$

Sand particle in each size group (%) $\dfrac{W(s) \times 100}{W_{oven\ dry}}$

where,

$W_{over\ dry}$ I the oven dried weight of the aggregates (ag) and Sand (a), and " I" is the size group

Soil Moisture

Soils are equilibrated with water at different pressures, and moisture content is assessed.

Low Range at 0-100 kpa (0-10 bar) Pressure

Apparatus

- ✰ One-bar pressure plate extractor and one-bar ceramic plates.
- ✰ Rubber rings (5cm diameter, 1cm height).
- ✰ Compressed air source with a manifold, regulator, and gauge.
- ✰ Balance and drying oven.
- ✰ Disposable aluminum dishes and desiccator.

Procedure

- ✰ Submerge the ceramic plates in water for 24 hours to achieve saturation, then position the plates on a workstation.
- ✰ Arrange the labeled rubber rings sequentially on the plate, which can hold 12 samples. Utilize a spatula to fill the ring with 2mm air-dry soil, approximately 20 grams of the sample. To prevent particle-size segregation, insert the complete soil sample onto the ring.
- ✰ Ensure the sample is level, but do not saturate it by adding water beneath the ring and cover plate. Introduce water between the rings until a surplus of at least 3 mm in depth accumulates on the plate.
- ✰ Encase samples with wax paper or a plastic sheet and permit them to rest overnight.
- ✰ The following morning, extract surplus water from the plate using a syringe, disposable pipette, or siphon.
- ✰ Position the triangular support at the base of the extractor vessel.
- ✰ Position the sample plate at the lowest point within the extractor. Subsequently, install the middle and upper plates, ensuring that the plastic spacers are positioned between the plates, and connect the outflow tubes.
- ✰ Secure the extractor and tighten it, ensuring that the "O" ring is properly positioned and all nuts are evenly tightened. Exert the required pressure within the range of 0 to 100 kPa (0 to 1 bar). Incrementally increase the pressure within the vessel.
- ✰ Position a beaker to gather water from the outflow tubes. Maintain pressure until water ceases to be released, often for 18 to 20 hours, although certain soils may require 48 hours or longer.

- Depressurize the extractor by detaching the outflow tubes from the water prior to powering down the instrument.
- Activate the extractor. Promptly transfer the moist soil sample from the ring to a tarred dish using a wide-bladed spatula. It is unnecessary to execute a quantitative transfer of the complete soil.
- Weigh the wet sample immediately (accuracy 0.01 g) and place it in a drying oven at 105 °C for 24 hours. Subsequently, transfer the sample to a desiccator, allow it to cool, and weigh it again.

Calculation

$$\% \, Moisture \, (\theta) = \frac{Wet \; soil \, (g) - Dry \; Soil \, (g)}{Dry \; soil \, (g)} x \; 100$$

High Range Moisture at 100-1500 kpa (0-15 bar) Pressure

Apparatus

- Fifteen-bar ceramic plate extractor.
- Fifteen-bar ceramic plates and rubber rings.
- Balance and Weighing dishes
- Drying oven, burette and desiccator.

Procedure

- PUtilize 15-bar ceramic plates and adhere to steps 1 through 12 of the preceding procedure (6.4.1). Applying pressure ranging from 1 to 15 bar (100 to 1500 kPa).
- Position the beaker to gather water from the outflow tubes. Allow it to remain overnight.
- Attach the outflow tube to a burette that is partially filled with tap water.
- Samples must remain in the extractor until flow has terminated from all samples on the plate and the soils have attained equilibrium (often 24 to 48 hours for most soils; however, certain fine-textured and organic soils may require up to 120 hours). A stable reading on the burette signifies that the flow from all samples has ceased and balance has been reached.
- Detachment of the burette is necessary to avert the reverse flow of tap water.
- Relieve pressure from the extractor.
- Execute the final two steps of the preceding technique.

Calculation

$$\% \, Moisture \, (\theta) = \frac{Wet \; soil \, (g) - Dry \; Soil \, (g)}{Dry \; soil \, (g)} x \; 100$$

Available Water Capacity Estimation

- The available water capacity refers to the volume of water held in the soil reservoir that can be extracted by plants.
- The estimation is based on the disparity in soil water content between Field Capacity and Permanent Wilting Capacity.

$$\% AWC = FC\ (\%) - PWD\ (\%)$$

Water Holding Capacity

Field capacity, or water holding capacity, refers to the volume of water retained in the soil following the drainage of excess gravitational water and the significant reduction of downward water movement.

Apparatus

- Polythene sheets and Spade.
- Soil auger and Moisture moisture boxes/cans.
- Balance and Oven.
- Oven - dry weight of soil.

Procedure

- Select a homogenous plot of 5 meters by 5 meters. Eliminate weeds, pebbles, and other debris, and construct bunds around the allotment.
- Introduce an adequate volume of water into the area to thoroughly soak the soil.
- Encase the plot area with a polyethylene sheet to monitor evaporation.
- Extract a soil sample from the center of the plot at the specified layer, commencing 24 hours post-saturation, and assess moisture content daily until the values of consecutive days approximate equality.
- Record the weight as below:

  Weight of empty moisture box = X

  Weight of moisture box + moist soil = Y

  Weight of moisture box + oven dry soil = Z
- Repeat above on next day and so on till a constant Z value is reached.

Calculation

Moisture content in soil = Y – Z

Weight of oven dry soil = Z – X

Percentage of moisture in soil (1st day) = $\left(\frac{Y-Z}{Z-X}\right) \times 1000 = a$

Percentage of moisture on succeeding days = A1, A2, *etc.*

Graph the daily measurements on graph paper. The minimum measurement is considered the soil's field capacity value.

Chapter 13

Soil Chemical Properties Analysis

Cation Exchange Capacity (CEC)

The cumulative quantity of exchangeable cations that a soil may retain is referred to as cation exchange capacity (CEC).

Apparatus

Centrifuge, 50 ml round bottom centrifuge tubes, Mechanical shaker, Flame Photometer and accessories which include Propane, Lithium and Sodium standards.

Reagents

- Sodium acetate (NaOAc) 1.0M: Dissolve 136.08 g of sodium acetate trihydrate in distilled water and bring the volume to 1 litre. Adjust the pH to about 8.2.
- Ethanol 95 per cent.
- Ammonium acetate (NH_4OAc) 1.0M: Dissolve 77.09 g of ammonium acetate in distilled water and dilute to approx. 900 ml. Adjust pH to 7.0 with dilute ammonium hydroxide or acetic acid as required and make up the volume to 1 litre.
- Standard solution of NaCl: Dissolve 5.845 g of AR grade NaCl in 1.0M ammonium acetate and make the volume to 1 litre. It will give 100 meq/litre of sodium in stock solution. From this solution take 0, 1, 2, 5, 7.5 and 10 ml and make up the volume to 100 ml each with the ammonium acetate. It will give 0, 1, 2, 5, 7.5 and 10 meq/litre of sodium.

Procedure

- Accurately weigh 5 g of soil and transfer the sample to a 50 ml centrifuge tube. 2. Introduce 25 ml of 1.0M sodium acetate solution into the tube stopper and agitate on a mechanical shaker for 5 minutes.
- Centrifuge at 2000 rpm for 5 minutes or until the supernatant is transparent. Decant the liquid entirely and perform the extraction three further times. Eliminate the decants.
- Reiterate steps 2 to 4 utilizing ethanol or isopropyl alcohol until the electrical conductivity of the decant measures below 40 mS/cm (often requiring 4 to 5 wash cycles). 6. To displace the adsorbed sodium, continue steps 2 to 4 utilizing the ammonium acetate solution. Gather the decant in a 100 ml volumetric flask equipped with a funnel and filter paper. Adjust the volume using ammonium acetate solution.
- To ascertain sodium concentration via flame photometry, prepare a series of sodium standard solutions within the range of 0 to 10 meq/litre of sodium. Construct a standard curve by graphing sodium concentration on the X-axis and flame photometric measurements on the Y-axis. An unknown sample extract is introduced into the flame photometer, and the reading is obtained, from which the sodium concentration is determined using the standard curve. To achieve optimal results, include LiCl into each standard to attain a final concentration of approximately 5 meq/litre of LiCl.

Calculation

Ammonium acetate extractable Na which is exchangeable Na in meq/100 g soil =

$$\frac{\text{Na. Conc. of extract in meq/litre(Y)}}{\text{Wt. of soil in g(5)}} \times \frac{100 \times \text{Vol. of extract in ml (100)}}{1000} = \frac{Y \times 10}{5} = 2Y$$

This displaced Na is actually a measure of the Cation Exchange Capacity (CEC) of the soil. So, meq Na/100 g soil is actually meq exchangeable cations (Ca, Mg, Na, and K)/100 g soil.

pH

It is the measure of soil alkalinity, acidity or neutrality.

Apparatus

pH meter with a range of 0-14 pH, pipette/dispenser, beaker and glass rod.

Reagent

- Buffer solutions of pH 4, 7 and 9: Dissolve separately one tablet of each buffer in water and make the volume to 100ml.

- Calcium chloride solution (0.01M): Dissolve 1.47 g $CaCl_2.2H_2O$ in 1 litre of water to obtain 0.01M solution.

Procedure

- Calibrate the pH meter utilizing two buffer solutions: one with a neutral pH of 7.0 and the other selected according to the soil's pH range. Obtain the buffer solution from the beaker. Alternately insert the electrode into the beakers holding two buffer solutions and calibrate the pH. The pH meter calibrated with buffers is prepared to analyze the samples.
- Measure 10 g of soil sample into a 50 or 100 ml beaker; introduce 20 ml of $CaCl_2$ solution (substitute water for $CaCl_2$ solution throughout the experiment if water serves as the suspension medium).
- Permit the soil to assimilate the $CaCl_2$ solution undisturbed, then agitate thoroughly for 10 seconds with a glass rod.
- Agitate the suspension for 30 minutes and document the pH using the calibrated pH meter.

Based on soil pH values, the following types of soil reactions are identified:

Soil Reaction	*Rating*
<4.6	Extremely acid
4.6-5.5	Strongly acid
5.6-6.5	Moderately acid
6.6-6.9	Slightly acid
7.0	Neutral
7.1-8.5	Moderately alkaline
>8.5	Strongly alkaline

Electrical Conductivity (EC)

Electrical conductivity (EC) quantifies ionic transport inside a solution between the anode and cathode. The EC is typically regarded as an indicator of the concentration of dissolved salts in a solution.

Apparatus

EC meter, Beakers (25 ml), Erlenmeyer flasks (250 ml), pipettes and filter paper.

Reagent

0.01M Potassium chloride solution: Dry a small quantity of AR grade potassium chloride at 60 °C for two hours. Weight 0.7456 g of it and dissolve in freshly prepared distilled water and make the volume to one litre. This solution gives an electrical conductivity of 1411.8x10-3 *i.e.* 1.412 mS/cm at 25 °C. For best result, select a conductivity standard (KCl^- solution) close to the sample value.

Procedure

- Transfer 40 g of soil into a 250 ml Erlenmeyer flask, add 80 ml of distilled water, seal the flask, and agitate on a reciprocating shaker for one hour. Utilize Whatman No. 1 filter paper for filtration. The filtrate is prepared for conductivity measurement.
- Clean the conductivity electrode with distilled water and thereafter rinse it with standard KCl solution. Introduce KCl solution into a 25 ml beaker and immerse the electrode in the solution. Calibrate the conductivity meter to display 1.412 mS/cm, adjusted to 25 °C.
- Clean the electrode and immerse it in the dirt extract. Document the digital display adjusted to 25 °C. The electrical conductivity reading in mS/cm quantifies the soluble salt content in the extract and indicates the salinity state of the soil. Conductivity may also be represented as mmhos/cm.

Organic Carbon/Organic Matter

Organic matter contributes to soil fertility and overall soil health by improving the physical, chemical, and biological aspects of the soil. The estimation of organic matter in soil can be accomplished using many approaches. The weight loss upon igniting serves as a clear indicator of the organic matter present in the soil.

Loss of Weight on Ignition

Apparatus

Sieve, Beaker, Oven and Muffle furnace.

Procedure

- Weigh 5.0 to 10.0 g (to the nearest 0.01 g) of sieved (2 mm) soil into an ashing container.
- Position the ashing jar with dirt in a drying oven calibrated to 105 °C and dry for a duration of 4 hours. Extract the ashing vessel from the drying oven and position it in a desiccated environment. Upon cooling, measure the weight to the closest 0.01 g. Insert the ashing jar with soil into a muffle furnace and elevate the temperature to 400 °C. Ash subjected to furnace conditions for four hours. Extract the ashing vessel from the muffle furnace, allow it to cool in a desiccated environment, and measure its weight to the nearest 0.01 g.

Calculation

Percent organic matter (OM)= $(W_1 - W_2) \times 100/ W_1$

Percent organic C = % OM × 0.58

where,

W_1 is the weight of Soil at 105°C and W_2 is the weight of soil at 400°C

Reagents

- Phosphoric acid – 85 per cent
- Sodium fluoride solution (chemically pure)
- Sulphuric acid (sp.gr.1.84) containing 1.25 per cent Ag_2SO_4 (silver sulphate)
- Standard 1N potassium dichromate: Dissolve 49.04 g of $K_2Cr_2O_7$ in water and dilute to 1 litre.
- Standard 0.5 N Ferrone ammonium sulphate solution: Dissolve 196g ferrone ammonium sulphate in 800 ml water, add 20 ml concentrated H_2SO_4 and make up the volume to 1 litre.
- Diphenylamine indicator: Dissolve 0.5 g reagent grade diphenylamine in a mixture of 20 ml water and 100 ml of concentrated H_2SO_4.

Procedure

- Weigh 1.0 g of the produced soil sample into a 500 ml conical flask, then introduce 10 ml of 1N $K_2Cr_2O_7$ solution and 20 ml of concentrated H_2SO_4 containing Ag_2SO_4.
- Thoroughly combine and permit the reaction to proceed for 30 minutes, then dilute the reaction mixture with 200 ml of water and add 10 ml of H_3PO_4.
- Incorporate 0.5 g of NaF solution and 1 ml of diphenylamine indicator. Titrate the solution with a normal 0.5 N $FeSO_4$ solution until the color transitions from blue-violet to a vivid green.
- A blank is concurrently executed without soil.

Calculation

Percent organic Caron (X) $\frac{10(S-T)\times 0.003}{S}\times\frac{100}{\text{wt.of soil}}$

Since one gram of soil is used, this equation simplifies to:

$$\frac{3(S-T)}{S}$$

where, S = ml $FeSO_4$ solution required for blank

T = ml $FeSO_4$ solution required for soil sample

3 = Eq Wt of C (weight of C is 12, valency is 4, hence Eq Wt is 12÷4 = 3.0)

0.003 = weight of C (1000 ml 1N $K_2Cr_2O_7$ = 3 g C. Thus, 1 ml 1N $K_2Cr_2O_7$ = 0.003 g C) Organic Carbon recovery is estimated to be about 77 per cent. Therefore, actual amount of organic carbon (Y) will be:

Percent value of organic carbon obtained x 100/77

Or percentage value of organic carbon x 1.3

Percent Organic matter = Y x 1.724 (organic matter contains 58 per cent organic carbon, hence 100/58 = 1.724)

Total Nitrogen (Kjeldahl Method)

Total N includes all forms of inorganic N, like NH_4–N, NO_3-N and also NH_2 (Urea) –N, and the organic N compounds like proteins, amino acids and other derivatives.

Apparatus

Kjeldahl digestion and distillation unit, Conical flasks, Burettes and Pipettes.

Reagents

- Sulphuric acid – H_2SO_4 (93-98 per cent)
- Copper sulphate – $CuSO_4.H_2O$ (AR grade)
- Potassium sulphate or anhydrous sodium sulphate (AR grade)
- 35 per cent sodium hydroxide solution: Dissolve 350 g solid NaOH in water and dilute to one litre
- 0.1M NaOH: Prepare 0.1M NaOH by dissolving 4.0 g NaOH in water and make volume to 1 litre. Standardize against 0.1N potassium hydrogen phthalate or standard H_2SO_4.
- 0.1M HCl or 0.1M H_2SO_4: Prepare approximately 0.1M acid solution and standardize against 0.1M sodium carbonate.
- Methyl red indicator
- Salicyclic acid for reducing NO_3 to NH_4, if present in the sample.
- Devarda's alloy for reducing NO_3 to NH_4, if present in the sample.

Procedure

- Weigh a 1 g sample of dirt and deposit it in a Kjeldahl flask. Incorporate 0.7 g of copper sulfate, 1.5 g of potassium sulfate, and 30 ml of sulfuric acid.
- Warm slowly until bubbling stops. If required, incorporate a minimal quantity of paraffin or glass beads to mitigate foaming.
- Boil vigorously until the solution becomes clear, then maintain digestion for a minimum of 30 minutes. Detach the flask from the heat source and allow it to cool, then incorporate 50 ml of water and transfer the mixture to a distillation flask.
- Precisely measure 20–25 ml of standard acid (0.1M HCl or 0.1M H2SO4) into the receiving conical flask to ensure a surplus of at least 5 ml of acid. Introduce 2-3 drops of methyl red indicator. Introduce sufficient water to submerge the terminal ends of the condenser output tubes.

- Introduce 30 ml of 35 percent NaOH into the distillation flask, ensuring that the contents remain unmixed, and heat the mixture to distill ammonia for approximately 30 to 40 minutes.
- Detach the receiving flask and cleanse the exit tube by rinsing it into the receiving flask with a small quantity of distilled water. Neutralize the surplus acid in the distillate using 0.1M NaOH.
- Ascertain the blank on reagents by employing an equivalent volume of standard acid in a receiving conical flask.

$$\text{Percent N} = \frac{1.401(V_1 M_1 - V_2M_2)-(V_3M_1 - V_4M_2)}{W} \times df$$

where,

V_1 - ml of standard acid taken in receiving flask for samples

V_2 - ml of standard NaOH used in titration

V_3 - ml of standard acid taken to receiving flask for blank

V_4 - ml of standard NaOH used in titrating blank

M_1 - Molarity of standard acid

M_2 - Molarity of standard NaOH

W - Weight of sample taken (1 g)

df - Dilution factor of sample (if 1 g was taken for estimation, the dilution factor will be 100).

Note: 1000 ml OF 0.1 M HCL or 0.1 M H_2SO_4 = 1.401 g nitrogen.

Nitrate by Phenoldisulfonic Acid Method

- One of the major difficulties in estimating NO_3 in soils by colorimetric methods is obtaining a clear colourless extract with low contents of organic and inorganic substances which interfere with the colorimetric method.
- In arid and salt affected soils, chloride (Cl) is the major anion which interferes with colour development of the phenoldisulfonic acid method. Therefore, if chloride concentration is more than 15mg/g soil, it should be removed before analysis by the use of Ag_2SO_4 to precipitate chloride as AgCl.
- The Ag_2SO_4 is added to the extract or to the reagent used for extraction, and the AgCl is removed by filtration or centrifugation after precipitation of the excess Ag_2SO_4 by basic reagent such as $Ca(OH)_2$ or $MgCO_3$.
- It is necessary to remove the excess Ag^+ before analysis of the extract because it also interferes with the phenoldisulfonic acid method of determining NO_3.

Apparatus

Reciprocating shaker, Heavy-duty hot plate, Spectrophotometer, Dispenser Erlenmeyer flask, Beakers and Glass rod.

Reagents

- Phenoldisulfonic acid (phenol 2, 4-disulfonic acid): Transfer 70 ml pure liquid phenol (carbolic acid) to an 800 ml Kjeldahl flask. Add 450 ml concentrated H_2SO_4 while shaking. Add 225 ml fuming H_2SO_4 (13-15 per cent SO_3). Mix well. Place Kjeldahl flask (loosely stoppered) in a beaker containing boiling water and heat for 2 hours. Store resulting phenoldisulfonic acid $[C_6H_3OH(H_SO_3)_2]$ solution in a glass-stoppered bottle.
- Dilute ammonium hydroxide solution (about 7.5M NH_4OH): Mix one part NH_4OH (specific gravity 0.90) with one part H_2O.
- Copper sulfate solution (0.5M): Dissolve 125 g $CuSO_4.5H_2O$ in 1 litre of distilled water.
- Silver sulfate solution (0.6 per cent): Dissolve 6.0 g Ag_2SO_4 in 1 litre of distilled water. Heat or shake well until all salt is dissolved.
- Nitrate-extracting solution: Mix 200 ml of 0.5M copper sulfate solution and 1 litre 0.6 per cent silver sulfate solution and dilute to 10 litres with water. Mix well.
- Standard nitrate solution (100 mg NO_3-N/ml, stock solution): Dissolve 0.7221 g KNO_3 (oven dried at 105 °C) in water and dilute to 1 litre. Mix thoroughly.
- Standard nitrate solution (10 mg NO_3-N/ml; working solution): Dilute 100 ml of 100 mg NO_3-N/ml stock solution to 1 litre with water. Mix well.
- Calcium hydroxide, AR grade powder (free of NO_3).
- Magnesium carbonate, AR grade powder (free of NO_3).

Procedure

- Introduce approximately 5 g of soil into an Erlenmeyer flask and incorporate 25 ml of nitrate-extracting solution.
- Agitate the contents for 10 minutes, then include approximately 0.2 g of $Ca(OH)_2$ and agitate for an additional 5 minutes.
- Incorporate approximately 0.5 g of $MgCO_3$, agitate for 10-15 minutes, then permit settling for several minutes.
- Filter using Whatman filter paper No. 42. Transfer 10 ml of clear filtrate into a 100 ml beaker. Evaporate to dryness on a hotplate at low temperature within a fume hood devoid of HNO_3 vapors. Cease heating at achieving dryness.

- Upon completion drying, chill the residue and swiftly add 2 ml of phenoldisulfonic acid (using a burette with the tip removed) to cover the residue promptly. Rotate the beaker to ensure the reagent contacts all remaining salt. Permit to rest for 10-15 minutes.
- Incorporate 16.5 ml of cold water. Rotate the beaker to dissolve the residue or stir with a glass rod until the residue is completely in solution.
- Once the beaker has cooled, gradually introduce 15 ml of dilute NH_4OH until the solution exhibits a pronounced alkaline nature, as evidenced by the emergence of a stable yellow hue.
- Incorporate 16.5 ml of water, resulting in a total amount of 50 ml. Combine comprehensively. Determine the concentration of NO_3-N at 415 nm with the standard curve.
- Preparation of the standard curve: Measure 0, 2, 5, 8, and 10 ml of the 10 mg NO_3/ml working solution into separate 100 ml beakers, add 10 ml of the NO_3 extraction solution, and evaporate to dryness. Execute steps 9 to 13, with the standard solutions containing 0, 0.4, 1.0, 1.6, and 2.0 mg NO_3-N/ml. Construct a standard curve for the quantification of NO_3 in the sample.

Calculation

NO_3 - in test solution (mg/ml) =

$$\frac{\text{Vol. after colour development(ml)}}{\text{Vol/ evaporated(ml)}} \times \frac{\text{Vol. of extracting soln. (ml)}}{\text{Wt of oven-dried soil(g)}}$$

Ammonium by Indophenol Blue Method

Apparatus

Erlenmeyer flask, Volumetric flask, Shaker and Spectrophotometer.

Reagents

- Potassium chloride (KCl) solution, 2M: Dissolve 150 g AR grade KCl in 1 litre distilled water.
- Standard ammonium (NH_4^+) solution: Dissolve 0.4717 g of ammonium sulfate $(NH_4)_2SO_4$ in water, and dilute to a volume of 1 litre. If pure dry $(NH_4)_2SO_4$ is used, the solution contains 100 mg of NH_4 -N/ml. Store the solution in a refrigerator. Immediately before use, dilute 4 ml of this stock NH_4^+ solution to 200 ml. The resulting working solution contains 2 mg of NH_4 -N/ml. Accordingly various concentrations of standard solution to be made for the standard curve.
- Phenol-nitroprusside reagent: Dissolve 7 g of phenol and 34 mg of sodium nitroprusside [disodium pentacyanonitrosylferrate, $Na_2Fe(CN)_5NO.2H_2O$]

in 80ml of NH_4^+-free water and dilute to 100 ml. Mix well and store in a dark-colored bottle in the refrigerator.

- Buffered hypochlorite reagent: Dissolve 1.480 g of sodium hydroxide (NaOH) in 70 ml of NH_4^+-free water, add 4.98 g of sodium monohydrogen phosphate (Na_2HPO_4) and 20 ml of sodium hypochlorite (NaOCl) solution (5-5.25 per cent NaOCl). Use less or more hypochlorite solution if the concentration is higher or lower, respectively than that is indicated above. Check the pH to ensure a value between 11.4 and 12.2. Add a small amount of additional NaOH if required to raise the pH. Dilute to a final volume of 100 ml.
- Ethylene diaminetetraacetic acid (EDTA) reagent: Dissolve 6 g of ethylene diamine tetraacetic acid disodium salt (EDTA disodium) in 80 ml of deionized water, adjust to pH 7, mix well, and dilute to a final volume of 100 ml.

Extraction

- Place 10 g of soil in a 250 ml wide-mouth Erlenmeyer flask and add 100 ml of 2M KCl.
- Put stopper and shake the flask on a mechanical shaker for 1 hour.
- Allow the soil-KCl suspension to settle (about 30 min) till the supernatant is clear.
- If the KCl extract cannot be analyzed within 24 hours, then filter the soil KCl suspension (Whatman no. 42 filter paper) and store in the refrigerator. Aliquots from this extract are used for the NH_4 estimation.

Estimation

- Transfer an aliquot (maximum 5 ml) of the filtered 2M KCl extract, containing between 0.5 and 12 mg of NH_4-N, into a 25 ml volumetric flask. Introduce 1 ml of the EDTA reagent and thoroughly mix the contents of the flask.
- Permit the content to rest for 1 minute, thereafter introduce 2 ml of the phenolnitroprusside reagent, followed by 4 ml of the buffered hypochlorite reagent, and promptly dilute the flask to a total volume of 25 ml with NH_4^+-free water, ensuring thorough mixing.
- Position the flask in a water bath regulated at 40 °C, and permit it to remain for 30 minutes. Extract the flask from the bath, allow it to cool to room temperature, and measure the absorbance of the colored complex at a wavelength of 636 nm relative to a reagent blank solution.
- Ascertain the NH_4-N content of the sample by consulting a calibration curve derived from the results of 25 ml standard samples containing 0, 2, 4, 6, 8, 10, and 12 mg of NH_4-N/ml.

- To construct this curve, introduce a suitable volume of 2M KCl solution (equivalent to that utilized for soil extract aliquots, approximately 5 ml) into a series of 25-ml volumetric flasks. Introduce 0, 1, 2, 3, 4, 5, and 6 ml of the 2 mg NH_4 -N/ml solution into the flasks, and assess the intensity of the blue color produced with these standards according to the aforementioned protocol for analyzing unknown extracts.
- (same volume as that used for aliquots of soil extract, *i.e.* about 5 ml)) to a series of 25-ml volumetric flasks. Add 0, 1, 2, 3, 4, 5, and 6 ml of the 2 mg NH_4 -N/ml solution to the flasks, and measure the intensity of blue colour developed with these standards by the procedure described above for the analysis of unknown extracts.

Calculation

NH_4 -N in the sample as noted from the standard curve= A (mg/ml)

$$\mu g \text{ of } NH_4^{-N} \text{ in 1 g soil} = \frac{A \times 100(\text{total vol. of extract})}{5(\text{vol. of extract estimted})} \times \frac{1}{10(\text{wt. of soil})}$$

where,

Weight of the soil taken for estimation = 10 g

Total volume of extract = 100 ml

Volume of extract taken for estimation = 5 ml.

Available Phosphorus

Two predominant approaches for assessing accessible phosphorus in soils are Bray's Method No. 1 for acidic soils and Olsen's Method for neutral and alkaline soils.ls.

Bray's Method No. 1 (Bray and Kurtz, 1945) for Acid Soils

Apparatus

Spectrophotometer, Pipette - 2 ml, 5 ml, 10 ml and 20 ml, Beakers/flasks - 25 ml, 50 ml, 100 ml and 500 ml.

Reagents

- Bray and Kurtz extracting solution (0.03N NH_4 F in 0.025N HCl): Dissolve 11.1 g of AR grade NH_4 F in 100 ml of distilled water, filter, and add to the filtrate 1 litre of water containing 20 ml of concentrated HCl, make up the volume to 10 litres with distilled water. It can be stored (one year).
- 1.5 per cent Ammonium Molybdate reagent: Dissolve 15g $(NH_4)_6MO_7O_{24}.4H_2O$ in 300 ml warm distilled water. Add the solution to 342 ml of concentrated HCl solution gradually with stirring. Dilute to 1 litre with distilled water.

- ✰ 40 per cent Stannous chloride solution (Stock Solution): Dissolve 10 g $SnCl_2H_2O$ in 25 ml of concentrated HCl. Store the solution in a glass stoppered bottle.
- ✰ Stannous chloride solution (Working Solution): Dilute 0.5 ml of the stock solution of stannous chloride to 66.0 ml with distilled water just before use. Prepare fresh dilute solution every working day.

Procedure

Preparation of the Standard Curve

Dissolve 0.1916 g of anhydrous KH_2PO_4 in 1 liter of distilled water. This solution has 0.10 mg P_2O_5 per milliliter. Maintain this as a normal stock solution of phosphate. Measure 10 ml of this solution and dilute it to a total volume of 1 litre using purified water. This solution comprises 1 mg P2O5 per milliliter (0.001 mg P_2O_5 per milliliter). Measure 1, 2, 4, 6, and 10 ml of this solution into individual 25 ml flasks. Incorporate 5 ml of the extractant solution and 5 ml of the molybdate reagent into each sample, then dilute with distilled water to a final volume of approximately 20 ml. Introduce 1 ml of diluted $SnCl_2$ solution, agitate once more, and adjust the volume to the 25 ml mark. After 10 minutes, measure the blue color of the solution using the spectrophotometer at a wavelength of 660 nm. Graph the absorbance readings versus mg P_2O_5 and connect the data points.

Extraction

Add 50 ml of the Bray's extractant No. 1 to the 100 ml conical flask containing 5 g soil sample. Shake for 5 minutes and filter.

Development of Colour

Measure 5 ml of the filtered soil extract using a bulb pipette into a 25 ml volumetric flask; add 5 ml of the molybdate reagent using an automated pipette, dilute to approximately 20 ml with distilled water, mix thoroughly, and incorporate 1 ml of the diluted $SnCl_2$ solution using a bulb pipette. Fill to the 25 ml line and shake vigorously. After 10 minutes, measure the blue color on the spectrophotometer at a wavelength of 660 nm, having calibrated the instrument to zero using a blank made in the same manner but excluding the dirt.

Calculation

$$P(kg / ha) = \frac{A}{10000000} \times \frac{50}{5} \times \frac{20000000}{5} = 4A$$

where,

Weight of the soil taken = 5 g

Volume of the extract = 50 ml

Volume of the extract taken for estimation = 5 ml

Volume made for estimation (dilution = 5 times) = 25 ml

Amount of P observed in the sample on the standard curve = A (mg).

Wt. of 1 ha of soil upto a depth of 22 cm is taken as 2 million kg.

Sodium Bicarbonate Extractable P or Olsen's Method (Olsen *et al.*, 1954)

Apparatus

Spectrophotometer, Pipette - 2 ml, 5 ml, 10 ml and 20 ml, Beakers/flasks - 25 ml, 50 ml, 100 ml and 500 ml.

Reagents

- Bicarbonate extractant (0.5M $NaHCO_3$): Dissolve 42 g Sodium bicarbonate in 1 litre of distilled water and adjust the pH to 8.5 by addition of dilute NaOH or HCl. Filter if necessary. Activated carbon – Darco G 60.
- Molybdate reagent: Dissolve 15g $(NH_4)_6MO_7O_{24}.4H_2O$ in 300 ml warm distilled water. Add the solution to 342 ml of concentrated HCl solution gradually with stirring. Dilute to 1 litre with distilled water.
- Stannous chloride solution: Dilute 0.5 ml of the stock solution of stannous chloride to 66.0 ml with distilled water just before use. Prepare fresh dilute solution every working day.

Procedure

- Preparation of the standard curve: The procedure is identical to that of Bray's Method.
- Extraction: Introduce 50 ml of the bicarbonate extractant into a 100 ml conical flask holding 2.5 g of soil sample. Incorporate 1 g of activated carbon. Agitate for 30 minutes using the mechanical shaker and subsequently filter.
- Color Development: Measure 5 ml of the filtered soil extract using a bulb pipette into a 25 ml volumetric flask; add 5 ml of the molybdate reagent using an automated pipette, dilute to approximately 20 ml with distilled water, mix thoroughly, and introduce 1 ml of the diluted $SnCl_2$ solution using a bulb pipette. Fill to the 25 ml mark and agitate thoroughly. After 10 minutes, measure the blue color on the spectrophotometer at a wavelength of 660 nm, having calibrated the instrument to zero using a blank produced identically but devoid of soil.

Calculation

Same as described in the Bray's Method.

Available Potassium

Flame Photometric Method (Toth and Prince, 1949)

Potassium present in the soil is extracted with neutral normal ammonium acetate solution. This is considered as plant available K in the soils. It is estimated with the help of flame photometer.

Apparatus

Multiple Dispenser– 25 ml, Flasks and beakers - 100 ml and Flame Photometer.

Reagents

- ☆ Molar neutral ammonium acetate solution: Dissolve 77 g of ammonium acetate ($NH_4C_2H_3O_2$) in 1 litre of water. Check the pH with bromothymol blue or with a pH meter. If not neutral, add either ammonium hydroxide or acetic acid as per the need to neutralize it to pH 7.0.
- ☆ Standard potassium solution: Dissolve 1.908 g pure KCl in 1 litre of distilled water. This solution contains 1 mg K/ml. Take 100 ml of this solution and dilute to 1 litre with ammonium acetate solution. This gives 0.1 mg K/ml as stock solution.
- ☆ Working potassium standard solutions: Take 0, 5, 10, 15 and 20 ml of the stock solution separately and dilute each to 100 ml with the N ammonium acetate solution. These solutions contain 0, 5, 10, 15 and 20 mg K/ml, respectively.

Procedure

- ☆ Preparation of the Standard Curve: Configure the flame photometer by alternately atomizing 0 and 20 mg K/ml solutions to achieve readings of 0 and 100, respectively. Atomize intermediate working standard solutions and document the measurements.
- ☆ Graph these measurements against the corresponding potassium concentrations and connect the points with a linear line to establish a standard curve.
- ☆ Extraction: Introduce 25 ml of ammonium acetate extractant into a conical flask secured in a wooden rack containing a 5 g soil sample. Agitate for five minutes and then strain.
- ☆ Quantify potash in the filtrate using the flame photometer.

Calculation

$$K(kg / ha) = A \times 25 / 5 \times \frac{20000000}{1000000} = 10A$$

where,

A = content of K (mg) in the sample, as read from the standard curve: Weight of 1 ha of soil upto a plough depth of 22 cm is approx. 2 million kg.

Available Sulphur

Available sulphur in mineral soils occurs mainly as adsorbed SO_4 ions.

Barium Sulphate Precipitation Method

Apparatus

Spectrophotometer, Mechanical shaker and Volumetric flask.

Reagents

- Mono-calcium phosphate extracting solution (500 mg P/litre): Dissolve 2.035 g of $Ca(H_2PO_4)_2.H_2O$ in 1 litre of water.
- Gum acacia-acetic acid solution: Dissolve 5g of chemically pure gum acacia powder in 500 ml of hot water and filter in hot condition through Whatman No.42 filter paper. Cool and dilute to one litre with dilute acetic acid.
- Barium chloride: Pass AR grade $BaCl_2$ salt through 1 mm sieve and store for use.
- Standard stock solution (2000 mg S/litre): Dissolve 10.89g of oven-dried AR grade potassium sulphate in 1 litre water.
- Standard working solution (10 mg S/litre): Measure exactly 2.5 ml of the stock solution and dilute to 500 ml.
- Barium sulphate seed suspension: Dissolve 18 g of AR grade $BaCl_2$ in 44 ml of hot water and add 0.5 ml of the standard stock solution. Heat the content to boiling and then cool quickly. Add 4 ml of gum
- Dilute nitric acid (approx. 25 per cent): Dilute 250 ml of AR grade conc. HNO_3 to one litre.
- Acetic-phosphoric acid: Mix 900 ml of AR grade glacial acetic acid with 300 ml of H_3PO_4 (AR grade).

Procedure

- Measure 20 g of soil sample into a 250 ml conical flask. Introduce 100 ml of the monocalcium phosphate extraction solution (500 mg P/litre) and agitate for one hour. Utilize Whatman No. 42 filter paper for filtration.
- Transfer 10 ml of the clear filtrate into a 25 ml volumetric flask.
- Incorporate 2.5 ml of 25 percent HNO_3 and 2 ml of acetic-phosphoric acid. Dilute to approximately 22 ml, seal the flask, and shake well if necessary.
- Agitate the $BaSO_4$ seed suspension, then include 0.5 ml of it along with 0.2 g of $BaCl_2$ crystals. Seal the flask, flip it three times, and set it aside.
- After 10 minutes, perform 10 inversions. After 5 minutes, invert 5 times.
- Let it rest for 15 minutes, then incorporate 1 cc of the gum acacia-acetic acid solution.

- Adjust the volume to 25 ml, invert three times, and let aside for 90 minutes.
- Invert ten times and measure the turbidity intensity at 440 nm (using a blue filter), while concurrently running a blank for comparison.
- Preparation of standard curve:
  - Put 2.5, 5.0, 7.5, 10.0, 12.5, 15.0 ml of the working standard solution (10 mg S/litre) into a series of 25 ml volumetric flasks to obtain 25, 50, 75, 100, 125 and 150 mg S.
  - Proceed to develop turbidity as described above for sample aliquots.
  - Read the turbidity intensity and prepare the curve by plotting readings against sulphur concentrations (in mg in the final volume of 25 ml).

Calculation

Available Sulphur (SO_4^{2-} -S) in soil (mg/kg) = W X 100/10 x 20 =W/2

where,

W stands for the quantity of sulphur in "mg" as obtained on X-axis against an absorbance reading (Y-axis) on standard curve "20" is the weight of the soil sample in g, "100" is the volume of the extractant in ml "10" is the volume of extractant in "ml" in which turbidity is developed.

Determination of Exchangeable Calcium and Magnesium

- Exchangeable cations are typically assessed using a neutral normal ammonium acetate extract of soil. Extraction is performed by agitating the soil-extractant combination, followed by filtration or centrifugation. Calcium and magnesium are quantified using either the EDTA titration method or atomic absorption spectrophotometry following the elimination of ammonium acetate and organic debris.
- It is noteworthy that soils may contain significant quantities of soluble calcium and magnesium. Consequently, these water-soluble cations are quantified in the 1:2 soil water extract and subtracted from the ammonium acetate-extractable calcium and magnesium (since ammonium acetate also extracts water-soluble cations) to determine the exchangeable calcium and magnesium. To prepare a soil water extract, typically 25g of soil is combined with 50ml of water, shaken for 30 minutes on a mechanical shaker, and subsequently filtered. The estimate procedure for water extract (water-soluble cations) and ammonium acetate extract (exchangeable cations) is same.
- The EDTA titration technique established by Chang and Bray (1951) is favored because to its precision, simplicity, and efficiency. The approach relies on the premise that calcium, magnesium, and several other cations create stable complexes with versenate (disodium ethylene diamine tetraacetic acid) at varying pH levels. The interference of copper, zinc, iron,

and manganese is mitigated by employing a 2 percent sodium cyanide solution or carbamate. Typically, the concentrations of interfering ions in irrigation water and soil extracts are minimal and can be disregarded. A specified volume of standard calcium solution is titrated with a 0.01N standard versenate solution, utilizing murexide (ammonium purpurate) as an indicator in the presence of NaOH solution. The endpoint is shown by a color transition from orange-red to purple at pH 12, when all calcium ions form a compound with EDTA.

Calcium by Versenate (EDTA) Method

Apparatus

Shaker, Porcelain dish, Beakers and Volumetric/conical flask.

Reagents

- Ammonium chloride – ammonium hydroxide buffer solution: Dissolve 67.5 g
- Standard 0.01N calcium solution: Take accurately 0.5 g of pure calcium carbonate and dissolve it in 10 ml of 3N HCl. Boil to expel CO_2 and then make the volume to 1
- EDTA solution (0.01N): Take 2.0 g of versenate, dissolve in distilled water and make the volume to 1 litre. Titrate it with 0.01N calcium solution and make necessary dilution so that its normality is exactly equal to 0.01N.
- Murexide indicator powder: Take 0.2 g of murexide (also known as ammonium purpurate) and mix it with 40 g of powdered potassium sulphate. Sodium diethyl dithiocarbamate crystals: It is used to remove the interference of Cu, Zn, Fe and Mn.
- Sodium hydroxide 4N: Prepare 16 per cent soda solution by dissolving 160 g of pure sodium hydroxide in water and make the volume to 1 litre. This will give pH 12.

Procedure

- Weigh 5 g of air-dried soil sample into a 150 ml conical flask and introduce 25 ml of neutral normal ammonium acetate. Agitate on a mechanical shaker for 5 minutes and filter using Whatman filter paper No. 1.
- Measure an appropriate quantity (5 or 10 ml) and introduce 2-3 crystals of carbamate together with 5 ml of 16 percent NaOH solution.
- Incorporate 40-50 milligrams of the indicator powder. Titrate with a 0.01N EDTA solution until the color transitions from orange-red to reddish violet (purple). It is recommended to introduce a drop of EDTA solution every 5 to 10 seconds, as the color change is not immediate.
- The endpoint must be compared to a blank reading. If the solution is excessively titrated, it should be returned titrated using a standard

calcium solution to determine the precise volume utilized. Record the volume of EDTA utilized for titration.

Calculation

If N_1 is normality of Ca^{2+} and V_1 is volume of aliquot taken and N2V2 are the normality and volume of EDTA used, respectively, then, $N_1V_1 = N_2V_2$

Calcium Plus Magnesium by Versenate (EDTA) Method

- Magnesium in solution can be titrated with 0.01N EDTA with Eriochrome Black T dye as an indicator at pH 10, in the presence of an ammonium chloride and ammonium hydroxide buffer. At the endpoint, the color transitions from wine red to blue or green.
- This titration will estimate both calcium and magnesium when calcium is present in the solution. Above pH 10, magnesium does not bind effectively to the Eriochrome Black T indicator, resulting in an indistinct endpoint.

Apparatus

Shaker, Porcelain dish, Beaker and Volumetric/conical flask.

Reagents

- EDTA or Versenate solution (0.01N): Same as in calcium determination.
- Ammonium chloride-ammonium hydroxide buffer solution: Same as in calcium determination.
- Eriochrome black T indicator: Take 100 ml of ethanol and dissolve 4.5 g of hydroxyl amine hydrochloride in it. Add 0.5 g of the indicator and prepare solution. Hydroxylamine hydrochloride removes the interference of manganese by keeping it in lower valency state (Mn^{2+}). Or mix thoroughly 0.5 g of the indicator with 50 g of ammonium chloride.

Procedure

- Weigh 5 g of air-dried soil into a 150 ml flask, add 25 ml of neutral normal ammonium acetate solution, mix on a mechanical shaker for 5 minutes, and filter through Whatman No. 1 filter paper.
- Transfer 5 ml of aliquot containing no more than 0.1 meq of Ca and Mg. If the solution exhibits a larger concentration, it need dilution.
- Incorporate 2 to 5 crystals of carbamate and 5 cc of ammonium chloride.
- Titrate this solution with 0.01N versenate until the color transitions to vivid blue.
- Sodium cyanide solution (2%) or sodium diethyl dithiocarbamate crystals. This is utilized to eliminate the interference of copper, cobalt, and nickel.

Calculation

If N_1 and V_1 are normality (concentration of Ca^{2+}/Mg^{2+}) and volume of aliquot taken and N2V2 are the normality and volume of EDTA used respectively, then

$N_1V_1 = N_2V_{2,}$

Ca^{2+}/Mg^{2+} meq/100 g soil = 100/wt of soil x extract volume/1000 x Ca^{2+} + Mg meq/litre.

Micro-Nutrients

- For the assessment of micronutrients, the bioavailable form is crucial, rather than the overall concentration. The primary aim of soil testing for micronutrients, akin to macronutrients, is to ascertain whether the soil can provide sufficient micronutrients for optimal crop yield or if deficits are anticipated in crops cultivated on such soils.
- The micronutrients most frequently examined are Zn, Cu, Fe, Mn, B, and Mo, which are addressed herein.

Available Zinc, Copper, Iron and Manganese

- Ethylene diamine tetraacetic acid (EDTA) combined with ammonium acetate is frequently employed for the extraction of several elements. DTPA (Diethylene triamine pentaacetic acid) is a prevalent extractant utilized for the concurrent extraction of elements such as Zn, Cu, Fe, and Mn.
- This extractant was developed by Lindsay and Norvell (1978). While a specific extractant that exhibits a stronger correlation with plant availability may be advantageous, a universal extractant is more cost-effective and time-efficient, particularly in a service laboratory where numerous samples must be analyzed rapidly.
- Consequently, DTPA, as a specific extractant, has been delineated in this paper. The quantification of components in the extract is conducted using an Atomic Absorption Spectrophotometer (AAS). Proposed critical limits for DTPA-extractable micronutrient elements by Lindsay and Norvell (1978).

Principle of Extraction

- DTPA is an important and widely used chelating agent, which combines with free metal ions in the solution to form soluble complexes of elements.
- To avoid excessive dissolution of $CaCO_3$ which may release occluded micronutrients that are not available to crops in calcareous soils and may give erroneous results, the extractant is buffered in slightly alkaline pH. Triethanolamine (TEA) is used as buffer because it burns cleanly during atomization of extractant while estimating in AAS.
- The DTPA has a capacity to complex each of the micronutrient cations as 10 times of its atomic weight. The capacity ranges from 550 to 650 mg/kg depending upon the micronutrient cations.

Extracting Solution (DTPA)

- DTPA 0.005M, 0.01M $CaCl_2.2H_2O$ and 0.1M TEA extractant: Add 1.967 g DTPA and 13.3 ml TEA in 400 ml distilled water in a 500 ml flask. Take 1.47 g $CaCl_2.2H_2O$ in a separate 1000 ml flask.
- Add 500 ml distilled water and shake to dissolve. Add DTPA+TEA mixture in $CaCl_2$ solution and make the volume to 1 litre. The pH should be adjusted to 7.3 by using 1M HCl before making the volume.

Principle of Estimation

- The extracted elements can be estimated by various methods, which include volumetric analysis, spectrometry and atomic absorption spectroscopy. Volumetric methods such as EDTA and $KMnO_4$ titrations are used for estimation of zinc and Mn, and iron, respectively. Copper can be estimated by titration with $Na_2S_2O_3$.
- Spectrometric methods are deployed in estimation of specific colour developed due to the presence of an element, which forms coloured compound in the presence of specific chemicals under definite set of conditions. The colour intensity has to be linear with the concentration of the element in question. The interference due to any other element has to be eliminated. Such methods are dithiozone method for estimation of zinc, orthophenanthroline method for iron, potassium periodate method for manganese, carbamate method for copper.
- The chemical methods are generally cumbersome and time taking. Hence the most commonly employed method is atomic absorption spectrometry. Here, the interference by other elements is almost nil or negligible because the estimation is carried out for an element at a specific emission spectra line.
- In fact in AAS, traces of one element can be accurately determined in the presence of a high concentration of other elements - IV.

Principle of Atomic Absorption Spectrophotometry

- The procedure is based on flame absorption rather than flame emission and upon the fact that metal atoms absorb strongly at discrete characteristic wavelengths which coincides with the emission spectra lines of a particular element. The liquid sample is atomized. The hollow cathode lamp which precedes the atomiser emits the spectrum of the metal used to make the cathode. This beam traverses the flame and is focused on the entrance slit of a monochromator, which is set to read the intensity of the chosen spectral line. Light with this wavelength is absorbed by the metal in the flame and the degree of absorption being the function of the concentration of the metal in the flame, the concentration of the atoms in the dissolved material is determined.

- For elemental analysis, a working curve or a standard curve is prepared by measuring the signal or absorbance of a series of standards of known concentration of the element under estimation. From such curve, concentration of the element in unknown sample is estimated. Atomic Absorption Spectroscopy can be successfully applied for estimation of Zn, Cu, Fe and Mn.
- For specific estimation on AAS, hollow cathode lamps, specific to specific elements are used. The specifications of relevant hollow cathode lamps are given in appendices. Running parameters which are specific to a particular model are given in the software provided with the equipment manual. Accordingly, the current supply, wave length of hollow cathode lamp, integration time and anticipated estimation ranges are fixed. Hollow cathode and Deuterium lamps are required to be properly aligned before starting the equipment.
- After proper alignment and adjustment, standard curves are prepared to ensure that the concentration of the elements in solutions perfectly relates to the absorbance.

Preparation of Standard Solutions

- Readymade standard solutions 1000 mg/ml or 1 mg/ml (1000 ppm) of dependable accuracy are supplied with the AAS and are also available with the suppliers of chemical reagents. If the standard solutions are to be prepared in the laboratory, either metal element foils of 100 per cent purity or the standard chemical salts can be used.
- In case of Zn, Cu and Fe, 1000 mg/ml (1000 ppm) standard solution are preferably prepared by dissolving 1.0 g pure metal wire and volume made to 1 litre as per the method described under each element. It is diluted to obtain the required concentration. In case of Mn, MnSO4.H2O is preferred.

Preparation of Standard Curves

Zinc

Reagents

- Standard Zinc Solution: Weigh 1.0 g of pure zinc metal in a beaker. Add 20 ml HCl (1:1). Keep for few hours allowing the metal to dissolve completely. Transfer the solution to 1 litre volumetric flask. Make up the volume with glass-distilled water. This is 1000 mg/ml zinc solution. For preparation of standard curve, refer 1000 mg/ml solution as solution A. Dilute 1 ml of standard A to 100 ml to get 10 mg/ml solution to be designated as standard B.
- Glass-distilled or demineralized acidified water of pH 2.5 + 0.5: Dilute 1 ml of 10 per cent sulphuric acid to one litre with glass-distilled or

mineralized water and adjust the pH to 2.5 with a pH meter using 10 per cent H_2SO_4 or NaOH. This solution is called acidified water.

- Working Zn standard solutions: Pipette 1, 2, 4, 6, 8 and 10 ml of standard B solution in 50 ml numbered volumetric flask and make the volume with DTPA solution to obtain 0.2, 0.4, 0.8, 1.2, 1.6 and 2.0 mg/ml zinc. Stopper the flasks and shake them well. Fresh standards should be prepared every time when a fresh lot of acidified water is prepared.

Procedure

- Prepare the solutions: Dilute the standards for atomic absorption spectrophotometry at a wavelength of 213.8 nm (zinc line of the instrument).
- Construct a standard curve of known zinc solution concentrations by graphing absorbance values on the Y-axis against their corresponding zinc concentrations on the X-axis.

Copper

Reagents

- Standard copper solution: Weigh 1 g of pure copper wire on a clean watch glass and transfer it to one litre flask. Add 30 ml of HNO_3 (1:1) and make up the mark. Stopper the flask and shake the solution well. This is 1000 mg/ml Cu solution and should be stored in a clean bottle for further use. Dilute 1 ml of 1000 mg/ml solution of copper to 100 ml to get 10 mg/ml standard copper solutions.
- Glass-distilled or demineralized acidified water of pH 2.5 + 0.5: Same as that done for Zn.
- Working Cu standard solutions: Pipette 2, 3, 4, 5, 6 and 7 ml of 10 mg/ml standard Cu solution in 50 ml numbered volumetric flasks and make the volume with DTPA solution to get 0.4, 0.6, 0.8, 1.0, 1.2 and 1.4 mg/ml copper. Stopper the flasks and shake them well. Prepare fresh standards every fortnight.

Procedure

- Calibrate the standards on an atomic absorption spectrophotometer at a wavelength of 324.8 nm (copper line of the instrument).
- Construct the standard curve by graphing the known concentration of copper on the X-axis against the corresponding absorbance values on the Y-axis.

Iron

Reagents

- ***Standard iron solution:*** Weigh accurately 1 g pure iron wire and put it in a beaker and add approximately 30 ml of 6M HCl and boil. Transfer it to one litre volumetric flask through the funnel giving several washings to the beaker and funnel with glass-distilled water. Make the volume up to the mark. Stopper the flask and shake the solution well. This is 1.000 mg/ml iron solution.

- Glass-distilled or demineralized acidified water of pH 2.5 ± 0.5: Same as that done for Zn.
- ***Working Fe standard solutions:*** Pipette 10 ml of iron stock solution in 100 ml volumetric flask and dilute to volume with DTPA solution. This is 100 mg/ml iron solution. Take 2, 4, 8, 12 and 16 ml of 100 mg/ml solution and dilute each to 100 ml to obtain 2, 3, 8, 12 and 16 mg/ml of Fe solution.

Procedure

- Calibrate the standards on an atomic absorption spectrophotometer at a wavelength of 248.3 nm (iron line of the instrument).
- Construct the standard curve by graphing the known concentrations of copper on the X-axis against the corresponding absorbance values on the Y-axis.

Manganese

Reagents

- ***Standard Mn solution:*** Weigh 3.0751 g of AR grade manganese sulphate ($MnSO_4H_2O$) on a clean watch glass and transfer it to one litre flask through the funnel giving several washings to watch glass and funnel with acidified water and make the volume up to the mark. This solution will be 1 000 mg/ml Mn. A secondary dilution of 5 ml to 100 ml with acidified water gives a 50 mg/ml solution.
- Glass-distilled or de-mineralized acidified water of pH 2.5 + 0.2: Same as that for Zn.
- ***Working Mn standard solutions:*** Standard curve is prepared by taking lower concentrations of Mn in the range of 0-10 mg/ml Take 1, 2, 4, 6 and 8 ml of 50 mg/ml solution and make up the volume with DTPA solution to 50 ml to obtain 1, 2, 4, 6 and 8 mg/ml working standards.

Procedure

- Calibrate the standards on an atomic absorption spectrophotometer at a wavelength of 279.5 nm (manganese line of the instrument).
- Construct the standard curve by graphing the known concentration of Mn on the X-axis against the corresponding absorbance values on the Y-axis.

Procedure for Extraction by DTPA

Once standard curves have been prepared, proceed for extraction by DTPA.

- Take 10 g of soil sample in 100 ml narrow-mouth polypropylene bottle.

- Add 20 ml of DTPA extracting solution.
- Filter the content using filter paper No.1 or 42 and collect the filtrate in polypropylene bottles 5. Prepare a blank following all steps except taking a soil sample.

Note: The resulting extract is utilized for the quantification of several micronutrients. Element-specific extractants are utilized to provide a more precise quantification of an element that exhibits a stronger association with plant availability. An extractant that is standardized or suggested for a certain context within a country may be utilized. The estimating process for AAS, however, remains unaltered.

Estimation on Atomic Absorption Spectrophotometer (AAS)

- Select a certain hollow cathode lamp for an element and install it on the AAS. Start the flame and set the instrument at zero by using blank solution.
- Sequentially aspirate the standard solutions of varying concentrations and document the readings. Construct a standard curve by graphing the concentration of the relevant element against the associated absorbance in various standard samples, as previously outlined.
- When executed precisely, a linear relationship is established between the element's concentration and absorbance in AAS, with a correlation coefficient approaching 1.0.
- Extract the soil solution for the assessment of nutrient elements in the specified soil sample and record the measurements.
- Determine the substance present in the soil extract by analyzing its concentration on the standard curve in relation to its absorbance.

Calculation

Content of micronutrient in the sample (mg/kg) = C mg/ml x 2 (dilution factor)

where,

Dilution factor = 2.0 (Soil sample taken = 10.0 g and DTPA used = 20 ml) Absorbance reading on AAS of the soil extract being estimated for a particular element = X

Concentration of micronutrient as read from the standard curve for the given absorbance (X) = C mg/ml.

Available Boron

The predominant technique for assessing available boron is hot water extraction of soil, as established by Berger and Truog in 1939. Several modified iterations of this method have been suggested, although the fundamental procedure persists unchanged.

Extraction Procedure

Weigh 25 grams of soil in a quartz flask or beaker. Introduce approximately 50 ml of double-distilled water and around 0.5 g of activated charcoal. The solution is heated to boiling for approximately 5 minutes and subsequently filtered using Whatman Filter Paper No. 42.

Estimation by AAS

The specs of the pertinent hollow cathode lamp are provided below:

Lamp current (m Ao)	5
Wave length (nm)	249.7
Linear range (µg/ml)	1-4
Slit width (nm)	0.02
Integration time (sec)	2.0

Flame Acetylene Nitrous Oxide

- The software accompanying the equipment instructions contains the running parameters relevant to each model.
- The present supply, wavelength of the hollow cathode lamp, integration time, and projected estimation ranges are established. Hollow cathode and deuterium lamps must be accurately aligned prior to initiating the device.
- Following appropriate alignment and calibration, standard curves are established to guarantee that the concentration of the element in solutions correlates precisely with the absorbance.

Reagents

- Standard Boron Solution: Dissolve 8.819 g $Na_2B_4O_7.10H_2O$ in warm water. Dilute to 1 litre to get 1000 mg/ml boron stock solution. Dilute 1 ml of standard to 100ml to get 10 mg/ml boron.
- Working standards: Take 1, 2, 3, 4, 5, 6, 7 and 10 ml of 10 ìg/ml solution and dilute each to 50 ml to get 0.2, 0.4, 0.6, 0.8, 1.0, 1.2, 1.4 and 2.0 mg/ml B.

Procedure

- The software accompanying the equipment manual contains model-specific running settings.
- The present supply, hollow cathode lamp wavelength, integration duration, and expected estimation ranges are established. Hollow cathode and Deuterium lamps must be accurately aligned prior to initiating the device.

- Following appropriate alignment and calibration, standard curves are established to guarantee that the concentration of the element in solutions correlates precisely with the absorbance.

Calculation

Content of B in the soil (mg/g or mg/kg) = C x dilution factor (10)

where,

C (mg/ml) = concentration of B as read from the standard curve against the absorbance reading of the soil solution on the spectrophotometer;

Dilution factor = 10, which is calculated as follows:

Weight of the soil taken = 25 g.

Volume of extractant (water) added = 50 ml.

First dilution = 2 times

Volume of the filtrate taken = 5 ml.

Final volume of filtrate after colour development = 25 ml.

Second dilution = 5 times.

Total dilution = 2 x 5 = 10 times.

Estimation by Colorimetric Method

The concentration of B in the filtered extract is quantified using the azomethine-H colorimetric technique.

Apparatus

Analytical balance, Flask or beaker, Volumetric flask, Funnels, Whatman No. 42 filter paper and Spectrophotometer.

Reagents

- *Azomethine-H:* Dissolve 0.45 g azomethine-H and 1.0 g L-ascorbic acid in about100 ml deionized or double-distilled water. If solution is not clear, it should be heated gently in a water bath or under a hot water tap at about 300 °C till it dissolves. Every week a fresh solution should be prepared and kept in a refrigerator.
- *Buffer solution:* Dissolve 250 g ammonium acetate in 500 ml deionized or double-distilled water and adjust the pH to about 5.5 by slowly adding approximately 100 ml glacial acetic acid, with constant stirring.
- *EDTA solution (0.025 M):* Dissolve 9.3 g EDTA in deionized or double distilled water.
- Standard stock solution: Dissolve 0.8819g $Na_2B_4O_710H_2O$ AR grade in a small volume of deionized water and make volume to 1000 ml to obtain a stock solution

- *Working standard solution:* Take 5 ml of stock solution in a 100 ml volumetric flask and dilute it to the mark. This solution contains 5 mg B/ml.

Procedure

- Take 5 ml of the clear filtered extract in a 25 ml volumetric flask and add 2 ml buffer solution, 2 ml EDTA solution and 2 ml azomethine-H solution. Mix the contents thoroughly after the addition of each reagent.
- Let the solution stand for 1 hour to allow colour development. Then, the volume is made to the mark. Intensity of colour is measured at 420 nm. The colour thus developed has been found to be stable upto 3-4 hours.
- Preparation of standard curve: Take 0, 0.25, 0.50, 1.0, 2.0 and 4.0 ml of 5 mg B/ml solution (working standard) to a series of 25 ml volumetric flasks. Add 2 ml each of buffer reagent, EDTA solution and azomethine-H solution. Mix the contents after each addition and allow to stand at room temperature for 30 minutes. Make the volume to 25 ml with deionized or double-distilled water and measure absorbance at 420 nm. This will give reading for standard solution having B concentration 0, 0.05, 0.10, 0.20, 0.40 and 0.80 mg B/ml.

Calculation

Content of B in the soil (ìg/g or mg/kg) = C x Dilution factor (10)

where,

C (mg/ml) = Concentration of B as read from the standard curve against the absorbance reading of the soil solution on the spectrophotometer.

Dilution factor = 10 which is calculated as follows:

Weight of the soil taken = 25 g

Volume of extractant (water) added = 50 ml

First dilution = 2 times

Final volume of filtrate after colour development = 25 ml

Second dilution = 5 times

Total dilution = 2 x 5 = 10 times.

Available Molybdenum

- Molybdenum (Mo) is an infrequent element in soils, occurring only in minimal quantities in igneous and sedimentary rocks. In the soil solution Mo exists mainly as $HMoO_4$ ion under acidic condition, and as MoO_4^{2-} ion under neutral to alkaline conditions.
- A shortage of molybdenum (Mo) in plants is prevalent at soil concentrations of 0.1 mg/g or lower. Molybdenum poisoning (molybdenosis) frequently occurs in cattle who graze fodder plants containing 10-20 mg Mo/g.

Estimation by Colorimetric Method

Apparatus

Spectrophotometer, Hot plate, Refrigerator and Water bath.

Reagents

- ***50 per cent potassium iodide solution:*** Dissolve 50 g in 100 ml of double-distilled water (DDW).
- ***50 per cent ascorbic acid solution:*** Dissolve 50 g in 100 ml of DDW.
- ***10 per cent sodium hydroxide solution:*** Dissolve 10 g of NaOH in 100 ml of DDW.
- ***10 per cent thiourea solution:*** Dissolve 10 g in 100 ml of DDW and filter. Prepare fresh solution on the same day of use.
- ***Toluene-3, 4-dithiol solution:*** Weigh 1.0 g of AR grade melted dithiol (51 °C) in a 250 ml glass beaker. Add 100 ml of the 10 per cent NaOH solution and warm the content upto 510 °C with frequent stirring for 15 minutes. Add 1.8 ml of thioglycolic acid and store in a refrigerator.
- ***10 per cent tartaric acid:*** Dissolve 10 g in 100 ml of DDW.
- Iso-amyl acetate and Ethyl alcohol.
- ***Ferrous ammonium sulphate solution:*** Dissolve 63 g of the salt in about 500 ml of DDW and then make the volume to one litre.
- Nitric-perchloric acid mixture (4:1).
- ***Extracting reagent:*** Dissolve 24.9 g of AR grade ammonium oxalate and 12.6 g oxalic acid in water and make the volume to one litre. Standard stock solution (100 ìg/ml Mo): Dissolve 0.150 g of AR grade MoO_3 in100 ml of 0.1M NaOH, make slightly acidic with dilute HCl and make the volume to 1 litre.
- ***Working standard solution (1 mg/ml Mo):*** Dilute 10 ml of the stock solution to 1 litre.

Procedure

- Weigh 25 g of air-dry soil sample in a 500 ml conical flask. Add 250 ml of the extracting solution (1:10 ratio) and shake for 10 hours.
- Filter through Whatman No. 50 filter paper and Collect 200 ml of the clear filtrate in a 250 ml glass beaker and evaporate to dryness on a water bath.
- Heat the contents in the beaker at 500°C in a furnace for 5 hours to destroy organic matter and oxalates. Keep overnight.
- Digest the contents with 5 ml of HNO3 - $HClO_4$ mixture (4:1), then with 10 ml of 4M H_2SO_2 and H_2O_2, each time bringing to dryness.

- Add 10 ml of 0.1M HCl and filter. Wash the filter paper, first with 10 ml of 0.1M HCl and then with 10 ml of DDW until the volume of the filtrate is 100 ml.
- Run a blank side by side (without soil). Take 50 ml of the filtrate in 250 ml separatory funnels and add 0.25 ml of ferrous ammonium sulphate solution and 20 ml of DDW and shake vigorously.
- Add excess of potassium iodide (KI) solution and clear the liberated iodine by adding ascorbic acid drop by drop while shaking vigorously. Add one ml of tartaric acid and 2 ml of thiourea solution and shake vigorously.
- Add 10 ml of iso-amyl acetate and separate out the contents (green colour) in colorimeter tubes/cuvettes. Read the colour intensity at 680 nm (red filter).
- Preparation of standard curve: Measure 0, 2, 5, 10, 15 and 20 ml of the working standard Mo solution containing 1 mg/litre Mo in a series of 250 ml separatory funnels. Proceed for colour development as described above for sample aliquots. Read the colour intensity and prepare the standard curve by plotting Mo concentration against readings.

Calculation

Available Mo in soil (mg/g) = A x 250/100 x 1/25 = A/20

where,

A = Mo concentration in mg/ml as obtained on X-axis against a sample reading.

Chapter 14

Analysis of Soil Biological Properties

Introduction

- Soil serves as a habitat for a multitude of flora and wildlife. They conduct several metabolic interactions to enable essential functions that soil offers, hence rendering soil a living system. Soil microbes engage in organic matter recycling, nitrogen modification, soil detoxification, and serve as a reservoir of genetic material for various bioactive compounds. Soil microbial activity maintains favorable soil conditions for optimal plant establishment through aeration, aggregation, nutrient fixation and solubilization, and plant protection, hence ensuring food security.
- In contrast to soil physical and chemical features, soil biological activity responds swiftly to management approaches, making it a sensitive indicator of soil health. Thus, assessing soil microbial activity and abundance will provide early indications regarding the promotion or inhibition of management actions.

Quantification and Separation of Soil Microbial Communities by Dilution Plating Method

Materials

- Soil sample (Rhizosphere and Non-Rhizosphere)
- Sterile 1 ml/10 ml pipettes
- Sterile water blanks 95 ml (in conical flask)
- Sterile water blanks 9 ml (in cotton plugged test tubes)
- Sterile Petri plates

Reagents

I. Thornton's Medium (for bacteria)

- ☆ Potassium di-hydrogen phosphate 1.0 g
- ☆ Magnesium sulphate 0.2 g
- ☆ Calcium chloride 0.1 g
- ☆ Sodium chloride 0.1 g
- ☆ Potassium nitrate 0.5 g
- ☆ Aspargin 0.5 g
- ☆ Mannitol 1.0 g
- ☆ Ferric chloride: Traces
- ☆ Agar-agar 20g
- ☆ Volume to make 1 L
- ☆ pH : 6.8-7.2

II. Martin's Rose Bengal Agar Medium (for Fungi)

- ☆ Dextrose 10.0 g
- ☆ Peptone 5.0 g
- ☆ Potassium di-hydrogen phosphate 1.0 g
- ☆ Magnesium sulphate heptahydrate 0.5 g
- ☆ Rose Bengal (1 per cent solution) 3.3 ml
- ☆ Agar-agar 20.0 g
- ☆ Streptomycin: Traces*
- ☆ Volume to make 1 L
- ☆ pH: 6.5-7.2

*A concentrated stock solution of streptomycin is made, filter sterilized, and added post-autoclaving, immediately prior to planting. The final concentration of streptomycin should be approximately 30 ppm.

III. Kenknight and Munaier's Medium (for Actinomycetes)

- ☆ Dextrose 1.0 g
- ☆ Potassium di-hydrogen phosphate 0.1 g
- ☆ Sodium nitrate 0.1 g
- ☆ Potassium chloride 0.1 g
- ☆ Magnesium sulphate heptahydrate 0.1 g
- ☆ Agar-agar 15.0 g
- ☆ Volume to make 1 L
- ☆ pH: 6.8-7.5

Equipments

Laminar flow assembly/inoculation camber, Incubator and Autoclaves.

Procedure

- ☆ Uproot a target plant from field along with soil around its roots preferably on the day of analysis
- ☆ Separate soil particles closely adhering to the roots by gentle tapping/ shaking. This constitute rhizosphere soil sample. The other soil sample from non-cropped area is considered non-rhizosphere sample. Avoid long period of storage and do not freeze the soil sample. If stored, keep below ambient temperature in plastic pouch to prevent drying, and incubate at room temperature for 24 hours, before analysis.
- ☆ Prepare serial dilution in Laminar flow assembly by adding 10 g soil sample in 95 ml water blank or 0.85 per cent NaCl solution prepared in 250 ml conical flask.
- ☆ Close the mouth of the flask and vigorously shake the flask to disperse soil thoroughly by giving vertical strokes (at least 20 minutes). This constitutes 1:10 or 10^{-1} dilution. The shaking should be uniform from all soil samples.
- ☆ Transfer 1 ml soil suspension obtained in previous step into 9 ml sterile water blank in the test tubes. Shake uniformly by rolling the tube between the palm of your hands to provide horizontal shaking. Continue the series in similar manner to get up to 10^{-6} dilution level. Mark the dilutions properly on the test tubes.
- ☆ Transfer 1 ml of required dilution (10^{-4} for fungi and actinomycetes and 10^{-6} for bacteria) into sterile petri plates.
- ☆ Pour the required medium (at approx. 45 °C cooled) uniformly for specific organism into petri plates. Rotate clockwise and anti-clockwise to mix soil suspension with medium. Make the rotation uniform for all samples. Allow the medium to solidify. Mark details on the lid.
- ☆ Incubate plates in inverted position at 28 °C in incubator for 2 days for fungi; 4 days for bacteria and one week for actinomycetes. Take at least four replications for each sample.
- ☆ Examine nature of colonies developing on plates after the required incubation period. Count the number of colonies. Discard plates showing large spreaders and mold colonies. Number of colonies multiplied by the respective dilution factor will give number of viable cell g^{-1} soil.
- ☆ Record the per cent moisture in soil sample by gravimetric method. Express results on the basis of oven dry weight of soil sample.
- ☆ Count the number of colonies appeared on plate and express colony forming unit (CFU) as

  No. of CFU/g= No. of colony/(Dilution×Volume of inoculums plated)

- Calculate standard deviation and standard error of mean. Statistically compare the number of organisms in the rhizosphere and non-rhizosphere soil samples.
- Calculate rhizosphere effect in terms of R:S ratio (R/S).

Evaluation of Nematode Populations in Soil Utilizing Cobb's Decanting and Sieving Technique

Nematodes, often known as roundworms or threadworms, constitute one of the most abundant groups of soil fauna. They consume bacteria, fungi, and various species, and can behave as parasites, resulting in illnesses in plants and animals. Predatory nematodes devour other nematodes, rotifers, and protozoa, hence regulating soil microbial populations. They must be extracted from affected plants or soils for identification and enumeration. The predominant technique for enumerating soil nematodes is outlined below.

Materials

- Two plastic pans
- A range of sieves-20, 60, 100, 200 and 325 mesh size.
- 100 ml petri dishes and Tissue paper
- Aluminium range
- 250 ml and 400 ml beakers.
- Graduating counting dish and Baermann funnel.

Equipment

Microscope.

Procedure

- Collect soil samples from periphery of the patch where the plants show gradual decline in growth. For perennial crops, collect samples from feeder root zone. For annual crops samples from rhizosphere zone give better assessment.
- Keep samples moist and store under shade. If necessary, samples may be stored at 10-15 °C without allowing the soil to dry.
- Keep 100 g soil in pan I and soak in one litre of water for 5 minutes.
- Break soil crumbs by hand. Discard big stones or granules separated by hand, and mix suspension thoroughly. Tilt pan and wait for 10 seconds to settle heavier particles. Then pour the muddy mixture through 20 mesh sieve into Pan II.
- To obtain optimum yield of nematodes, add more water to the residue in Pan I, and repeat the process thrice. To ensure that even very long nematode pass through the sieve, rinse materials on sieve with gentle jet of water or partly immerse in Pan II with gentle shaking.

- Discard residue in Pan I and on 20 mesh sieve. Mix contents in Pan II thoroughly and pour the same way through 60 mesh sieve back to Pan I. Repeat the process two or three times and rinse the sieve as before.
- Wash the residue on the sieve into a 250 ml beaker to observe nematode cysts. Repeat the process of decanting and sieving with the rest three sieves 100, 200 and 325 mesh with the additional precaution to keep the sieve at 45o angle with ground while sieving. This will prevent the filamentous nematode to pass easily through the sieve.
- Collect the residue on each sieve separately through backwash. If the suspension of the residue is cloudy due to presence of colloidal clay, return to the respective sieve and rinse again before proceeding to the next stage.
- Discard materials coming through the 325 mesh. This is supposed not to carry any nematode.
- Allow the contents of beaker to settle for 2 h and pour out excess supernatant water. Pool contents of all the beakers (except those collected over 60 mesh sieve) in 100 ml beaker and again allow to settle for reduction of the volume.
- Pour suspension on the tissue paper in Baermann funnel assemble and leave for 24 to 48 h to allow the nematode to swim across the tissue paper in the water below leaving behind the other debris. This helps in separating the nematodes from suspension
- Measure the volume of contents in the petri dish or collecting vial. Spread 5 to 10 per cent of it over a graduated dish for counting by observation under microscope at 40-60x magnification.

Calculation

Express nematode population in number per unit volume of moist soil.

Isolation and Enumeration of Soil Protozoa

- Protozoa are the most basic unicellular eukaryotic organisms that are prevalent in soil. Their influence on the soil environment is limited; however, a decline in their population can significantly impact decomposition and transformation processes, perhaps due to their role in regulating soil microbial populations. Soil protozoa directly consume (prey upon) host bacteria. It will lead to the eradication of bacterial growth. A sequence of replicated soil dilutions is plated on petri dishes containing host bacterium growth. The elimination of host bacterium development owing to predation indicates the existence of at least one protozoan, which is classified as positive (+).
- When the host bacteria remains unaffected, the absence of protozoa (predator) is recorded as negative (-). By tallying both positive and

negative growth in each dilution and their replicates, the quantity of protozoa can be determined from the aggregate of negative responses utilizing Fisher's method of negative plates.

Materials

Soil Sample, Sterile 1 ml pipettes, Sterile test tubes, Glass rings (2 cm internal diameter and 1 cm depth; 1-2 mm thickness), *Aerobacter aerogen* and Sterilized petri plates.

Reagent

Sterile normal saline solution. Dissolve 8.5 g sodium chloride in 1000 ml distilled water. Sterilize for 15 minutes at 15 kg cm^2 pressure and 120 °C temperature.

Equipment

Incubator.

Procedure

- Take 10 g of soil and shake for 4 min with 50 ml sterile normal saline solution.
- From the soil suspension make a series of two fold dilution by adding 1 ml of each dilution to 1 ml sterile normal solution in each test tube. Make a series of 15 dilutions.
- Pour about 15 ml of molten agar containing 5 g L^{-1} NaCl into each of the sterilized petri plates of 10 cm diameter.
- Place eight sterile glass rings in each petri plate quickly before the agar solidifies.
- Make thick suspension of *Aerobacter aerogen* from a week-old growth slant. Place one drop of this suspension to each dilution in the eight rings.
- Place 0.5 ml suspension of each dilution in the eight rings in one petri plate. You will have thus 15 such plates with 8 rings in each.
- Incubate plates for 15 days at 28± 1 °C in an incubator.
- Examine each ring for bacterial growth. Count rings showing bacterial growth as (+) and those without bacterial growth (-) for the presence of protozoa.
- Count total number of (+) and (-) rings. Calculate (showing no protozoa) by applying Fisher's method of negative plates from the fisher's table.

Quantification of Soil Algae by Most Probable Number (MPN) Method

A fundamental prerequisite of this approach is that the organism must induce a discernible alteration in the media or exhibit distinctive identifiable growth, such as green or blue-green coloring resulting from algal proliferation. This method can also be employed for quantifying nitrifying bacteria (nitrate creation) and Rhizobium (nodule production).

Equipment

Autoclave and Laminar flow assembly/Inoculation chamber.

Reagent

Bristol's Medium

- Sodium nitrate: 1 g
- Potassium monohydrogen phosphate: 1 g
- Magnesium sulphate heptahydrate: 0.3 g
- Calcium chloride hexahydrate ' 0.1 g
- Sodium chloride: 0.1 g
- Ferric chloride hexahydrate: Traces
- Volume to make 1 L

Procedure

- Weigh 10 g soil and suspend it in 95 ml sterile water in a 250 ml conical flask. Shake vigorously and allow the coarse particle to settle. This establishes 1:10 or 10^{-1} dilution.
- Following serial dilution techniques, prepare dilution up to 10^{-6}.
- Take 10 ml of Bristol's medium in each of the five test tubes for each level of dilution (total 30 test tubes), plug with cotton and sterilize.
- Transfer 1 ml of the each dilution into 5 test tubes containing Bristol's medium.
- Incubate tubes in a growth chamber (at 22 °C) or green house or near a window exposed to light for a period of one month.
- Observe for development of green/blue-green pigmented growth and at the end count the number of tubes showing algal growth for each dilution. These are the positive tubes.
- Determine moisture content of the soil gravimetrically.
- Calculate the most probable number of algae in the soil sample as described below. Express the results on the basis of oven dry weight of the soil.

Calculation

- Identify the lowest dilution in which all tubes are positive (*i.e.*, showing algal growth) and select as P1 the number of positive tubes in that dilution (P1 = 5 in this case).

- Count P2 and P3 which represent the number of positive tubes in the next two higher dilutions. Obviously, P2 and P3 values will be less than P1 (*i.e.*, 5).
- Refer the table of most probable number(MPN) for use with 10-fold dilution and 5 tubes per dilution as given in Cochran, W.G. (1950).
- Find the row of numbers across the table to column headed by the observed value of P3. The figure at the intersection is the most probable number of organism in the original sample represented in the inoculum added in second dilution.
- Multiply this figure by appropriate dilution factor (*i.e.* second dilution among the three) to obtain value of original sample.

Isolation of Rhizobium from Root Nodules

Rhizobium establishes a symbiotic relationship with legume roots, resulting in nodule formation; hence, legume root nodules serve as optimal specimens for the isolation of rhizobia. The newly excised nodule is surface sterilized, crushed in a minimal volume of sterilized water, and subsequently plated on an appropriate growth medium, which can be purified for single-cell isolation through strain identification.

Materials

Test tubes 5 ml sterile water, Petriplates (Sterilized), Glass rod, Forceps (Sterilized), Sterile tap water and Sterile gloves.

Reagents

Mercuric chloride (0.1 per cent), Ethyl alcohol (70 per cent) and Yeast Extract Mannitol Agar (YEMA) medium. To distinguish rhizobia from other contaminanting bacteria, the YEMA media can be a supplemented with Congo red dye (10 ml of 1/400 aqueous solution). The rhizobia appears whitish coloured on congo red supplemented YEMA medium (CRYEMA) whereas other bacteria may take the stain and form red colony.

Equipment

Laminar Flow Assembly, Incubator and Autoclave.

Procedure

- Gently uproot about one month old healthy leguminous plant and thoroughly wash the root system with tap water.
- Select a plump, pink nodule on the tap root. Remove nodule with help of scalpel or a sharp razor blade along with a portion of the root at the point of its attachment.
- Wash the nodule with the tap water several times to remove any debris adhering to it and transfer into a sterile beaker (50 ml).

- ✰ Add a sufficient amount of 0.1 per cent mercuric chloride solution so as to dip the nodule in it and shake by hand for 3 min. Decant and then add 70 per cent ethyl alcohol to dip the nodule in it for 20-30 sec. and decant the alcohol.
- ✰ Wash the nodule with sterilized water thoroughly for 5-6 times to remove the sterilizing chemicals.
- ✰ Place the surface-sterilized nodule in a 5 ml sterilized vial containing approx. 1 ml water. Crush with sterilized glass rod.
- ✰ Plate the suspension of crushed nodule on Congo Red YEMA medium and incubate at 28 °C till small, round, slimy colourless/white colonies with red dot in the centre appears.

Transfer single isolated colony to fresh YEMA plates for purification and maintain pure culture on YEMA slant pending strain identification.

Chapter 15

Chemical Fertilizers

Introduction

- A fertilizer is a product utilized to enhance soil fertility, including manure or a combination of nutrient elements in the form of organic or inorganic chemical compounds. It is administered to soils or plant tissues (often leaves) to provide one or more critical nutrients for plant growth. Fertilizers are chemical substances that directly contribute to the significant growth of plants and crops by supplying essential nutrients such as nitrogen, phosphorus, and potassium, hence maintaining soil fertility. Chemical fertilizers are substances with a high concentration of essential nutrients for plant growth. In addition to the three primary elements—carbon, oxygen, and hydrogen—plants necessitate a significant amount of nutrients. Nutrients are categorized into primary nutrients, secondary nutrients, and micronutrients.
- The components that supply these nutrients are enumerated in Table 15.1. The fertilizer industry primarily produces essential plant nutrients for soil application. Nitrogenous fertilizers contain nitrogen in the form of ammoniacal nitrogen (e.g., ammonium sulfate, ammonium chloride), nitrate nitrogen (e.g., calcium ammonium nitrate, which contains both ammoniacal and nitrate nitrogen), and amide nitrogen (e.g., urea).

Table 15.1: Essential Elements for Plant Growth

Primary Nutrients	*Secondary Nutrients*	*Micro-Nutrients*
Nitrogen	Calcium	Boron
Phosphorus	Magnesium	Chlorine
Potassium	Sulphur	Copper
		Iron
		Manganese
		Molybdenum
		Zinc
		Nickel

- In phosphatic fertilizers, phosphate exists as accessible phosphate, including single super phosphate (SSP) and triple super phosphate (TSP). PThe potassium need can be satisfied using muriate of potash (potassium chloride) and sulphate of potash.
- Fertilizers are often categorized according to the type of nutrition they contain. A fertilizer containing a single nutrient is referred to as a straight fertilizer, while a fertilizer containing many nutrients is classified as a complex or mixed fertilizer.
- The production of chemical fertilizers requires substantial energy; for instance, the production of 1 kilogram of nitrogenous, phosphatic, and potassic fertilizers demands 80 MJ. Energy consumption amounts to 12 MJ and 8 MJ, respectively.

Classification of Fertilizers

Straight Fertilizers

A straight fertilizer is a single compound containing one primary (N, P, K) nutrient only

For example

Nitrogenous Straight Fertilizer

- Urea
- Ammonium sulphate
- Calcium ammonium nitrate (CAN)

Phosphatic Straight Fertilizer

- Single super phosphate (SSP)
- Triple Superphosphate (TSP)
- Dicalcium phosphate

Potassic Straight Fertilizer

- Muriate of Potash
- Sulphate of Potash
- Schoenite

Compound or Complex Fertilizer

A compound or complex fertilizer constitutes a singular compound. However, it comprises two or more primary nutrients that are in chemically mixed form.

For example

- Diammonium phosphate (DAP)
- Nitrophosphate (NP)
- Ammonium phosphate sulphate

Mixed Fertilizers

Mixed fertilizers are formulated by the physical amalgamation of two or more fertilizers. NPK fertilizers are sophisticated formulations intended to provide the three essential nutrients, denoted by various grades that indicate the corresponding percentages of nitrogen (N), phosphorus (P_2O_5), and potash (K_2O).

For example:

NPK 15-15-15 grade means that the fertiliser contains 15 percent each of N, P_2O_5 and K_2O

Nitrogenous Fertilizers

Nitrogen is the primary macronutrient typically utilized in commercial fertilizers. Nitrogen is a crucial nutrient for plants and appears to exert the most rapid and significant impact. A nitrogenous fertilizer contains nitrogen in the form of nitrate (NO_3^-), ammonium (NH_4^+), amide ($-NH_3$), or cyanamide ($-CN_2$), or a mix thereof.

- Nitrate-containing nitrogenous fertilizers are favored by most plants due to their high solubility in water and rapid availability to plants. In saturated and soggy conditions, these substances are leached or transformed into gaseous form, rendering them inaccessible to plants. Sodium nitrate (16% N), calcium nitrate (15.5% N), and potassium nitrate (13% N) are utilized globally.
- Ammonical fertilizers contain nitrogen in ammoniacal form and exhibit significant resistance to leaching, making them suitable for usage in waterlogged situations. Despite being readily soluble in water, they are not as rapidly accessible to plants as nitrate fertilizers. A period is necessary to transform NH_4^+ into NO_3^-; therefore, it is appropriate for slow-growing, long-duration crops. Ammonium sulfate (20 percent N), ammonium chloride (24-26 percent N), and diammonium phosphate (18 percent N) are extensively utilized in India.

- Ammonical-nitrate fertilizers have nitrogen in two forms: ammonical and nitrate, such as ammonium nitrate (33 percent N) and calcium ammonium nitrate (25 percent N).
- Amide fertilizers exhibit high solubility in water and are readily degraded by soil microbes. These are converted into ammoniacal and subsequently nitrate forms to become accessible to plants. Urea is the most economical nitrogen fertilizer. Urea in soil is degraded by the enzyme urease produced by soil microbes. It exhibits resistance to leaching.

Some Commonly Used Nitrogenous Fertilizers

- **Urea (H_2NCONH_2):** Urea is presented in the form of white prills and exhibits significant hygroscopic properties. This is achieved by heating a combination of liquid ammonia and gaseous carbon dioxide at approximately 185°C and 200 atmospheres of pressure. Urea contains approximately 46 percent nitrogen and is utilized during seed incorporation into the soil; however, it should not come into touch with the seed. Water is administered 3-4 days subsequent to the application of urea on the soil.
- **Ammonium sulphate [$(NH_4)_2$ SO_4]:** It is a non-hygroscopic crystalline salt that exhibits white, grey, brown, red, or yellow coloration and possesses exceptional physical qualities. This is a nitrogenous fertilizer utilized to enhance the yield of crops such as paddy and potato. This fertilizer contains around 20.6 percent nitrogen, all in ammoniacal form. The ammonia in this fertilizer is converted into nitrate by denitrifying bacteria in alkaline soil. Consequently, these nitrates are readily assimilated by crops and plants. This fertilizer is manufactured in India on a huge scale in Sindri, Jharkhand.
- **Calcium Ammonium Nitrate [Ca $(NO_3)_2$ NH_4NO_3):** It is a composite of ammonium nitrate and powdered chalk, appearing as gray or light brown granules coated with soapstone, and is highly hygroscopic. This fertilizer contains 25 percent nitrogen, with half in ammoniacal form and the other half in nitrate form. When incorporated into the soil, it exhibits no adverse effects, and its high solubility in water facilitates its seamless integration into the soil.
- **Ammonium chloride (NH_4Cl):** Ammonium chloride is a byproduct of soda ash production and is a white crystalline salt. It contains 25 percent nitrogen in the form of ammonium nitrogen.

Phosphatetic Fertilizers

Phosphorus is the second most deficient major nutritional element in Indian soils, following nitrogen. It is the second fertilizer element and a vital component of all living cells, required for the sustenance of both plants and animals. It plays an active role in all forms of plant metabolism. Phosphate in phosphatic fertilizers may exist in either water-soluble or citrate-soluble forms. The phosphate that dissolves in water is termed water-soluble phosphate, while the phosphate that does not dissolve in water but is soluble in a 2 percent neutral ammonium citrate solution is referred to as citrate-soluble phosphate. The aggregate of water-soluble and citrate-soluble values is referred to as accessible phosphates.

- Water-soluble phosphatic fertilizers include phosphate in a highly soluble form ($H_2PO^-_4$) in water, making it readily accessible to plants. Single super phosphate (SSP: 16-18% phosphate); Double super phosphate (DSP: 32% phosphate); Triple super phosphate (TSP: 46% phosphate).
- Citric acid soluble phosphatic fertilizers, which dissolve easily in acidic water or weak acids, include phosphorus in an accessible form (H_3PO_4). Basic slag, containing 18 percent phosphate, is a byproduct of the iron and steel industries. Dicalcium phosphate abundant in phosphorus.
- Insoluble phosphatic fertilizers are entirely insoluble in water and exhibit moderate solubility in weak acids, such as citric acid; for example, rock phosphates contain 20-40 percent P_2O_5. Bone food.

Some Commonly Used Phosphatic Fertilizers Single Super Phosphate

- It is a gray or brownish powder or granular substance including monocalcium phosphate and gypsum. It exhibits minor hygroscopicity and contains 16 percent water-soluble P_2O_5.
- Ammonium phosphate, specifically diammonium phosphate, is the most prevalent phosphate fertilizer. It is water-soluble and possesses favorable physical properties. It contains 18 percent nitrogen and 46 percent P2O5.

Potassic Fertilizers

Potassium is the third essential nutrient in fertilizers. Potassium functions as a chemical regulator, root enhancer, stalk fortifier, nutrition synthesizer, sugar and starch transporter, protein synthesizer, respiration modulator, water facilitator, and disease inhibitor; however, its efficacy is contingent upon the presence of co-nutrients such as nitrogen and phosphorus.

- Over 90 percent of potassic fertilizer utilized is in the form of KCl, while the remainder consists of sulfate and a minor quantity of mixed salts. Potassic fertilizers are not manufactured in India; all potassic fertilizer utilized in India is imported.
- The primary commercial potassic fertilizers are muriate of potash (KCl) containing 60 percent K_2O and potassium sulfate (K_2SO_4) containing 50 percent K_2O.
- Muriate of potash, a crystalline substance ranging from white to red, is extremely soluble in water, allowing plants to directly utilize its potassium content in KCl form. It may be applied prior to or during planting and is appropriate for all crops except tobacco and potatoes, owing to its

chloride content. The potassium concentration is greater in KCl than in K_2SO_4. K_2SO_4 possesses the advantage of containing sulfur and is suited to tobacco and potatoes.

Secondary Nutrients

- Calcium, magnesium and sulphur are called the secondary nutrients. It is required for plant growth in relatively smaller quantities than primary nutrients. Any deficiency of the secondary nutrients and other essential elements reduce the efficiency of primary nutrient by restricting the yield to low levels. Therefore, to obtain optimum results, crops have to be supplied with secondary nutrient in addition to primary nutrients.
- Calcium and magnesium are the elements that are largely responsible for the acidity or alkalinity of soil. They are moderately soluble, so in regions of high rainfall they are washed from the soil and carried away. Soils in these regions are usually acid, and lime (calcium carbonate) is often added every year to make the soil less acid. In areas of low rainfall, and where the soil is derived from limestone, the soil is high in calcium and magnesium and is usually alkaline.
- Sulfur is present in many commercial fertilizers, even though it is not listed on the label. As sulfate, it is highly acid and is often used to make alkaline soils acid.

Sources of Fertilizers with Secondary Nutrients

- ***Calcium sources:*** Calcium is mostly utilized in liming agents to ameliorate soil acidity. In cases of inadequacy, calcium may be administered as calcium chloride or incorporated into foliar sprays.
- ***Magnesium sources:*** The magnesium requirements of crops are significant. Deficiencies are rather prevalent, particularly in light, sandy soils. Dolomitic limestone is the predominant source of magnesium, supplying both calcium and magnesium while mitigating soil acidity. Additional sources comprise potassium-magnesium sulfate, magnesium oxide, and basic slag.
- **Sulphur sources:** Soil organic matter is the principal source of sulfur in soil. Over 95 percent of the sulfur present in the soil is bound inside organic materials. Additional natural sources comprise animal excrement, irrigation water, and the atmosphere. Animal manures possess sulfur concentrations varying from below 0.02 to approximately 0.3 percent. The composition significantly changes based on species, storage methods, and applications, among other factors. Sulfur dioxide and other atmospheric gases dissolved in precipitation can contribute up to 22 kg of sulfur per hectare annually, with higher amounts in certain industrialized regions. Irrigation water may possess elevated concentrations of sulfur. When the sulfate-S concentration in irrigation water surpasses 5 ppm, a sulfur

deficit is improbable. Nonetheless, the application of starting fertilizer containing sulfur for new crops may prove advantageous due to the mobility of sulfate during winter precipitation. Table 15.2 presents common sources of calcium, magnesium, and sulfur.

Table 15.2: Prevalent Sources of Secondary Nutrients

Material	*Per cent Ca*	*Per cent Mg*	*Per cent S*
Calcitic limestone	32		
Burnet lime	60		
Hydrated lime	46		
Mari	24		
Gypsum	22		12-18
Dolomitic limestone	29	3-12	
Basic slag	22	3	
Magnesia (Mg oxide)		55-60	
Magnesium chloride		7.5	
Magnesium sulphate		9-20	14
Potassium-magnesium sulphate		11	22
Ammonium sulphate			24
Ammonium thiosulphate			26
Ammonium polysulphide			40-50
Potassium sulphate			18
Elemental sulphur			>85

Micronutrients

- Micronutrients are critical nutrients for plant growth, required by plants in minimal quantities. Intensive cropping rapidly depletes all nutrients, including micronutrients, from the soil. Consequently, the judicious application of micronutrients is essential for enhancing agricultural output.
- Micronutrients are frequently integrated into NPK fertilizers. They may also be applied in solution, either directly or sprayed onto bulk mixtures. In situations necessitating prompt intervention, aqueous solutions of micronutrient salts are applied to crop foliage via spraying.
- Iron, zinc, manganese, copper, boron, molybdenum, nickel, and chlorine are classified within this category.

Table 15.3: Prevalent Sources of Micronutrients

Source	*Micronutrient (per cent)*
Boron: Borax	B, 11.3
Sodium pentaborate	B, 18.0
Fertilizer Borate 46	B,14.0
Fertilizer Borate 65	B, 20.0
Boric acid	B,17.0
Colemanite	B,10.0
Solubor	B,20.0
Boronated Single Superphosphate	B, 0.18
Copper:Copper sulphate	Cu, 22.5-24
Copper ammonium Phosphate	Cu,30.0
Iron: Iron sulphate	Fe, 19-23
Iron oxide	Fe, 69-73
Iron ammonium sulphate	Fe, 14
Iron ammonium polyphosphate	Fe, 22
Iron chelates	Fe, 5-14
Manganese: Manganese sulphate	Mn, 26-30.5
Manganese oxide	Mn, 41-68
Manganese chelate	Mn, 12
Manganese carbonate	Mn, 31
Manganese chloride	Mn, 17
Molybdenum: Ammonium molybdate	Mo, 54
Sodium molybdate	Mo, 39-41
Molybdic acid	Mo, 47.5
Zinc : Zinc sulphate (hydrated)	Zn, 23-36
Zinc Oxide	Zn, 78
Basic Zinc sulphate	Zn, 55
Zinc-ammonia complexes	Zn, 10
Zinc chelates	Zn, 9-14

Complex Fertilizers

Complex NPK fertilizers has the benefit of containing each nutrient within every granule. Their cost exceeds that of the corresponding nutrient quantities obtained by the application of straight fertilizers; however, the quality of the product is assured. Sulfur and/or other nutrients can be readily integrated if necessary. By utilizing the suitable complex fertilizer, farmers are providing at least an estimation of each necessary nutrient. Exact precision is unnecessary due to soil variations, even within a single area.

Complex fertilizers are effective products; nonetheless, the trends in fertilizer usage over the past thirty years have been dictated by commercial factors rather than agricultural efficiency or sustainability.

Table 15.4: Common Complex Fertilizers

Material	*Total Nitrogen (N)*	*Neutral Ammonium Citrate Soluble Phosphate (P_2O_5)*	*Water Soluble Phosphate (P_2O_5)*	*Water Soluble Potash(K_2O)*
Ammonium Phosphate				
Mono Ammonium Phosphate 11-52-0	11.0	52.0	44.2	
DAP 18-46-0	18.0	46.0	41.0	
Ammonium Phosphate Sulphate				
16-20-0	16.0	20.0	19.5	
20-20-0	20.0	20.0	17.0	
18-9-0	18.0	9.0	8.5	
Ammonium Phosphate Sulphate Nitrate				
20-20-0	20.0	20.0	17.0	
Nitrophosphate				
20-20-0	20.0	23.0	12.0	
23-23-0	23.0	23.0	18.5	
Ammonium Nitrate Phosphate				
23-23-0	23.0	23.0	20.5	
Urea Ammonium Phosphate				
28-28-0	28.0	28.0	25.2	
24-24-0	24.0	24.0	20.4	
20-20-0	20.0	20.0	17.0	
Potassium Nitrate (cystalline/prilled)				
13-0-45	13.0			45.0
Mono Potassium Phosphate				
0-52-34			52.0	34.0
NPK fertilizers				
15-15-15	15.0	15.0	4.0	15.0
10-26-26	10.0	26.0	22.1	26.0
12-32-16	12.0	32.0	27.2	16.0

Material	*Total Nitrogen (N)*	*Neutral Ammonium Citrate Soluble Phosphate (P_2O_5)*	*Water Soluble Phosphate (P_2O_5)*	*Water Soluble Potash(K_2O)*
22-22-11	22.0	22.0	18.7	11.0
14-35-14		35.0	29.0	14.0
17-17-17	17.0	17.0	14.5	17.0
14-28-14	14.0	28.0	23.8	14.0
19-19-19	19.0	19.0	16.2	19.0
17-17-17	17.0	17.0	13.6	17.0
20-10-10	20.0	10.0	8.5	10.0

Application of Fertilizers

Fertilizers are frequently utilized for cultivating various crops, with application rates contingent upon soil fertility, typically assessed through a soil test, and tailored to the specific crop. Legumes, for instance, assimilate atmospheric nitrogen and typically do not necessitate nitrogen fertilizer.

Liquid vs Solid

Fertilizers are utilized on crops in both solid and liquid forms. Approximately 90 percent of fertilizers are utilized in solid form. The most commonly utilized organic fertilizers are urea, diammonium phosphate, and potassium chloride. Solid fertilizer is generally in granulated or powdered form. Solids are frequently present in the form of prills, which are solid globules. Liquid fertilizers consist of anhydrous ammonia, aqueous ammonia solutions, and aqueous solutions of ammonium nitrate or urea. These concentrated materials can be diluted with water to create a concentrated liquid fertilizer (e.g., UAN). The benefits of liquid fertilizer are its expedited efficacy and enhanced coverage. The incorporation of fertilizer into irrigation water is termed "fertigation." Furthermore, the utilization of slow-release fertilizer enhanced the efficacy of administered nutrients.

Diverse compounds are employed to augment the efficacy of nitrogen-based fertilizers. Farmers can mitigate the detrimental impacts of nitrogen runoff. Nitrification inhibitors, or nitrogen stabilizers, impede the transformation of ammonia into nitrate, an anion that is more susceptible to leaching. 1-Carbamoyl-3-methylpyrazole (CMP), dicyandiamide, nitrapyrin (2-chloro-6-trichloromethylpyridine), and 3,4-Dimethylpyrazole phosphate (DMPP) are widely utilized. Urease inhibitors are employed to impede the hydrolytic transformation of urea into ammonia, which is susceptible to evaporation and nitrification. The enzymatic conversion of urea to ammonia is facilitated by urease. N-(n-butyl) thiophosphoric triamide (NBPT) is a widely recognized urease inhibitor.

Foliar Application

Foliar fertilizers are administered directly onto leaves. This approach is predominantly employed to administer water-soluble straight nitrogen fertilizers and micronutrients, particularly for high-value crops like fruits.

Overfertilization

Meticulous fertilization techniques are crucial, as surplus nutrients might be harmful. Fertilizer burn may transpire when excessive fertilizer is administered, leading to harm or potential mortality of the plant. Fertilizers differ in their propensity to cause burning, generally correlating with their salt index.

Synthetic vs. Organic Fertilizers

- Organic fertilizers are substances obtained from plant and animal components or byproducts. Examples include blood meal, compost, bat guano, manure, seaweed, and worm castings..
- Synthetic fertilizers are artificially produced inorganic substances, typically originating from petroleum industry by-products. Examples are Ammonium Nitrate, Ammonium Phosphate, Superphosphate, and Potassium Sulfate.
- Plants demand 17 essential nutrients. The three principal macronutrients are nitrogen (N), phosphorus (P), and potassium (K). These nutrients are utilized in substantial quantities by developing plants, necessitating their frequent replenishment to maintain output.
- The secondary nutrients include calcium (Ca), magnesium (Mg), and sulfur (S). Typically, the soil contains sufficient levels of these nutrients, rendering additional fertilization unnecessary. The micronutrients utilized in minimal amounts are boron (B), copper (Cu), iron (Fe), chloride (Cl), manganese (Mn), molybdenum (Mo), zinc (Zn), and nickel (Ni).
- Plants are unable to differentiate between organic and synthetic fertilizers; the nutrients are metabolized identically. Nonetheless, the resemblance ceases at that point.
- Chemical fertilizers contribute nutrients to the soil, but they do not provide any additional substances. Plants require more than mere nutrition for survival. They also require organic substances and living organisms. Synthetic fertilizers do not promote microbial life in the soil. The utilization of synthetic fertilizers effectively eradicates a substantial proportion of beneficial bacteria. These little organisms facilitate the decomposition of organic materials into a stable amendment that enhances soil quality and fertility. Certain organisms transform atmospheric nitrogen into a form used by plants.
- Compost and organic matter introduce advantageous microbes. Microorganisms prevalent in soil and compost transform organic nitrogen

into inorganic nitrogen through a process known as mineralization. Plants may subsequently absorb the nutrients produced by them. Composts harbor a remarkable diversity of microorganisms, many of which may be advantageous in pathogen management. Beneficial microorganisms assist in regulating plant diseases. Enhancing organic matter enhances soil structure, yielding a crumb-like formation. It also augments water retention and boosts soil fertility. Microorganisms can decompose pollutants in soil and water into less environmentally hazardous components.

While there are certain drawbacks to utilizing organic fertilizers, the positives surpass the limits. Constraints of organic amendments:

- The formulation of organic fertilizers can exhibit significant variability. Organic resources provide a less concentrated source of nutrients in comparison to inorganic fertilizers.
- Organic fertilizers may be economically unfeasible for large-scale operations.
- Organic fertilizers can be cumbersome and may necessitate additional effort for application.
- The release of nutrients is highly variable and correlates with the extent of microbial activity, which typically fluctuates with soil temperature.

Limitations of Chemical Fertilizers

- Most chemical fertilizers lack micronutrients.
- Synthetic fertilizers do not promote microbial activity in the soil.
- Chemical fertilizers do not contribute organic matter to the soil.
- Synthetic chemicals can be excessively applied, potentially damaging roots or resulting in dangerous salt concentrations.
- Chemical fertilizers may release nutrients excessively rapidly, resulting in substantial top growth before the roots can adequately develop. This type of growth frequently results in weaker, disease-prone plants with reduced fruit production.
- Synthetic fertilizers frequently leach due to their solubility, releasing nutrients more rapidly than plants can assimilate them.

Chapter 16

Pesticides

Introduction

Pesticides are chemical agents designed to exterminate pests. A pesticide is a chemical or biological substance, including viruses, bacteria, antimicrobials, or disinfectants, that repels, incapacitates, or exterminates pests. The prevalence of pesticide usage has become the term synonymous with plant protection product. It is frequently employed to eradicate or manage various agricultural pests that may harm crops and livestock, hence diminishing farm productivity. The most often utilized pesticides include insecticides for exterminating insects, herbicides for eradicating weeds, rodenticides for eliminating rodents, and fungicides for managing fungus, mold, and mildew. The Food and Agriculture Organization (FAO) defines pesticide as any substance or combination of substances designed to prevent, eliminate, or manage pests, including vectors of human or animal diseases, undesirable plant or animal species, and those that cause damage or disrupt the production, processing, storage, transport, or marketing of food, agricultural products, wood and wood products, or animal feed, as well as substances administered to animals for pest control on or within their bodies. Conventional agriculture employs insecticides for plant protection and fertilizers to promote growth and fertility. In organic agriculture, their utilization is significantly limited. Pesticides are frequently referred to as agricultural chemicals. While these chemicals assist in eliminating undesirable pests and diseases, they can also impact human health and the environment.

Brief History

Pesticides are not novel innovations! Numerous ancient civilizations employed insecticides to save their crops from insects and pests. The ancient Sumerians employed elemental sulfur to save their crops from insects. Medieval farmers utilized poisons such as arsenic and lead on prevalent crops. The Chinese employed arsenic and mercury compounds for the management of body lice and other vermin. The Greeks and Romans employed oil, ash, sulfur, and other substances to safeguard themselves, their livestock, and their crops from numerous pests. During the nineteenth century, researchers concentrated on natural methods utilizing chemicals derived from the roots of tropical crops and chrysanthemums. In 1939, Dichloro-Diphenyl-Trichloroethane (DDT) was discovered and subsequently became an exceedingly efficient and widely utilized pesticide globally. Nevertheless, two decades later, because to biological impacts and human safety concerns, DDT has been prohibited in nearly 86 nations.

Benefits of Pesticides

Pesticides enable farmers to yield more crops on reduced area, enhancing crop output by 20 to 50 percent. Pesticides enable farmers to optimize the advantages of other essential agricultural resources, like high-quality seeds, fertilizers, and water supplies. Pesticides are thus an essential instrument for the sustainable cultivation of high-quality food and fibers.

- **Promote Abundant Harvests:** Multiple scientific studies indicate that the frequent consumption of fruits and vegetables diminishes the risk of various malignancies, hypertension, cardiovascular disease, diabetes, stroke, and other chronic ailments. Pesticides have been employed to augment production in response to the rising food demand. The production of main crops has more than tripled over the past six decades due to the utilization of pesticides.
- **Maintain Food Affordability:** Farmers increase food production on the same acreage through the utilization of pesticides. Research indicates that organic vegetable cultivators incur substantially higher expenses on manual weeding than those employing pesticides. This elucidates the rationale behind the higher cost of organic food compared to conventionally cultivated food.
- **Mitigate Waterborne and Vector-Borne Diseases:** Conditions including malaria, Lyme disease, and West Nile virus. Pesticides promote human health by mitigating disease outbreaks via the regulation of rodent and insect populations.
- **Environmental Conservation:** They allow farmers to yield greater crop production per unit area with less tillage, thereby diminishing deforestation, preserving natural resources, and mitigating soil erosion. Pesticides are essential for managing invasive species and harmful weeds.
- The reduction of herbicides has alleviated the burden of manual weeding, enabling agricultural families globally to seek education and possibilities beyond agriculture, so enhancing their quality of life and living standards.

- Pesticides have converted developing nations into food producers: Crop protection products have enabled farmers in these regions to cultivate two or three crops annually, positioning these countries as potential 'breadbaskets' for the global market. Food exports advantage individuals in temperate nations characterized by abbreviated growing seasons.
- **Securing Stored Goods:** Even post-harvest, crops remain vulnerable to insect infestations. Insects, fungi, and rodents can damage valuable grains. Pesticides employed in stored items can extend the shelf life of produce, mitigate significant post-harvest losses due to pests and illnesses, and ensure the grain is safe for consumption.

Adverse Effects of Pesticides

- The harmful compounds in these are intentionally released into the environment. Although each pesticide is designed to eliminate a specific insect, a significant proportion of pesticides inadvertently reach locations other than their intended target. Instead, they infiltrate the atmosphere, aquatic systems, sediments, and ultimately contaminate our food supply.
- Pesticides are associated with human health risks, ranging from acute effects such as headaches and nausea to chronic consequences including cancer and reproductive damage.
- Their utilization also diminishes overall soil biodiversity. The absence of chemicals in the soil enhances its quality, facilitating improved water retention essential for plant growth.

Types of Pesticides

Grouped by Types of Pests They Kill

- Insecticides - insects
- Herbicides - plants
- Rodenticides - rodents (rats and mice)
- Bactericides - bacteria
- Fungicides - fungi
- Larvicides - larvae

Based on How Biodegradable they are:

- **Biodegradable:** This type can be decomposed by microorganisms and other organisms into innocuous molecules.
- **Persistent:** The persistent substances are those that may require months or years to decompose.

Classification of Pesticides According to Chemical Forms

- **Organophosphate:** The majority of organophosphates function as insecticides, impacting the nervous system by interfering with the enzyme that modulates a neurotransmitter.
- **Carbamate:** Like organophosphorus pesticides, carbamate pesticides impact the brain system by inhibiting an enzyme that modulates neurotransmitter activity. Nevertheless, the effects of enzymes are typically reversible.
- **Organochlorine insecticides:** Organochlorine pesticides, such as DDT, chlordane, and toxaphene, were frequently utilized in the past; however, most countries have since eliminated them from their markets due to their detrimental health and environmental impacts, as well as their tenacity.
- **Pyrethroid:** These are synthetic derivatives of pyrethrin, a naturally occurring pesticide derived from chrysanthemums (flowers). They were designed to optimize their environmental stability.
- **Sulfonylurea Herbicides:** Sulfonylurea herbicides have been marketed for weed management, including pyrithiobac-sodium, cyclosulfamuron, bispyribac-sodium, terbacil, and sulfometuron-methyl. Sulfosulfuron, rimsulfuron, pyrazosulfuron-ethyl, imazosulfuron, nicosulfuron, oxasulfuron, flazasulfuron, primisulfuron-methyl, halosulfuron-methyl, flupyrsulfuron-methyl-sodium, ethoxysulfuron, chlorimuron-ethyl, bensulfuron-methyl, azimsulfuron, and amidosulfuron.
- Biopesticides are specific insecticides generated from natural materials, including animals, plants, microbes, and certain minerals.

Pesticide Ingredients

Pesticide formulations comprise both "active" and "inert" components. The phrases "active ingredient" and "inert ingredient" have been delineated by federal legislation, namely the Federal Insecticide, Fungicide, and Rodenticide Act (FIFRA), since 1947.

- ***Active Ingredients:*** An active component is defined as a substance that prevents, eradicates, repels, or alleviates a pest, or functions as a plant regulator, defoliant, desiccant, or nitrogen stabilizer. The label must legally identify the active ingredient by name along with its weight proportion.
- ***Inert Ingredients:*** An inert ingredient is any component of the product that is not designed to impact a target pest. Isopropyl alcohol may serve as an active ingredient and antimicrobial pesticide in certain products; conversely, in other formulations, it functions as a solvent and may be classified as an inert ingredient. The law does not mandate the identification of inert substances by name and percentage on the label; rather, the aggregate percentage of these ingredients must be disclosed.

Classifiction of Pesticides

The Classification According to Usage

- ☆ **Acaricides:** These are compounds utilized to exterminate mites and ticks or to impede their growth or development. Examples include DDT, dicofol, carbofuran, methiocarb, propoxur, abamectin, milbemectin, flufenoxuron, chlorpyrifos, oxydemeton methyl, phorate, phosalone, fenpyroximate, fipronil, bifenthrin, cyhalothrin, fluvalinate, permethrin, and chlorfenapyr.
- ☆ **Algicide:** These are compounds employed to eradicate or suppress algae. Examples include copper sulfate, diuron, isoproturon, oxyfluorfen, and simazine.
- ☆ **Antifeedants:** These are substances that inhibit an insect or other pest from feeding. Examples include chlordimeform, fentin, and azadirachtin.
- ☆ **Avicides:** These are the chemicals employed to exterminate avian species. The list includes fenthion and strychnine.
- ☆ **Bactericides:** These are substances derived from or synthesized by microorganisms (e.g., bacteria or fungi), or analogous chemicals created intentionally, that are employed to eradicate or suppress bacteria in plants or soil. Examples include copper hydroxide, kasugamycin, streptomycin, and tetracycline.
- ☆ **Bird repellents:** Chemicals that function as bird repellents include copper oxychloride, diazinon, methiocarb, thiram, and ziram.
- ☆ **Chemosterillant:** These are the substances that render an insect sterile, hence inhibiting its reproduction. Certain insects that engage in single mating can be managed or eliminated by the release of substantial quantities of sterilized individuals, which serve as sterilizing agents for the population. All of these function in one of three manners: (a) they impede the formation of ova or sperm. If it fails, proceed to the second stage. (b) induce mortality in the spam or eggs. (c) If these stages are completely unsuccessful, they result in deadly mutations in the sperm or egg material, causing significant harm to the genetic and chromatin material of both sperm and eggs. This produces a zygote, but the resulting progeny will completely lose their reproductive capability. For example: diflubenzuron
- ☆ **Fungicides:** These are the chemicals employed to prevent, treat, and eliminate fungus. Examples include cymoxanil, carpropamid, metalaxyl, metalaxyl-M, carboxin, aureofungin, kasugamycin, streptomycin, validamycin, carbendazim, thiabendazole, thiophanate-methyl, cyproconazole, difenoconazole, flusilazole, tebuconazole, triadimefon, Bordeaux mixture, copper oxychloride, iprodione, captan, ferbam, thiram, and ziram. mancozeb, maneb, metiram, propineb, zineb, isoprothiolane, tridemorph, edifenphos, fosetyl-Al, fenarimol, tricyclazole.

- **Herbicide softeners:** A chemical that safeguards crops from damage caused by herbicides, while allowing the herbicide to eliminate weeds. Examples include benoxacor, cloquintocet, cyometrinil, and cyprosulfamide.
- **Herbicides:** These are compounds utilized to eradicate plants or to impede their growth or development. Examples include alachlor, butachlor, metolachlor, pretilachlor, methabenzthiazuron, pendimethalin, oxyfluorfen, imazethapyr, anilofos, glyphosate, oxadiargyl, oxadiazon, 2,4-D, clodinafop, cyhalofop, quizalofop, paraquat, atrazine, isoproturon, linuron, metoxuron, chlorimuron, and sulfosulfuron.
- **Insect attractant:** A chemical that attracts pests to a trap, thereby eliminating them from crops, livestock, or stored goods. Ex Gossyplure, Gyplure, Muscalure (the nomenclature concludes with 'lure' as they attract the bugs)
- **Insect repellents:** A substance that repels insects from alighting on humans or animals. Examples include Citronella oil and Permethrin.
- **Insect Growth regulator:** A compound that inhibits the growth or development of an insect. Examples include. Diflubenzuron, buprofezin
- **Insecticides:** A pesticide utilized to exterminate insects or to impede their growth or development. Examples include azadirachtin, pyrethrins, carbofuran, carbosulfan, methomyl, buprofezin, diflubenzuron, fenoxycarb, abamectin, emamectin, milbemectin, spinosad, cartap, clothianidin, imidacloprid, thiamethoxam, acetamiprid, thiacloprid, DDT, lindane, endosulfan, dichlorvos, monocrotophos, phosphamidon, demeton-O-methyl, ethion, malathion, phorate, dimethoate, phosalone, azinphos-methyl, chlorpyrifos, pirimiphos-methyl, quinalphos, triazophos, cyfluthrin, cyhalothrin, lambda-cyhalothrin, cypermethrin, alpha-cypermethrin, cyphenothrin, deltamethrin, fenpropathrin, esfenvalerate, fluvalinate, imiprothrin, tofenprox, chlorfenapyr, clothianidin, thiamethoxam, thiacloprid, isoprothiolane.
- **Mammal repellents:** A substance that repels mammals from nearing or consuming crops or stored goods.
- **Mating disrupters:** Chemicals that disrupt the mechanisms by which male and female insects identify one another through airborne pheromones, thereby inhibiting reproduction.
- **Molluscicides:** These are the agents employed to exterminate slugs and snails. Examples include copper sulfate, metaldehyde, thiacloprid, and thiodicarb.
- **Nematicides:** Nematicides are the chemicals employed to manage nematodes. Examples include abamectin, benomyl, carbofuran, carbosulfan, methyl bromide, fenamiphos, phosphamidon, chlorpyrifos, dimethoate, phorate, and triazophos.

- ☆ **Plant growth regulators:** Plant growth regulators are compounds that modify the anticipated growth, blooming, or reproductive rates of plants. This definition excludes fertilizers and other plant nutrients. Examples include 2,4-D, α-naphthaleneacetic acid, ethephon, metoxuron, gibberellic acid, chlormequat, paclobutrazol, and triacontanol.
- ☆ **Rodenticides:** These are substances employed to exterminate rats and similar animals. Examples include strychnine, bromadiolone, coumachlor, coumatetralyl, warfarin, zinc phosphide, lindane, and aluminium phosphide.
- ☆ **Synergists:** A substance that amplifies the toxicity of a pesticide to a bug, yet is not inherently poisonous to the pest itself. For instance: piperonyl butoxide.
- ☆ **Virucide:** An agent has the ability to eradicate or neutralize viruses. Ribavirin (unavailable in India).
- ☆ **Miscellaneous:** Aluminium phosphide, sodium cyanide.
- ☆ **Biologicals:** Viruses, bacteria, fungus, and flora Nematodes, insects, and other parasites or predators.

Insecticides

Substances utilized to exterminate insects are termed insecticides. Insecticides are extensively utilized in medicine, agriculture, and industry. They possess the capacity to significantly modify ecosystem components and are harmful to both animals and humans. Certain pesticides accumulate as they propagate through the food chain.

Types of Insecticides

There are three different types of insecticides. They are

- ☆ ***Systemic*** – This insecticide is applied to the soil for absorption by the plant roots. Upon infiltration into the roots, the insecticide disseminates to exterior regions including leaves, fruits, twigs, and branches. It creates a coating on the plant's surface and serves as a toxin to any insect that attempts to consume the plant.
- ☆ ***Ingested*** – Examples of swallowed pesticides include rodenticides and insecticides.
- ☆ ***Contact*** – These insecticides function like precision projectiles, specifically targeting and exterminating insects upon administration. Typically, domestic insect spray functions as a contact insecticide, requiring direct application to the insect.

Classification of Insecticide

- ☆ It is categorized as organic and inorganic based on its chemical composition.

- Insects are divided according to their mechanism of introduction into four categories: contact poisons, fumigant poisons, stomach poisons, and systemic poisons.
- It is categorized according to its mechanism of action as physical poisons, nerve poisons, respiratory poisons, protoplasmic poisons, general poisons, and chitin inhibitors.
- It is categorized into four categories based on toxicity.
    - Extremely toxic - Color: red, symbol: skull and poison, oral LD50: 1-50
    - Moderately toxic - Color: blue, symbol: danger, oral LD50: 501 - 5000
    - Highly toxic - Color: yellow, symbol: poison, oral LD50: 51 - 500
    - Less toxic - Color: green, symbol: caution, oral LD50: >5000
- It is categorized as ovicides, pupicides, larvicides, and adulticides according to the level of specificity.

Insecticides are categorized into four classes based on their chemical composition:

- Organic insecticides
- Synthetic insecticides
- Inorganic insecticides
- Miscellaneous compounds

Organochlorine Pesticides

This group includes polychlorinated derivatives of cyclohexane (Lindane), polychlorinated biphenyls (DDT, dicofol), and polychlorinated cyclodiene (Endosulfan).

Effect of Organochlorine Pesticides in the Environment

Insecticides can exterminate bees, leading to a reduction in pollination, the depletion of bee populations that pollinate flora, and colony collapse disorder (CCD). Several organochlorine pesticides have been prohibited from most use globally, and they are regulated internationally through the Stockholm Convention on Persistent Organic Pollutants. (POP's). The substances comprise aldrin, chlordane, DDT, dieldrin, endrin, heptachlor, mirex, and toxaphene.

Organophosphorous Pesticides

These are the esters of derivatized phosphoric acid, thiophosphoric acid, and dithiophosphoric acid, referred to as phosphates, thiophosphates, and dithiophosphates, respectively. Examples of each kind of pesticides include the following:

Group	Example
i. Phosphates	Monocrotophos, Phosphamidon, DDVP
ii. Thiophosphates or	Methyl parathion, fenitrothion, Phosphoro-thiatesoxy demeton methyl
iii. Dithiophosphates or phosporodithioates	Dimethoate, Phorate, Phosalone

The classification of these groups, based on the organic moiety linked to the phosphoric acid, includes aliphatic, phenyl, and heterocyclic derivatives.

Carbamates

Carbamates are esters derived from carbamic acids or thiocarbamic acids. The Carbamates can be further categorized into three sub-groups as outlined below.

Group	Example
i. Aryl N methyl carbamate	Carbaryl, Propoxur
ii. Hetero cyclic mono or dimethyl Carbamates	Carbofuran
iii. carbamoylatedoximes	Methomyl
iv. Thiocarbamates	Cartap hydrochloride (neriestoxin group of insecticide

Pyrethroids

- Living organisms inherently possess a multitude of compounds, some of which provide defense against alien invasions. Numerous such molecules have been isolated, identified, and assessed for their biological activity.
- The flowers of chrysanthemum contain chemicals known as pyrethrins, which have significant pesticidal efficacy but demonstrate reduced environmental stability.
- Pyrethrins are chemically the esters of chrysanthemic acid and pyrethric acid, which includes a dimethyl cyclopropane group, combined with alcohols such as pyrethrolone, cinerolone, and jasmolone. Consequently, there exists a total of six esters as illustrated below.

Ester Extract	*Acid Portion*	*Alcohol Portion*	*Per cent Content in Pyrethrum*
Pyrethrin -I	Chrysanthemic Acid	Pyrethrolone	35
Cinerin I	Chrysanthemic Acid	Cinerolone	10
Jasmolin-I	Chrysanthemic Acid	Jasmolone	5
Pyrethrin -II	Pyrethric acid	Pyrethrolone	32
Cinerin I	Pyrethric acid	Cinerolone	14
Jasmolin-I	Pyrethric acid	Jasmolone	4

Synthetic Pyrethroids

- ☆ Allethrin was the inaugural synthetic pyrethroid created in 1949, succeeded by resemethrin. However, they have not maintained the expected characteristics and have demonstrated significant photolability. Permethrin was the inaugural photo-stable pyrethroid produced. Subsequently, cypermethrin, deltamethrin, and fenvalerate were administered. Numerous synthetic pyrethroids have been developed and utilized globally for plant protection.
- ☆ The most recent installment of the series in India is...Synthetic pyrethroids comprise halogenated derivatives of dimethyl cyclopropane carboxylic acid and cyano phenoxy benzyl alcohol. Fenvalerate is an anomaly, as its acid component is p-chlorophenyl isopropyl acetic acid rather than cyclopropane carboxylic acid. The alcohol component of permethrin lacks a cyano group and is only phenoxy benzyl alcohol.

Neonicotinoids

Neonicotinoids are modified derivatives of nicotine that have been developed with enhanced bioefficacy.

- ☆ Acetamiprid
- ☆ Clothianidin
- ☆ Imidacloprid
- ☆ Thiamethoxam
- ☆ (2nd generation- thianicotinyl)

Miscellaneous Pesticides

- ☆ The following are new arrivals in India from several classes of chemical groups, with significant potential for biological action. Spinosad and abamectin are of biological origin.
- ☆ Numerous pesticides necessitate the application of minimal doses of active ingredients to attain pest control and do not bioaccumulate.
    - ❖ Spinosyns (spinosad- Contact and stomach poison) neriestoxin (cartap)
    - ❖ Fiproles (or Phenylpyrazoles)(Fipronil- Broad spectrum insecticide with contact action)
    - ❖ Pyrroles (chlorfenapyr-Used against insects and mites stomach and stomach contact action)
    - ❖ Quinazolines (fenazaquin-Contact acaricide)
    - ❖ Benzoylureas(diflubenzuron a Contact insect growth regulator IGR)
    - ❖ Antibiotics (abamectin - Contact and stomach poison-insecticide and acaricide)

Classification of Herbicides

- Herbicides can be categorized based on their application as foliar and soil-applied. Based on their activity upon application, these compounds can be categorized as contact and systemic herbicides.
- Foliar-applied herbicides can be classified as either contact or systemic, whereas soil-applied herbicides are predominantly systemic, absorbed by the plant through the roots and translocated throughout the organism.
- Another classification based on their action is selective and non-selective. Non-selective herbicides eradicate all plants indiscriminately, whereas selective herbicides target specific species inside certain crops.
- Additionally, these herbicides are categorized according to the timing of application and their efficacy as pre-emergent and post-emergent. Pre-emergent herbicides are administered to the soil prior to weed germination, and post-emergent herbicides are sprayed immediately after germination occurs.

Herbicides can additionally be categorized based on their method of action as follows:

- Growth regulators
- Amino acid synthesis regulators
- Lipid synthesis inhibitors
- Seedling growth inhibitors
- Photosynthetic inhibitors
- Cell membrane disrupters
- Pigment Inhibitors

Growth regulators: These substances disturb hormonal equilibrium and protein synthesis, leading to the plant's demise. Phenoxyacetic herbicides, aryloxyphenoxypropionic herbicides. These are predominantly herbicides targeting broadleaf plants. For example, 2,4-D.

Amino acid synthesis regulators: These chemicals inhibit the specific enzyme responsible for amino acid synthesis, namely Acetolactate synthase. For example, imidazolinone herbicides (Imazethapyr), pyrimidinylsulfonylurea herbicides (Bensulfuron Methyl), and amino acid derivatives that block the EPSPS enzyme (Glyphosate).

Lipid synthesis inhibitors: Lipid synthesis inhibitors impede the creation of fatty acids, which are crucial for lipid biosynthesis. Aryloxy phenoxypropionic herbicides block ACCase, which is essential for fatty acid synthesis (e.g., Clodinafop-propargyl).

Seedling growth inhibitors: Seedling growth inhibitors are agents that impede shoot and root development.These substances hinder cell division and the formation of lipids or proteins in the seedling. For example, acetanilide herbicides such as Butachlor.

Photosynthetic inhibitors: Photosynthetic inhibitors impede electron transfer during photosynthesis, hence obstructing the conversion of sunlight into chemical energy, such as Triazines (Atrazine).

Cell membrane disrupters: This group of compounds disrupts cell membranes. For instance, by pyridyliums or quaternary ammonium compounds, such as Paraquat.

Pigment Inhibitors: These compounds inhibit the development of pigments essential for photosynthesis. Leaves undergo bleaching or whitening. For example, Isoxazolidones (Clomazone).

The herbicides now utilized in India belong to the following class of chemicals:

- Anilide herbicides (flufenacet)
- Chloroacetanilide herbicides (metolachlor)
- Chlorotriazine herbicides (atrazine)
- Pyrimidinyl thiobenzoic acid herbicides (pyrithiobac sodium)
- Phenylurea herbicides (isoproturon)
- Pyrimidinyl sulfonylurea herbicides (bensulfuron methyl)
- Dinitroaniline herbicides (fluchloralin pendimethalin)
- Imidazolinone herbicides (imazethapyr)
- Organophosphorus herbicides(anilofos)
- Oxadiazolone herbicides (oxadiargyl)
- Benzothiazole herbicides (methabenzthiazuron)
- Phenoxyacetic herbicides (2,4-d)
- Aryl oxyphenoxy propionic herbicides (clodinafop-propargyl)
- Quaternary ammonium herbicides (paraquat)
- Unclassified herbicide
- Halogenated aliphatic herbicides
- Oxadiazolone herbicides
- Nitrophenyl ether (oxyflourfen)
- Isoxazolidones (clomazone)

Fungicides

A fungicide is a chemical agent employed to eradicate fungi, fungal spores, and fungal diseases. Natural and synthetic fungicides are both available. Commercial fungicides, mostly synthetic, are widely employed in horticulture to eradicate fungal diseases that afflict plants. Fungal infections are the primary cause of global crop loss, inflicting significant harm in agriculture and leading to substantial reductions in yield, quality, and profitability for both plants and animals.

- A fungicide, or antimycotic, is a poisonous agent employed to eradicate or suppress fungal development. Fungicides are typically employed to manage parasitic fungi that inflict economic harm on crops or decorative plants or pose a threat to the health of domestic animals or humans.
- The majority of agricultural and horticultural fungicides are administered as sprays or powders.
- Seed fungicides are administered as a protective coating prior to germination. Systemic fungicides, or chemotherapeutants, are administered to plants, where they disseminate throughout the tissue to eliminate current diseases or to safeguard against potential infections.
- In human and veterinary medicine, pharmaceutical fungicides are frequently utilized as topical antifungal creams or administered as oral drugs.
- Fungicides target fungal pathogens, whereas bactericides target bacterial pathogens.
- Fungicides can be fundamentally categorized into inorganic fungicides, organometallic fungicides, and organic fungicides.
- Examples of inorganic fungicides include sulfur and copper compounds. Methoxy ethyl mercuric chloride, classified as a restricted-use pesticide, exemplifies an organometallic fungicide. All other synthetic compounds, such as alachlor and oxyfluorfen, are classified as organic fungicides.

The classification of these compounds is based on their field behavior as contact and systemic fungicides.

- Non-therapeutic or non-eradicant or protectant and therapeutic or eradicant.
- All systemic fungicides are classified as both protective and curative fungicides.

Non therapeutic Or Non eradicant or Protectant: These compounds have restricted mobility and are applied before the commencement of the infection. Comprehensive coverage is crucial and serves to mitigate the further dissemination of the disease when implemented post-infection. These exhibit somewhat lower persistence. Copper fungicides, sulfur, dithiocarbamates.

Therapeutic or Eradicant: Therapeutic properties are demonstrated solely at optimal or elevated concentrations, whereas the same drug displays only protective effects at lower quantities. Lime Sulphur, Edifenphos, Captan, Dinocap, Chlorothalonil.

Systemic: These compounds function as both protective and therapeutic fungicides. They are assimilated by the plant and primarily translocated in an upward direction, resulting in enhanced control and longevity inside the plant, so providing prolonged protection. Certain compounds are recognized for their ability to activate defensive mechanisms in plants. For example, TPMs that exhibit no fungicidal activity in vitro may yet serve as highly effective fungicides.Examples of this group include Carbendazim, Fenarimol, TPM, Carboxin, Triazoles, and Antibiotics.

Copper compounds: Copper sulfate (acidic) is soluble in water, whereas all other compounds are insoluble in water.Copper oxychloride (alkaline) is water-insoluble, available in 50 percent wettable powder and 56 percent oil-based formulations. The Bordeaux mixture, consisting of 1 percent cuprous oxide derived from a 1:1 ratio of copper sulfate and lime in water, is alkaline and water-insoluble; it must not include any free copper.

Sulphur compounds: Sulphur functions as a fungicide and acaricide, available in wettable powder (WP) at 80 percent and dispersible powder (DP) at 85 percent, as well as in 40 percent controlled release (CS) formulation. Lime sulphur (22 percent solution) is produced by heating lime with a red-colored char solution. Highly alkaline and unstable, transformed into $CaSO_4$, $Ca(OH)_2$, $CaCO_3$, H_2S, and sulfur. It is fundamentally a calcium polysulfide ($CaS.Sx + CaS_2O_3$). H_2O:Sulfur: lime is at a ratio of 17:2:1.

Phthalimide fungicides: The captan fungicide belongs to the phthalimide group and its mode of action is multi-site contact activity. This is a protective and curative (contact) fungicide.

Dithio thiocarbamates: They are utilized as seed dressings and applied foliarly as a protective fungicide. Mancozeb is offered in conjunction with metalaxyl. somewhat poisonous. somewhat enduring. Thiram exhibits environmental stability and persists in plants for 1 to 1.5 months and in soil for 6 to 24 months.

Benzimidazole fungicides: This category of compounds slows mitosis by attaching to beta-tubulin. These compounds exhibit systemic properties with both protective and curative effects, possessing low toxicity. Unstable in alkaline environments and stable in acidic situations. Benomyl and thiophanate methyl are metabolized into carbendazim inside the plant system under the influence of water, sunshine, and enzymatic activity. Benomyl exhibits more persistence than carbendazim and TPM.

Dinitrophenol fungicides: These compounds operate as contact fungicides, demonstrating both protective and therapeutic effects. Their mode of action entails the uncoupling of mitochondrial oxidative phosphorylation. Illustration: Dinocap.

Organophosphorus fungicides: Include Edifenphos, Fosetyl-Al, and Iprobenfos (IBP) (Kitazin). Edifenphos is classified as a contact fungicide. Extremely hazardous substance. It undergoes enzymatic hydrolysis in the plant and exhibits moderate persistence.

Fosetyl-Al (a true systemic): Inhibits mycelial development and sporulation, while iprobenfos, a phospholipid biosynthesis inhibitor, is a systemic fungicide with considerable persistence.

Conazole fungicides (triazoles): These chemicals are mostly inhibitors of sterol production. These exist as non-volatile solids with low solubility in water. These are systemic fungicides exhibiting both protective and curative properties. Triazole fungicides are hydrolyzed in plants and soil, resulting in the formation of corresponding hydroxy compounds.

Imidazole, oxazole and strobillurin fungicides: Imidazole and oxazole fungicides impair mitochondrial respiration. Fenamidone and femoxadone serve as examples of imidazole and oxazole, respectively, both of which are systemic fungicides that undergo photodegradation in aquatic environments. Azoxystrobin is classified as a strobilurin, and its mechanism of action parallels that of imidazoles and oxazoles.

Pyrimidine fungicides: Pyrimidine fungicides limit the biosynthesis of ergosterol and are classified as systemic fungicides. Fenarimol is available as Fenarimol 12 percent EC.

Anilide fungicides: Anllide fungicides are systemic agents. A few examples belonging to this group are as follows.

- Carpropamid at 27.8 percent Supreme Court: Mechanism of action: Inhibition of nucleic acid synthesis (suppression of melanin). Augments crop resilience by elevating phytoalexin synthesis subsequent to rice blast infection.
- **Thifluzamide**: Inhibits succinate dehydrogenase in the tricarboxylic acid cycle. Thifluzamide 24 percent SC is utilized in rice to combat sheath blight caused by Rhizoctonia solani.

Dithiolane fungicides: Isoprothiolane is classified as a fungicide and plant growth regulator. It is a systemic fungicide that obstructs the penetration and elongation of infection hyphae by preventing the production of infection pegs or the secretion of cellulase. Utilized for combating blast in rice.

Strobilurin fungicides: This category of compounds comprises systemic fungicides that impede mitochondrial respiration. Azoxystrobin and Pyraclostrobin are currently registered for use in India.

Antibiotic fungicides: Streptomycin, kasugamycin, and validamycin are classified as antibiotic fungicides utilized in agriculture. Their mechanism of action mostly involves the suppression of protein synthesis.These compounds are systemic and have considerable persistence. Examples of this group of compounds include Kasugamycin, Streptomycin sulphate (sesqui salt), Validamycin, and Streptomycin Sulphate (9 percent) combined with Tetracycline Hydrochloride (1 percent) SP.

The market availability of pesticides for application or usage leads to an additional classification based on the kind of formulation. Solid formulations, such as WP, SP, DP, and granules, and liquid formulations, including EC, SC, AF, and ULV. Domestic pesticide formulations are available in several forms, including liquid volume (LV), aerosol, gels, mats, and coils.

Biofungicides

Biofungicide is a naturally derived microbial or biochemical substance. There exist three categories of biopesticides:

- Microbial biopesticides contain an active ingredient that serves as a biological control agent, which is an organism capable of attacking or competing with a pathogen or pest.
- Plant biopesticides, or plant-incorporated protectants, are defined by the EPA as "pesticidal substances produced by plants from genetic material that has been introduced into the plant."
- Biochemical biopesticides comprise naturally occurring chemicals. Certain biochemicals may also result from fermentation. Biochemicals can directly influence pathogenic organisms or may induce systemic acquired resistance (SAR). Biopesticides are often narrow-spectrum, exhibit low toxicity, disintegrate rapidly, and are therefore regarded as having less potential for adverse environmental effects.

Fungicide Resistance

- Fungicide resistance is a permanent, heritable characteristic that leads to diminished sensitivity to a fungicide in an individual fungus. This capability is acquired via evolutionary mechanisms. Fungicides exhibiting a single-site mode of action are at a comparatively elevated risk for the development of resistance, in contrast to those employing a multi-site mode of action.
- Currently, the majority of fungicides in development employ a single-site mode of action, as this is linked to a reduced likelihood of adverse effects on the environment, including non-target organisms.
- Fungicide resistance arising from the alteration of a singular main gene leads to pathogen subpopulations that are either susceptible or markedly resistant to the pesticide. In this context, resistance is characterized by a total failure of disease management that cannot be restored with increased dosages or more frequent applications of fungicides. This form of resistance is typically termed "qualitative resistance."
- Fungicide resistance arising from alterations in several interacting genes leads to varying susceptibility among pathogen isolates, contingent upon the extent of genetic modifications. The sensitivity variation within the population is continuous. Resistance, in this context, is perceived as a diminishment of disease control that can be restored through increased dosages or more frequent treatments. Prolonged selection for pathogen resistance by repeated treatments may ultimately lead to the highest

recommended rates and/or the shortest application intervals failing to effectively manage the disease. This form of antifungal resistance is frequently termed "quantitative resistance."

- Fungal isolates resistant to one fungicide frequently exhibit resistance to other closely related fungicides, even without prior exposure, due to their analogous modes of action. This phenomenon is referred to as cross-resistance. Fungicides sharing the same Group Code are likely to demonstrate cross-resistance. Occasionally, negative cross-resistance arises across unrelated fungicides, since the genetic alteration that imparts resistance to one fungicide renders the resistant isolate more susceptible to another fungicide.
- Effectively managing fungicide resistance is essential to prolong the efficacy of at-risk fungicides. The principal objective of resistance management is to postpone its emergence rather than to address resistant fungal strains once they have been established. Consequently, resistance management plans must be instituted upon the initial availability of at-risk fungicides for commercial application. The aim of resistance management is to reduce the application of the vulnerable fungicide while maintaining effective disease control. This is achieved by employing the at-risk fungicide with other fungicides and non-chemical control strategies, such as disease-resistant cultivars, within an integrated disease management framework.
- Utilizing an efficient disease control program is essential to postpone the emergence of resistant strains. At-risk fungicides must be utilized at the manufacturer's prescribed dosage (full rate) and treatment frequency. Employing full rates is anticipated to reduce the selection of strains exhibiting intermediate fungicide sensitivity when resistance is governed by many genes (quantitative resistance). At-risk fungicides must be alternated with other at-risk fungicides that possess distinct modes of action or belong to other chemical groups, and should be either mixed or alternated with fungicides that exhibit a low resistance risk.
- When a crop can act as a source of inoculum for a subsequent crop, the alternation strategy among susceptible fungicides should persist between consecutive crops, ensuring that the initial fungicide applied to a crop is from a distinct cross-resistance group than the final fungicide used on the preceding crop. Certain at-risk fungicides are manufactured as premix products in conjunction with other fungicides to mitigate resistance. At-risk fungicides should be employed solely when absolutely necessary. The optimal period for employing them in resistance management is at the onset of an epidemic when the pathogen population is minimal. Multi-site contact fungicides should be utilized independently during the late growing season, as they have demonstrated adequate disease control to safeguard yield. An essential aspect of resistance management involves

evaluating disease control and documenting any reduction in efficacy that may be attributable to resistance.

Herbicides

- Herbicides are substances that eradicate or suppress plant development. They are essential components of an Integrated Pest Management strategy.
- Selective herbicides target specific weed species without significantly affecting the desired crop, whereas non-selective herbicides, often referred to as total weedkillers in commercial products, are employed to eliminate all plant material in areas such as waste ground, industrial sites, construction zones, railways, and railway embankments.
- In addition to selective and non-selective classifications, other significant distinctions encompass persistence (or residual action: the duration the product remains effective), means of uptake (whether it is absorbed solely by above-ground foliage, through the roots, or via alternative methods), and mechanism of action (the operational process of the product).
- Historically, substances like common salt and various metal salts served as herbicides; however, their usage has declined over time, and in certain countries, many of these are prohibited due to concerns of soil persistence, toxicity, and groundwater contamination. Herbicides have been employed in warfare and conflict.
- Contemporary herbicides frequently serve as synthetic analogs of natural plant hormones that disrupt the growth of targeted flora. The term organic herbicide refers to herbicides designed for organic agriculture. Certain plants, like the genus Juglans (walnuts) and the tree of paradise, generate their own natural herbicides. This phenomenon, along with other associated chemical interactions, is referred to as allelopathy.
- In response to herbicide resistance, a significant issue in agriculture, some products integrate herbicides with several modes of action. Integrated pest management may incorporate herbicides in conjunction with alternative pest control strategies.

Types of Herbicides

- Translocated herbicides reach the site of action through the plant's transport systems, specifically the xylem and phloem. The xylem conveys water and nutrients from the soil to growth locations, whereas the phloem distributes photosynthetic products, such as sugars, to growth and storage areas. Symptoms on the target weeds may take up to two weeks to manifest, contingent upon the herbicide dosage, environmental circumstances, and species involved.
- Contact herbicides exhibit restricted mobility within the plant, thus necessitating thorough covering of the target. In contrast to translocated

herbicides, such as glyphosate, contact herbicides, including paraquat, oxyfluorfen, diquat, and bromoxynil, typically exhibit symptoms swiftly, generally within 24 hours.

- Selective herbicides eliminate target weeds while preserving intended plants (the crop or pasture) when administered at a set application rate.
- Non-selective herbicides, often known as knockdown herbicides, such as glyphosate or paraquat, will harm the majority of plants.
- Residual pesticides persist in the soil for an extended duration (months) and can influence subsequent weed germinations.
- Non-residual herbicides, such the non-selective paraquat and glyphosate, have minimal or negligible soil activity and are rapidly destroyed in the soil. They are either degraded or adhered to soil particles, rendering them less accessible to growing plants. They may possess minimal or no capacity for absorption by roots.
- Post-emergent and pre-emergent denote the objectives and time of herbicide treatment. Post-emergent denotes the foliar application of herbicide subsequent to the emergence of target weeds from the soil, whereas pre-emergent signifies the administration of herbicide to the soil prior to weed emergence.
- Herbicide combinations and sequential applications entail the use of multiple herbicides, typically to broaden the range of controlled weed species and to manage resistance. A mixture entails the simultaneous application of many products. When herbicides are incompatible and cannot be combined in a single tank, they are administered in succession.

Nomenclature of Commonly Used Herbicides

- Before the extensive application of chemical herbicides, mechanical and cultural treatments, including modifications of soil pH, salinity, or fertility, were employed to manage weeds.
- The inaugural extensively utilized herbicide was 2,4-dichlorophenoxyacetic acid, also referred to as 2,4-D, which eradicates several broadleaf plants while predominantly preserving grasses (elevated concentrations of 2,4-D during critical growth phases may adversely affect grass crops such as maize or cereals). The economical price of 2,4-D has resulted in its persistent utilization, making it one of the most widely employed herbicides globally.
- The triazine family of herbicides, including atrazine, was launched in the 1950s. Atrazine does not decompose fast (within a few weeks) following application to soils with a pH above neutral. Atrazine is reported to possess carryover, a typically unfavorable characteristic for herbicides. Glyphosate, often marketed as Roundup, was launched in 1974 for nonselective herbicide use. It is now a prominent herbicide for selective weed management in cultivated crops, owing to the advancement of crop varieties that exhibit resistance to it.

☆ Numerous contemporary agricultural chemical herbicides are explicitly designed to degrade rapidly upon application. This is advantageous since it permits the cultivation of crops potentially impacted by the herbicide in subsequent seasons. Nevertheless, herbicides with minimal residual activity (i.e., those that breakdown rapidly) sometimes fail to deliver prolonged weed control throughout the season.

Table 16.1: Catalog of Herbicides with Corresponding Common and Chemical Names

Sl.No.	*Common Name*	*Trade Name*	*Chemical Name*
Aliphatic carboxylic			
1	Acrolein	Aqualin	2-propenal
2	Allyl alcohol	Allyl alcohol	2-propen-1-ol
3	TCA	TCA	
4	Dalapon	Dalapon, Dowpan	
Phenoxy Carboxylic			
1	2,4,D	2,4,D; Fernoxone	2,4-dichlorophenoxyacetic acid
2	2,4,5 T	Weedar; Weedone	2,4,5-trichlorophenoxy acetic acid
Phenyl acetic acid			
1	Fenac	Fenac	2,3,6-trichlorophenyl acetic acid
Benzoic acid			
1	Dicamba	Banvel	2-methoxy - 3,6•dichloro benzoic acid
2	2,3,6 - TBA	Trysben; Benzac	2,3,6-trichlorobenzoic acid
Phthalic acid			
1	Naptalam	Aanap; Peach thin 322	N-1-naphthylph halamic acid
2	Endothol	Aquathal; Hycrothal	
Phenol			
1	PCP	P'emerge; DNBP	Pertachlorophenol
2	Dinoseb	Basanite	2-sec-butyl-4,6-dinitrophend
Dinitroaniline			
1	Fluchloralin	Basalin	N-propyl -N(2-chloroeth yl)- 4-(tricluoromethyl) - 2,6 dinitro-aniline
2	lsopropalin	Paarlon	N, Ai-dipropyl- 4-isopropyl -2,6-dintro anil ne
3	Pendimethalin	P-owl: S:omp	Herbadex: N-(1-ethylpropy1)-3.4-dimethy1-2.6- dini:roaniline
4	Trifluralin	Treflan	N,N-dipropy1-4-(trifluor omethyl) 2,6-dinitroaniline

Benzonitnle			
1	Dichlobenil	Casoron	2,6, - dinitrobenzonitrile
2	Bromoxynil	Brominol; Butril	3,5-dibromo-4-hydroxybenzonitrile
Triazines			
1	Atrazine	Aatres; Gesprim	2-chloro-4-ethylamino-6-isopropyl-amino-S -triaziie
2	Simazine	Princep; gesatop; Tapazine	2-chloro-4,6-bis(ethylamino) -S -triazine
Urea			
1	Diuron	Karmex	3-(3,4-dichloropheny1)-1,1-dimethylurea
2	Monuron	Telvar	3-(4-chloropheny1)-1,1-dimethylurea
3	Isoproturon	Tolkan; Arelon	3-(4-isopropylpheny1)-1,1-dimethylurea
Phenyl Carbamate			
1	Phenmedipham	Betanol	3-(methoxycarbonyl)amino]phenyl (3-methylphenyl)carbamate
2	Propham	Chem-hoe	Isopropyl phenylcarbamate
Thiocarbamate			
1	Butylate	Bolero; Saturn	S-ethyl diisobutyl thiocarbamate
2	Thiobencarb	Odram; Saturn	S-4-chlorobenzyl diethyl(thiocarbamate)
3	Metham	Vepam; VPM	Sodium methyl dithiocarbomate
Acid amide			
1	Alachlor	Lasso	2-chloro-2',6'-diethyl-Nmethoxymethyl-acetanilide
2	Butachlor	Machete; Delchlor	N-butoxymethy1-2-chloro-2',6'- diethyl-acetanilide
Organic arsenical			
1	Cacodylic acid	Phyto 138; Eras	Dimethyl arsonic acid
2	MAA	Ansar	Methane arsonic acid
Uracil			
1	Bromacil	Hyvar-X; Krovar -1	5-bromo-3-sec-butyl-6-methyluracil
Diphenyl ether			
1	Nitrofen	TOK E 25	2,4-dichlorophenyl p-nitrophenyl ether
2	Oxyfluoren	Goal	2- choloro – 1-(3-ethoxy-4nitrophenoxy) - 4-(trifluoro methyl) benzene
Bipyridllium			
1	Dicuat	Regone	6,7-dihydrodipyrido[1.2-a:2',1'- a] pyrazinedkm
2	Parocuat	Cramaxonc	1,1'-dimethy1-4,4'tipyriclinium
Unclassified			
1	Glyphosate	Round up; Glycel	N-(phosphanomethy0 glyc ne

2	Pidoram	Tnrcon	4-amino 3.5,E –trir:hlnro nicotinic. aniri
Inorganic herbicides			
1	Amonium sulphmate		
2	Sodium arscnite		

Herbicide Poiosoning and First Aid

Numerous contemporary pesticides selectively eradicate weeds by disrupting metabolic pathways specific to plant organisms. Consequently, their systemic toxicities in mammals are typically minimal. Nevertheless, several herbicides have a considerable risk of toxicity if mishandled, and many cause irritation to the eyes, skin, and mucous membranes (Table 16.2).

Table 16.2: Herbicides and Their Toxicological Manifestations

Sulfonyl urea	Chlorimuron-ethyl, Sulfosulfuron, Halosulfuron, Triasulfuron, Metsulfuron-methyl, Ethoxysulfuron, Sulfemetunm-methyl	Systemic toxicity is unlikely unless large amounts have been ingested
Urea derivatives	Chlorotoluron, Diuron, Flumeturon, Isoproturon, Linuron, Methabenzthiazuron	Many substituted ureas are irritating to eyes, skin, and mucous membranes
Bipyridylium	Paraquat, Diquat	Strongly irritant to eyes, skin, and respiratory tract

Table 16.3

Chemical Class	*Generic Name*	*Known or Suspected Adverse Effects*
Acetamides	Metolachlor	Irritant to eyes and skin
Aliphatic Acids	Trichloroacetic acid (TCA), Dichloropropionic acid (Dalapon)	Irritant to skin, eyes and respiratory tract
Chlorophenoxy aliphatic acid	2,4 Dichlorophenoxyacetic acid	Irritant to lung, stomach and intestinal lining, injure liver, kidney and nervous system
Anilides	Alachlor, Propachlor, Propanil	Mild irritant, Dermal irritant and sensitizer, Irritaing to skin, eyes and respiratory tract
Benzoic anisic acid derivatives	Trichlorobenzoic acid	Moderately Irritating to skin, eyes and respiratory tract
Benzonnitriles	Dichlobenil	Minimal toxic, irritant effect
Benzothiadiazinone dioxide	Asulam Terbucarb, Butylate, Cycloate, Pebulate, Vernolate, EPCT, Diallate, Triallate, Thiobencarb	Irritaing to skin, eyes and respiratory tract, Particularly in concentrated form. some may be weak inhibitors and cholinesterase

Chemical Class	*Generic Name*	*Known or Suspected Adverse Effects*
Cyclohexenone	Sethoxydim	Irritating to skin and eyes
Dinitroaminobenzene derivative	Butralin, Pendimethalin, oryzalin	May be moderately irritating. These herbicides do not uncouple oxidative phosphorylation or generate methemoglobin
Fluorodinitrotoluidine compounds	Benfluralin, Dinitromine, Fluchloralin, Profluralin, Trifluralin	May be mildly irritating, These herbicides do not uncouple oxidative phosphorylation or generate methemoglobin
Isoxazolidinone	Clomazone	May be moderately irritating
Nicotinic Acid isopropylamine derivative	Imazapyr	Irritant to eyes and skin
Phosphonates	Glyphosate	Irritating to eyes, skin and respiratory tract
Triazines	Ametryn, Atrazine, Cyanazine, Desmetryn, Metribuzin, Prometryn, Propazine, Simazine, Terbuthylazine, Tertutryn, Prometon	Systemic Toxicity is unlikely unless large amount have been ingested. Some triazines are moderately irritant to the eyes, skin and respiratory tract. Prometon is strongly irritating to eyes, skin and respiratory tract
Triazole	Amitrole, Aminotriazole	Minimal systematic toxicity. Slight irritant effect
Sulfonyl urea	Chlorimuron-ethyl, Sulfossulfuron, Halosulfuron, Triasulfuron, Metsulfuron. Ethoxysulfuron	Systemic toxicity is unlikely unless large amount has been ingested
Urea Derivatives	Diuron, Linuron, Isoproturon, Flumeturon, Chlorotoluron, Methabenzthiazuron	Many substituted ureas are irritating to eyes, skin and mucous membrane
Bipyridylium	Paraquat, Diquat	Strongly irritant to eyes, skin, and respiratory tract
Acetamides	Metolachlor	Irritant to eyes and skin
Aliphatic Acids	Trichloroacetic acid(TCA), Dichloropropionic acid (Dalapon)	Irritant to skin, eyes and respiratory tract
Chlorophenoxy aliphatic acid	2,4 Dichlorophenoxyacetic acid	Irritant to lung, stomach and intestinal lining, injure liver, kidney and nervous system
Anilides	Alachlor, Propachlor, Propanil	Mild irritant, Dermal irritant and sensitizer, Irritating to skin, eyes and respiratory tract
Benzoic anisic acid derivatives	Trichlorobenzoic acid	Moderately Irritating to skin, eyes and respiratory tract

Chemical Class	Generic Name	Known or Suspected Adverse Effects
Benzonnitriles	Dichlobenil	Minamial toxic, irritant effect
Benzothiadiazinone dioxide	Asulam Terbucarb, Butylate, Cycloate, Pebulate, Vernolate, EPCT, Diallate, Triallate, Thiobencarb	Irritating to skin, eyes and respiratory tract, Particularly in concentrated form. some may be weak inhibitors and cholinesterase
Cyclohexenone	Sethoxydim	Irritating to skin and eyes
Dinitroaminobenzene derivative	Butralin, Pendimethalin, oryzalin	May be moderately irritating. These herbicides do not uncouple oxidative phosphorylation or generate methemoglobin
Fluorodinitrotoluidine compounds	Benfluralin, Dinitromine, Fluchloralin, Profluralin, Trifluralin	May be mildly irritating, These herbicides do not uncouple oxidative phosphorylation or generate methemoglobin
Isoxazolidinone	Clomazone	May be moderately irritating
Nicotinic Acid isopropylamine derivative	Imazapyr	Irritant to eyes and skin
Phosphonates	Glyphosate	Irritating to eyes, skin and respiratory tract
Triazines	Ametryn, Atrazine, Cyanazine, Desmetryn, Metribuzin, Prometryn, Propazine, Simazine, Terbuthylazine, Tertutryn, Prometon	Systemic Toxicity is unlikely unless large amount have been ingested. Some triazines are moderately irritant to the eyes, skin and respiratory tract. Prometon is strongly irritating to eyes, skin and respiratory tract
Triazole	Amitrole, Aminotriazole	Minimal systematic toxicity. Slight irritant effect
Sulfonyl urea	Chlorimuron-ethyl, Sulfossulfuron, Halosulfuron, Triasulfuron, Metsulfuron. Ethoxysulfuron	Systemic toxicity is unlikely unless large amount has been ingested
Urea Derivatives	Diuron, Linuron, Isoproturon, Flumeturon, Chlorotoluron, Methabenzthiazuron	Many substituted ureas are irritating to eyes, skin and mucous membrane
Bipyridylium	Paraquat, Diquat	Strongly irritant to eyes, skin, and respiratory tract

Treatment

Pesticide poisoning can impact the body in two manners: it may induce a localized reaction upon contact with exposed skin or eyes, or it may be absorbed systemically, resulting in a broader physiological response. Skin contamination must be addressed immediately by cleansing with soap and water. Eye contamination necessitates immediate treatment through extensive irrigation with copious volumes of clean water. Immediate medical assistance should be sought if cutaneous or ocular discomfort continues.

Gastrointestinal Decontamination

Consumption of these herbicides is likely to result in vomiting and diarrhea due to their irritating characteristics. Management relies on: (1) the most accurate estimation of the consumed quantity, (2) the duration since ingestion, and (3) the subject's clinical condition. General precautions should be implemented to avert the absorption of the toxin. Activated charcoal, at a concentration of 50 g in 400 ml of water, may serve as an absorbent for residual pesticides in the intestinal tract when delivered at a dosage of 5 ml/kg body weight, either orally or by gastric lavage. This gastric lavage may be repeated as necessary. To eliminate hazardous compounds from the colon, a cathartic like sodium sulfate or magnesium sulfate can be taken orally at a dosage of 30 g in 250 mL of water. The volume of fluids administered must not surpass 50 ml per 10 kg of body weight. Dehydration and electrolyte imbalances may necessitate the administration of oral or intravenous fluids.

No specific antidotes exist for poisoning by these herbicides. In instances of suicidal ingestions, it is imperative to consider the potential for the ingestion of numerous harmful chemicals. If substantial quantities of herbicide have been consumed and the patient is evaluated within one hour of consumption, gastrointestinal cleaning should be contemplated. Administer activated charcoal and sorbitol orally if the quantity of ingested herbicides is minimal, if successful emesis has already transpired, or if therapy is postponed.

Intravenous Fluids

Severe vomiting and diarrhea can lead to significant dehydration and electrolyte deficiency. In such circumstances, it is imperative to monitor blood electrolytes and fluid balance, and to deliver intravenous infusions of glucose, normal saline, Ringer's solution, or Ringer's lactate to replenish extracellular fluid volume and electrolytes. Oral nutrition may be administered once fluids can be retained.

Biopesticides

Biopesticides are specific categories of pesticides obtained from natural sources, including animals, plants, microbes, and certain minerals. Canola oil and baking soda possess pesticidal properties and are classified as biopesticides. Canola oil and baking soda possess pesticidal properties and are classified as biopesticides.

Types of Biopesticides

- Microbial pesticides
- Plant pesticides
- Biochemical pesticides

Microbial Pesticides

- Microbial pesticides comprise a microorganism (e.g., bacterium, fungus, virus, or protozoan) as the active component. Microbial pesticides can manage various pests, while each distinct active ingredient is largely specialized to its target pest(s). Certain fungus regulate specific weeds, while others exterminate particular insects.

- The predominant microbial insecticides are sub-species and strains of *Bacillus thuringiensis*, commonly referred to as Bt. Each strain of this bacteria generates a distinct combination of proteins, targeting and lethally affecting one or several closely related species of insect larvae. Some *Bacillus thuringiensis* (Bt) strains target moth larvae on plants, whereas others are specialized to the larvae of flies and mosquitoes. The specific insect species targeted depend on the ability of the particular Bt to create a protein that binds to a larval gut receptor, resulting in the starvation of the insect larvae.
- Microbial biopesticides are of three kinds
    - *Bacterial Biopesticde:* Bacteria such as *Bacillus thuringiensis, Bacillus papillae*, and *Bacillus lentimorbus* has the capability to exterminate specific insect pests and are classified as entomopathogenic.
    - *Bacillus thuringiensis:* It is a gram-positive, spore-forming, rod-shaped bacterium found in soil. During sporulation, the bacteria synthesizes insecticidal proteins in the form of parasporal crystals. These are referred to as delta endotoxins, also known as Cry proteins. Cry proteins have particular toxicity against insects from the orders Lepidoptera, Coleoptera, and a limited number of other insect taxa.
    - *Fungal Biopesticides:* Entomopathogenic fungi infect insects, inducing illnesses that result in their mortality. Two notable fungi utilized as mycopesticides are *Beauveria bassiana*, responsible for white muscardine illness, and *Metarhizium anisopliae*, which induces green muscardine disease.
    - ***Viral Biopesticides:*** Viral insecticides are infections that target insects and other arthropods. Viral pesticides are employed to manage Lepidopteran larvae such as Helicoverpa and Spodoptera, affecting crops like cotton, corn, sorghum, and tomatoes. Baculoviruses are the prevalent viral biopesticides utilized. They are minuscule and consist of double-stranded DNA. The genus Baculoviridae comprises three subgroups:

        1. Nuclear Polyhedrosis viruses (NPVs),
        2. Granulosis viruses (GVs),
        3. Non occluded viruses

Plant Pesticides

- Plant pesticides are pesticidal compounds synthesized by plants through the incorporation of exogenous genetic material.

For instance, researchers can incorporate the gene encoding the Bt pesticidal protein into the plant's genetic code. The plant, rather than the Bt bacteria, produces the chemical that eradicates the pest. Both the protein and its genetic material are subject to regulation by the EPA; the plant itself is not controlled.

Biochemical Pesticides

These pesticides are naturally occurring agents that manage pests through non-toxic ways. In contrast, conventional insecticides are typically synthetic substances that directly eliminate or deactivate pests. Biochemical pesticides encompass compounds like insect sex pheromones that disrupt mating, along with diverse aromatic plant extracts that lure insect pests to traps. The EPA has formed a specialized committee to ascertain whether a compound qualifies for designation as a biochemical pesticide, given the challenges in making such determinations.

Biopesticides Technology

- The Biopesticide sector constitutes a little fraction of the extensive pesticide market in India. The previous decade has saw significant expansion in this sector, particularly in the standardization of production procedures for Trichoderma, Gliocladium, Paecilomyces, Pseudomonas, Trichogramma, NPV, and Bacillus for application against various insect pests and diseases.
- Currently, these biological control agents have been effectively utilized in India. Trichogramma, a stingless wasp that preys on the eggs of sugarcane borers, has been employed to combat borers in the states of Tamil Nadu, Rajasthan, Uttar Pradesh, Bihar, and Haryana. Likewise, Trichogramma, Bracon, Chelonus, and Chrysopa species are employed for the management of cotton bollworms.
- *Trichogramma* has also been employed against the rice stem borer and leaf folder. The sugarcane scale insect has been managed through the utilization of predatory Coccinellid beetles in Uttar Pradesh, West Bengal, Gujarat, and Karnataka. This is a compilation of various technologies employed in the production of biopesticides, along with a few samples.

Table 16.4

Bioagent	*Manufacturing Process*	*Utilization*
Trichogramma species (egg parasite)	Mass multiplication utilizing stored grain pests as hosts. Production entails the proliferation of host insects on sorghum grains, which are subsequently permitted to be parasitized by Trichogramma. Eggs are indicated on cards as 'tricho cards.'	Utilized for the management of sugarcane early shoot borer, cotton bollworms, and sorghum stem borer.
Crysoperla carnea (Chrysopid predetor)	Mass propagation conducted in the laboratory on the ova of the stored grain pest.	To manage larval pests in pulses, vegetables, and fruits.

Bioagent	*Manufacturing Process*	*Utilization*
Cryptolaemus montrouzieri (Ladybird beetle)	Mass proliferation of mealybugs facilitated by pumpkin under laboratory conditions.	To manage mealybugs, particularly on fruit.
NP Virus of *Helicoverpa armigera* and *Spodoptera litura*	Production commences with the emergence of pod borer and tobacco caterpillar larvae (host culture) on a semi-synthetic diet. Non-Player A virus is applied to cultured larvae. Subsequently, the infected larvae are gathered to extract viral solution using blending, filtration, and centrifugation.	Employed to combat bollworms in cotton and pod borers.
Trichoderma Fungal spp.	Produced in laboratories and manufactured as a powder utilizing a carrier material (talc powder).	To manage root rot and wilt infections, particularly in pulses.
Pheromone lures for *Helicoverpa armigera* and *Spodoptera litura*	Sex pheromones are dispensed into plastic traps at the specified concentration using micro-pipettes and subsequently inserted into rubber septa. The septum is affixed to the trap.	To capture the male of the gram pod borer and tobacco caterpillar

Table 16.5: Various bio-pesticides, accompanied with their trade names and target organisms.

Category	*Trade Name*	*Target*
Bacteriocide		
Azotobacter radiobacter k 84	Galltrol A	Crown Gall Disease
P agglomorans C9-1	Blightban C9-1	Fireblight
Fungicide		
Bacillus lichenoformis SB3086	Ecogaurd	Fungal Disease
Bacillus subtilis MB 1600	Histick N/T	Damping off
Fungicide/Bacteriocide		
Bacillus subtilis QST713		Foliar fungal and bacterial disease

Principal Categories of Bio-agents Available for Commercial Production

Various types of bio-agents can be economically mass-produced for extensive distribution to farmers for the management of insect pests. They are as follows:

Parasitoids	*Predators*	*Insect Pathogens*
Trichogramma chilonis, T. brasiliensis and *T. pretiosum* (egg parasites) - for tomato fruit borer	*Cryptolaemus montrouzieri* (Australian ladybird beetle) for the management of several mealybug and soft scale species.	**Virus**: Nuclear Polyhedrosis Virus (NPV) - for significant polyphagous pests such as *Helicoverpa armigera* (gram pod borer) and *Spodoptera litura* (tobacco caterpillar)

Parasitoids	*Predators*	*Insect Pathogens*
Trichogramma chilonis - for brinjal shoot and fruit borer, shoot borers of cotton, sugarcane, rice *etc.*	*Chrysopa* spp. (green lacewing bug) - for the control of aphids, white flies *etc.*	**Bacteria:** *Bacillus thuringiences* (*Bt*) - for control of lepidopterous pests
		Fungi: *Trichoderma viride* and *Trichoderma harziarum* against soil borne fungal diseases
		Namatodes: for control of soil-borne grubs, lepidopterans and some foliar pests

Effectiveness of Biopesticides in Field Applications

State Agricultural Universities and ICAR Research Institutes/Stations have undertaken field efficacy experiments to assess the effectiveness of pest control by biopesticide application. The proportion of pest control attained for designated bio-control agents is as follows:

Bio-agent	*Efficacy of Pest Control*
Trichogramma spp.	60-90 per cent
Cryptolaemous montrouzieri	100 per cent
NPV	70-80 per cent
Trichoderma viridae	60-90 per cent

The advancement of the biopesticides sector must be regarded as a strategic, holistic, and future-oriented endeavor. The increasing population and its escalating demands necessitate a greater supply of food and other things. The rising apprehension among customers and government regarding food safety has prompted growers to investigate alternative environmentally sustainable approaches to replace or, at a minimum, augment existing chemical-based procedures. The utilization of bio-pesticides has emerged as a possible alternative to chemical pesticides. Biopesticides hold a significant position in the future of Integrated Pest Management techniques.

Table 16.6: Bio-pesticides Registered under Section 9(3) of the Insecticides Act, 1968 for Use in India

Sl.No.	*Name of Biopesticide*
1	*Bacillus thuringiensis* var. *israelensis*
2	*Bacillus thuringiensis* var. *kurstaki*
3	*Bacillus thuringiensis* var. *galleriae*
4	*Bacillus sphaericus*
5	*Beauveria bassiana*

Sl.No.	Name of Biopesticide
6	*Trichoderma viride*
7	*Verticillum lecanii*
8	NPV of *Helicoverpa armigera*
9	NPV of *Spodoptera litura*
10	*Trichoderma harzianum*
11	*Metarrhizium anisopliae*
12	*Pseudomonas fluorescence*
13	*Verticillum chlamydosporium*
14	*Paecilomyces lilacinus*
15	*Ampelomyces quisqalis*
16	*Bacillus subtilis*

Advantages of Biopesticides

- Biopesticides are typically less detrimental than conventional pesticides.
- Biopesticides typically impact only the intended pest and closely related species, unlike broad-spectrum conventional pesticides, which can damage a diverse range of organisms, including birds, insects, and mammals.
- Biopesticides often are effective in very small quantities and often decompose quickly, thereby resulting in lower exposures and largely avoiding the pollution problems caused by conventional pesticides.
- Biopesticides, when incorporated into Integrated Pest Management (IPM) systems, can significantly reduce the reliance on conventional pesticides while maintaining good agricultural yields.

Disadvantages of Biopesticides

- High specificity necessitates precise identification of the pest or pathogen and the application of several products; nevertheless, this characteristic can also be advantageous as the biopesticide is less likely to adversely affect non-target species.
- Their often sluggish response renders them inappropriate for immediate pest outbreaks threatening crops.
- Efficacy is frequently inconsistent due to the effects of several biotic and abiotic conditions, since certain biopesticides consist of living organisms that achieve pest or pathogen control by proliferating within or next to the target organism.
- Living creatures undergo evolution and enhance their resistance to biological, chemical, physical, or any other type of control. Should the target population remain extant and capable of reproduction, the surviving individuals may develop a tolerance to imposed stressors, leading to an evolutionary arms race.

Table 16.7

Banned in India	*Banned for Import, Manufacture and Use*	*Banned for Use but Continued to Manufacture*	*Withdrawn and Accepted by Registration Committee (S.O 915 (E) dated 15th June, 2006*
Aldicarb	Carbofuran 50 per cent SP	Captafol 80 per cent Powdr	Dalapon
Aldrin	Methomyl 12.5 per cent L	Nicotin Sulfate	Ferbam
Benzene Hexachloride	Methomyl 24 per cent formulations	Formothion	
Calcium cyanide	Phosphamidon 85 per cent SL		Nickel Chloride
Chlrobenzilate			Paradichloroben
chlrodone			Simazine
chlorofenvinphos			Warfarin
Copperacetoarsenite			
Dibromochloropropane			
Dieldrin			
Endrin			
Ethyl Mercury Chloride			
Ethyl Parathion			
Ethylene Dibromide			
Heptachlor			
Lindane			
Maleic Hydrazide			
Menazon			
Metoxuron			
Nitrofen			
Paraquat Dimethyl Sulphate			
Pentachloro Nitrobenzene			
Pentachlorophenol			
Sodium Methnae Arsonate			
TCA (Trichloroacetic Acid)			
Tetradifon			
Toxaphene (Camphechlor)			

http://www.spices.res.in/pages/pesticides-formulations-banned-india.

References

American Society for Testing and Materials E 1519. 1999. Annual Book of ASTM Standards, Vol. 11.05. Designation E 1519-95, Standard Terminology Relating to Agricultural Tank Mix Adjuvants. p. 905–906.

Bhattacharyya, A. Barik, S.R. and Ganguly, P. 2009. New pesticide molecules, formulation technology and uses: Present status and future challenges. The Journal of Plant Protection Sciences 1(1) : 9-15.

Central Insecticide Board and Registration Committee, Directorate of Plant Protection, Quarantine and Storage, N.H. IV, Faridabad-121 001.

Cecil, P.F. 1986. Herbicidal warfare: The Ronch Hand Project in Vietnam. Praeger special studies, Praeger scientific, New York, USA.

Das, T.K. 2014. Weed Science: Basics and Applications. 2 Ed. Jain Brothers, New Delhi.

Davies, J. 2001. Herbicide safeners-commercial products and tools for agrochemical research. Pesticide Outlook. February: 11-15.

Directorate of Plant Protection Quarantine and Storage. 2015. Pesticide and documentation Unit. (http://ppqs.gov.in/IpmPesticides.htm).

Dixit, A. and Varshney, J.G. 2009. Handbook on Herbicide Recommendations. Directorate of Weed Science Research, Jabalpur.

FAO Stat. 2014. http://faostat3.fao.org

Food Safety and Standards (contaminants, toxins and residues) Regulations. 2011. Food Safety and Standards Authority of India, Ministry of Health and Family Welfare, GOI. (www.fssai.gov.in).

Green, J.M. and Green, J.H. 1993. Surfactant structure and concentration strongly affect rimsulfuron activity. Weed Technology 7(3): 633-640.

Green, J. 2000. Adjuvant outlook for pesticides. Pesticide Outlook. October: 196-199.

Holm, F.A. and Johnson, E.N. 2009. The history of herbicide use for weed management on the prairies. Prairie Soils and Crops Journal 9: 1-11.

Janaki, P., Sharma, N., Chinnusamy, C., Sakthivel, N. and Nithya, C. 2015. Herbicide residues and their management strategies. Indian Journal of Weed Science 47(3): 329–344.

Parmar, B.S. and Tomar, S.S. 2004. Pesticide formulation: Theory and practice. CBS Publishers and Distributors, New Delhi.

Pesticide Properties Database, 2015. University of Hertfordshire (http://sitem.herts.ac.uk/aeru/ppdb/en/Reports/1160.htm).

Pesticide Safety Education Program (PSEP), Cornelle University (http://psep.cce.cornell.edu/).

Chapter 17

Environmental Soil Chemistry

Introduction

- Soil chemistry has conventionally concentrated on the chemical interactions inside soils that influence plant development and nourishment. Beginning in the 1970s and particularly in the 1990s, as apprehensions regarding inorganic and organic contaminants in water and soil and their effects on plant, animal, and human health intensified, the focus of soil chemistry shifted to environmental soil chemistry.
- Environmental soil chemistry examines the chemical interactions between soils and significant environmental constituents such as plant nutrients, radionuclides, metals, metalloids, and organic compounds.

Soil Pollution

- Soil pollution is characterized by the accumulation of enduring poisonous substances, chemicals, salts, radioactive elements, or pathogenic agents in soils, which negatively impact plant development and animal health.
- Soil is the superficial layer of organic and inorganic substances that envelops the Earth's lithosphere. The organic component, originating from the decomposed remains of flora and fauna, is concentrated in the dark topmost layer of topsoil. The inorganic component, consisting of rock pieces, was created over millennia by the physical and chemical weathering of bedrock.
- Fertile soils are essential for agriculture to provide the global population with adequate food supplies. Soil can get contaminated through various means, including:

- Seepage from a landfill
- Indiscriminate discharge of industrial waste into the soil
- Percolation of contaminated water through the soil
- Rupture of underground storage tanks
- Excess application of pesticides, herbicides or fertilizer
- Solid waste dumping and seepage from it

☆ The predominant substances contributing to soil contamination are:

- Petroleum hydrocarbons
- Heavy metals
- Pesticides
- Solvents

Geogenic Sources of Pollution

Arsenic

Arsenic Toxicity in Soil-Plant-Animal-Human System

☆ Arsenic (As), a hazardous trace element, poses significant environmental issues owing to its occurrence in soil, water, flora, fauna, and the human body. The elevated toxicity and prevalence in the biosphere have elicited public and political apprehension. Among the 20 countries, including Argentina, Chile, Finland, Hungary, Mexico, Nepal, Taiwan, Bangladesh, India, and others, where groundwater arsenic contamination and its associated human suffering have been documented, Bangladesh exhibits the highest severity, followed by West Bengal, India (Sanyal et al., 2015a). The magnitude of the issue is severe and unparalleled, placing millions in the Bengal delta basin at risk.

☆ The extensive arsenic pollution of groundwater in many regions of West Bengal, particularly in five districts adjacent to the Bhagirathi River and neighboring districts in Bangladesh, is a significant concern. Arsenic contamination in groundwater exceeding the permissible limit of 50 μg As. L–1 (WHO, 2001) has been identified in various regions beyond the Bengal delta basin, including Chandigarh (1976), Nepal (2001), Bihar (2002), Uttar Pradesh (2003), Jharkhand (2003-2004) (Sanyal and Dhillon, 2005), Chhattisgarh, and Punjab (2006-2007).

Guideline Value of Maximum Arsenic Concentration

☆ The World Health Organization (WHO) has recommended a provisional guideline value for total arsenic (As) concentration in drinking water of 10 μg As L^{-1} since 1993, primarily due to the unreliability of measuring lower levels on a large scale for the protection of human health. The National Standard for Maximum Acceptable Concentration (MAC) of arsenic in drinking water is 50 µg As. L^{-1} in most countries, including India and Bangladesh, according to prior WHO guidance from 1971. The suggested new standard value of 5 µg As·L^{-1} is now being evaluated (WHO, 2001). This is primarily because inorganic arsenic compounds are included in Group 1 (carcinogenic to humans) based on substantial evidence of carcinogenicity in humans and little evidence in animals (IARC, 1987). Sufficient data on the carcinogenicity of organic arsenic have not been produced.

- ☆ The joint FAO/WHO Expert Committee on Food Additives (JECFA) established a provisional maximum tolerable daily intake (PMTDI) for inorganic arsenic in humans at 2.1 µg As. kg body weight^{-1} day^{-1} in 1983 and affirmed a Provisional Tolerable Weekly Intake (PTWI) of 15 µg As. kg body weight^{-1} day^{-1} in 1988 (FAO/WHO, 1989). Guideline values for soil, plant, and animal systems are unavailable.
- ☆ Two principal hypotheses of geogenic origin have been posited to explain the extensive arsenic pollution in the groundwater of certain regions in West Bengal and Bangladesh, situated inside the delta delineated by the Bhagirathi and Ganga-Padma rivers. Among the two hypotheses, the arsenopyrite oxidation hypothesis and the ferric oxyhydroxide reduction hypothesis, the latter aligns more closely with the experimental observations documented for the aquifer sediments and groundwater of the Bengal delta basin (Sanyal et al., 2012; Das et al., 2014). This concept posits that anoxic conditions in the aquifer facilitate the mobilization of arsenic from arsenic-bearing sediments into the groundwater. The preservation of anoxic conditions is believed to be supported by the prevalent practice of wetland paddy production in the impacted region.

Table 17.1: Arsenic Concentration in Rocks and Various Other Materials

Sl.No.	*Types of Rocks/Materials*	*Arsenic Content (mg As. kg^{-1})*
1.	Rocks	
	Igneous rocks	
	Ultrabasic:	
	Peridotite, Dunite, Serpentine	0.3-15.8
	Basic:	
	Basalt (extrusive)	0.18-113
	Gabbro (intrusive)	0.06-28
	Intermediate:	
	Latite, Andesite, Trachyte (extrusive)	0.5-5.8
	Diorite, Granodiorite, Syenite (intrusive)	0.09-13.4
	Acidic:	

Sl.No.	Types of Rocks/Materials	Arsenic Content (mg As. kg^{-1})
	Rhyolite (extrusive)	3.2-5.4
	Granite (intrusive)	0.18-15
	Metamorphic rocks	
	Quartzite	2.2-7.6
	Slate/Phyllite	0.5-143
	Schist/Gneiss	0.0-185
	Sedimentary rocks	
	Marine:	
	Shale/Claystone(near-shore)	4.0-25
	Shale/Claystone(off-shore)	3.0-490
	Carbonates	0.1-20.1
	Phosphorites	0.4-188
	Sandstone	0.6-9.0
	Non-marine:	
	Shales	3.0-12
	Claystone	3.0-10
2.	Coal	Up to 2000
3.	Crustal Average	2

Source: Sanyal (2005).

Natural Abundance

- Dissolved arsenic concentrations in natural waters, excluding groundwater, are typically low, particularly in regions associated with geothermal activity and/or mining operations. Sedimentary rocks often have a greater arsenic concentration (Table 8.2) compared to igneous and metamorphic rocks, but suspended and bottom sediments in the majority of aquatic systems possess higher arsenic levels than most natural waters.
- The ability to retain arsenic is mostly determined by sediment particle size and the presence of surface coatings made up of clays, clay-sized iron and manganese oxides, and organic materials.
- Arsenic retained in solid phases within sediments, particularly iron oxides, organic materials, and sulfides, may serve as the principal sources of arsenic in groundwater when conditions favor its release from these solid phases. These encompass abiotic events (oxidation/reduction, ion exchange, chemical transformations) and biotic reactions (microbial methylation) (Sanyal et al., 2012).

Table 17.2: Arsenic Levels in Water Excluding Groundwater

Source	*Arsenic Concentration (µg As. L^{-1})*
Rainwater and snow	< 0.002-0.59
Rivers	0.20-264
Lakes	0.38-1.00
Sea water	0.15-6.00
Ponds (West Bengal, India)	4-70
Canals (West Bengal, India)	40-150

Source: Sanyal *et al.* (2012).

Arsenic Poisoning and its Impact on Health

- Arsenic is a prevalent hazardous element found in natural habitats. As little as 0.1 g of arsenic trioxide can be fatal to humans (Jarup, 1992). Initial manifestations of arsenic poisoning encompass dermatological conditions, asthenia, fatigue, loss of appetite, nausea, and emesis accompanied by either diarrhea or constipation. As poisoning advances, the symptoms exhibit more distinctive characteristics, including acute diarrhea, edema (notably of the eyelids and ankles), skin pigmentation, arsenical melanosis, hyperkeratosis, liver enlargement, respiratory ailments, and skin cancer.
- Severe cases also exhibit gangrene in the extremities and malignant neoplasms (Sanyal et al., 2012). The "Bell Ville Disease," characterized by arsenic-induced cutaneous symptoms in the population of Bell Ville, Argentina, "Black Foot Disease" in Taiwan, and "Kai Dam" sickness in Thailand are recognized health problems resulting from arsenic poisoning (Sanyal et al., 2012). Numerous researchers have investigated hair, nail, skin-scale, and urine samples from a substantial population inhabiting in the impacted region of West Bengal (India) and Bangladesh. Numerous samples exhibited arsenic concentrations exceeding the allowable limits.

Chemistry of Arsenic in Groundwater-Soil Environment

- Arsenic (As) in groundwater and soil exists as dissolved oxyanions, namely arsenites ($As^{III}O_3^{3-}$; Hn $As^{III}O_3(3^{-n})^-$, where n = 1, 2) or arsenate ($As^VO_4^{3-}$, Hn $As^VO_4(3^{-n})^-$, where n = 1, 2), in addition to organic forms. The solubility, mobility, bioavailability, and hence toxicity of arsenic in soil-crop systems essentially depend on its chemical form, particularly the oxidation state of arsenic. The toxicity of arsenic compounds in groundwater and soil is primarily influenced by their oxidation state, which is affected by the redox conditions and pH, as well as the presence of organic compounds containing arsenic.

- The toxicity of arsenic compounds in groundwater and soil environments adheres to the following hierarchy:
    - Arsine [AsH_3; arsenic (As) valence state: 3] Organo-arsine compounds > arsenites (As^{3+} state) and oxides (As^{3+} state) > arsenates (As^{5+} state) > arsonium metals (+1) > native arsenic metal (0).
- Arsenites have significantly greater solubility, mobility, and toxicity compared to arsenates in aquatic and terrestrial ecosystems. Organic compounds, specifically dimethyl arsinic acid (DMA) or cacodylic acid, which upon reduction (e.g., in anoxic soil conditions) yield di- and trimethyl arsines, are also found in soil. Monomethyl arsonic acid (MMA) is another chemical compound found in groundwater and soil. At pH 6-8 and in an aerobic oxidized environment (redox potential, Eh = 0.2-0.5V), arsenic acid species and arsenate oxyanions, specifically $HnAs^{V}O_4(3^{-n})^-$ ions, with n = 1, 2 (pentavalent arsenic forms), are present in significant proportions in most aquatic systems. Conversely, under mildly reducing conditions, such as those found in flooded paddy soils with Eh = 0-0.1V, arsenous acid, $H_3As^{III}O_3$, and arsenite oxyanion species ($As^{III}O_3{}^{3-}$; $HnAs^{III}O_3(3^{-n})^-$, with n = 1, 2) (trivalent arsenic) are the predominant species.
- Moreover, As $^{(III)}$ is more commonly found in soils with a neutral pH range, particularly in the afflicted regions of West Bengal, India, and Bangladesh, as well as in most groundwater, which raises concerns. This is mainly due to the fact that As $^{(III)}$ predominantly exists as a neutral, uncharged species, specifically arsenous acid, $H_3As^{(III)}O_3$ (pKa = 9.2), at the pH levels of neutral soils and most natural groundwater, as anticipated from Henderson's equation, rendering it less susceptible to retention by the charged mineral surfaces in soils and sediments.
- Both direct and indirect data indicates that arsenic (and selenium) is retained in soils and sediments by oxides (e.g., of Fe, Al, Mn) through the creation of inner-sphere complexes via a ligand-exchange mechanism. This is seen below by the subsequent reaction scheme.

[M–OH] + H_2O Ì! [M–$OH_2{}^+$] + OH^- (17.1)

[M–$OH_2{}^+$] + [$As^{V}O_4{}^{3-}$] ⇌ [M-OH_2 +......$As^{V}O_4{}^{3-}$] ⇌
[M–$As^{V}O_4{}^{2-}$] + H_2O (17.2)

- Nonetheless, non-specific adsorption (by electrostatic mechanisms) of arsenic transpires at pH levels underneath the point of zero charge (ZPC) for a specific adsorbent (Sanyal et al., 2012). The aforementioned ligand exchange often enhances the negative charge of the soil colloidal fraction, such as iron oxides, hence lowering the zero point of charge (ZPC) of the arsenic-contaminated soil at reduced pH levels. This was demonstrated by the simultaneous increase in the negative magnitude of the variable-surface charge and the surface potential of the associated soil colloidal fraction (Sanyal et al., 2012). Nonetheless, non-specific adsorption (by electrostatic mechanisms) of arsenic transpires at pH levels beneath the point of zero charge (PZC) for a certain adsorbent.

- It is important to note that groundwater or soil solution, influenced by inflows and outflows, circulation, and anthropogenic disturbances due to extraction, cannot be anticipated to maintain thermodynamic equilibrium, as it is predominantly an open system in thermodynamic terms. Consequently, the concentration ratio of arsenic species, specifically [(As^{III})/(As^{V})], in field soils frequently does not align with the values calculated from the measured redox potential (Eh) and the application of Nernst's equation (at 25°C) to the equilibrium redox reaction.

$$As^{V}O_4{}^{3-} + 2H^{+} + 2e = As^{III}O_3{}^{3-} + H_2O \quad (17.3)$$

$$Eh \cong E0\,h - 0.0295 \log [(As^{III}O_3{}^{3-})]/(As^{V}O_4{}^{3-})] - 0.059\,pH \quad (17.4)$$

- The (III) terms denote the equilibrium concentrations of the corresponding ionic species in dilute soil solution, and Eo h represents the standard redox (reduction) potential of the ($As^{V}O_4{}^{3-}/As^{III}O_3{}^{3-}$) redox couple at 25°C. Equation 17.4 clearly indicates that the concentration of As^{III}, and consequently the soluble arsenic levels in soil, should significantly rise with decreasing Eh and increasing pH. Moreover, at elevated pH levels, the concentration of OH^- ions would rise, resulting in the displacement of As^{III} and As^{V} species from their binding sites via competitive ligand exchange processes. The relationship between arsenic sorption and the pH of the sorption medium is mostly determined by the characteristics of the soil colloidal fraction. A decline in arsenate adsorption was observed with rising pH, but this trend was evident only at lower arsenate concentrations, which was reversed at elevated arsenate equilibrium values. This pattern was elucidated by the fluctuating electrostatic potential of the variable-charge soil colloidal surfaces in relation to pH, solubility product principles, and the buffering capacity of the used arsenic salt (Majumder and Sanyal, 2003).

Selected Findings in Soil-Crop Systems

Arsenic Loading in Soils of West Bengal

Research studies conducted in the designated affected areas indicated that total and Olsen-extractable arsenic (i.e., 0.5M $NaHCO_3$, pH 8.5-extractable arsenic, which represents the soil arsenic pool accessible to plant uptake) ranged from 8.4 to 24.3 mg As. kg^{-1} and from 2.90 to 15.8 mg As. kg^{-1}, respectively (Sanyal et al., 2015a), in the impacted soils of West Bengal. The arsenic concentrations in the soil of these regions were typically greater than those documented for the soils of many other countries, including Argentina, China, Italy, Mexico, France, and Australia (Sanyal et al., 2012).

The inorganic arsenic fractions in the contaminated soils were categorized into distinct arsenic pools: water-soluble arsenic (Ws–As), aluminium-bound arsenic (Al–As), iron-bound arsenic (Fe–As), and calcium-bound arsenic (Ca–As), utilizing a sequential extraction methodology. The results indicated that the inorganic soil arsenic (As) pools ranked as follows: Ws–As < Al–As < Ca–As < Fe–As. The Fe–As fraction accounted for 45 percent to 74.7 percent of the total soil arsenic sequential sum (Sanyal et al., 2012).

Interaction of Arsenic with Organic Compounds in Soil Systems

- Soil serves as an efficient reservoir for arsenic found in contaminated groundwater utilized for crop irrigation. The organic components of soil, specifically humic acid (HA) and fulvic acid (FA), function as efficient accumulators of toxic heavy metals by creating metal-humate complexes (chelates) with varying stability (Datta et al., 2001; Mukhopadhyay and Sanyal, 2004; Sinha and Bhattacharyya, 2011; Ghosh et al., 2012; Sanyal et al., 2012; Sanyal et al., 2015a). Furthermore, soil clays, aluminum oxides, and iron oxides, particularly the amorphous forms of iron and aluminum oxides in soil, significantly affect arsenic absorption by soils, soil minerals, and sediments. The aforementioned organo-arsenic compounds exhibited considerable stability, even in the presence of competing oxyanions such as phosphate, sulfate, and nitrate (Sanyal et al., 2015a).
- The moderating effect of organic fractions from FYM, vermicompost, municipal sludge, mustard cake, and surface soil of West Bengal was evaluated regarding the stability of the associated arseno-humic/fulvic complexes formed in organic manure-treated contaminated soils (Sinha and Bhattacharyya, 2011; Ghosh et al., 2012). The application of organic manures as soil amendments significantly diminished the arsenic concentration in sesame seeds by up to 65.5% (vermicompost), 50% (phosphocompost), 42% (mustard cake), and 40% (FYM), in comparison to the control group (Sinha et al., 2011). The risk linked to dietary exposure to arsenic-contaminated sesame oil amounted to 15.6 percent of the proposed acceptable weekly intake for arsenic at the highest concentration of arsenic in sesame oil. Consequently, enhancing the soil organic matter content in the arsenic-affected tropical soils, which are relatively deficient in native organic matter, through the implementation of suitable management practices (such as the recycling of crop residues and the incorporation of appropriate organic manure) will promote arsenic retention in these soils.

Interaction of Arsenic with Phosphorus and Micronutrients

- Phosphorus (P) is a vital macronutrient necessary for plant growth. The placement of As and P in Group Vb of the Periodic Table renders their interaction within the soil-plant system a significant concern for arsenic mobilization. Evidence suggests that these oxyanions are unlikely to be adsorbed separately in mixtures; instead, they are likely to compete for similar adsorption sites (Sanyal et al., 2012; Das et al., 2014).

- Numerous researchers demonstrated that phosphate presence diminishes arsenate adsorption, with a significantly greater reduction observed in the competitive influence of arsenate on phosphate adsorption by soil minerals; however, considerable variability in the extent of competition between these two oxyanions has also been documented (Mukhopadhyay et al., 2002; Sanyal et al., 2015a).
- Conversely, the introduction of specific micronutrients, such as zinc and iron salts, to contaminated soil low in these elements aids in alleviating arsenic toxicity within the soil-crop system. As previously mentioned, arsenic is released more significantly into soil solutions during submergence, which can be mitigated by the application of graded amounts of zinc, likely through the creation of the comparatively less soluble zinc arsenate.
- The application of Zn to soil diminished arsenic deposition in rice crops cultivated under submerged conditions, concurrently resulting in increased yield (Das et al., 2016). The application of zinc in boro (summer) rice significantly diminished the plant's accumulation of arsenic and its translocation to biomass, while its residual effects greatly impacted the mitigation of soil arsenic, thus reducing its accumulation in subsequent crops. Analogous effects were noted when these studies were replicated using iron instead of zinc (Das et al., 2016).

Arsenic in Soil-Plant System

- Numerous workers have documented the uptake and transformation of arsenic by various plant species cultivated in arsenic-affected regions. These crops (including rice, elephant-foot yam, green gram, cowpea, sesame, peanut, etc.) exhibited a significant accumulation of arsenic in various plant sections. Pointed gourd, a creeping vegetable plant, has exhibited significant arsenic accumulation when grown in the contaminated soils of West Bengal. Several other crops, specifically cauliflower, tomato, and bitter gourd, were observed to accumulate arsenic in their commercially viable yield. The distribution of arsenic concentration in plant components often adhered to the sequence:

  root > stem > leaf > economic produce.
- The conversion of arsenate to the more hazardous arsenite is promoted by a decrease in redox potential (Eh) in anoxic soil conditions, with arsenite exhibiting greater solubility and mobility than arsenate. The rice plant is particularly vulnerable to arsenic toxicity due to its cultivation in submerged soil conditions (low Eh). The processing of rice, including

parboiling and milling, was observed to elevate arsenic levels in both traditional and high-yielding varieties (Sanyal et al., 2012). The toxicity of arsenic species in plants is claimed to follow this hierarchy: Arsine (AsH_3) > As_{3+} > As_{5+} > Monomethyl arsonic acid (MMA) > Dimethyl arsinic acid (DMMA) (NRCC, 1978).

- ☆ Additionally, the assessment of 200 rice varieties revealed significant variability in arsenic buildup within the grain. No substantial association was identified in the arsenic absorption patterns among the roots, shoots, and grains of the 200 rice lines analyzed. Preliminary investigation indicated that the arsenic concentration in grain is regulated by many genes (BCKV, Unpublished work). Research indicates that crops such as potato, pumpkin, and sesame acquire lower levels of arsenic compared to other varieties (Sanyal et al., 2012).
- ☆ Upon storing arsenic-contaminated groundwater from a shallow tube well in a pond, there was a gradual decrease in arsenic concentration in the pond water over time, accompanied with an accumulation of arsenic in the matching pond sediment samples. The reduction in arsenic concentration in the stored water may have resulted from the deposition of arsenic into the pond sediment, hence increasing the arsenic accumulation in the latter. The precipitation throughout the rainy season (July to September) considerably reduced the pollution of the stored water through dilution. This facilitates the conjunctive utilization of surface and groundwater in agriculture during the dry season (January to May, characterized by minimal or no precipitation) as a viable corrective strategy (Sanyal et al., 2012).

Detoxification of Arsenic by Plants and Microbial Species

- ☆ The documented hyperaccumulation of arsenic (As) from contaminated soils by the brake-fern, Pteris vittata, and its subsequent translocation into the aboveground biomass indicates that the arsenic accumulated by the plant predominantly exists in toxic inorganic forms, with the proportion of highly toxic As (III) significantly exceeding that of the less toxic As (V) form, in contrast to their distribution in the contaminated soil where the fern proliferates (Ma et al., 2001). It is important to note that the accumulation of arsenic does not inherently result in its detoxification inside the plant, unless the poison is efficiently neutralized or transformed into less dangerous forms by metabolic processes. A comprehensive search for phytoaccumulating or phytoexcluding plant species is essential.
- ☆ A review of literature identifies several plant and microbial species recognized for their arsenic accumulation capabilities or as bio-indicators, which can efficiently extract arsenic (and other heavy metals) from aquatic systems. For example, water hyacinth can absorb 170 μg As. g dry weight^{-1} in its stem and 340 μg As. g dry weight^{-1} in its leaves when cultivated in a pond with 10 mg As. dm^{-3} (Chigbo et al., 1982).

Nevertheless, the accumulated arsenic in water hyacinth (*Eichornia crassipes*) is prone to seeping into the water body, especially during the breakdown of this aquatic plant. Therefore, caution must be taken when assessing the arsenic levels in aquatic environments based on water hyacinth buildup.

- Several microbial species, including the bacterial species *Proteus* sp., *Escherichia coli*, *Flavobacterium* sp., *Corynebacterium* sp., and *Pseudomonas* sp.; the fungus *Candida humicola*; and the freshwater algae *Chlorella ovalis*, *Phaeodactylum tricornutum*, and *Oscillatoria rubescens*, have been documented to exhibit differing capacities for arsenic accumulation. Various weed species, typically associated with crops such as rice, potato, jute, and mustard, were observed to thrive in arsenic-contaminated soils (2-14 mg As. kg $soil^{-1}$) and were irrigated with arsenic-laden groundwater, resulting in significant arsenic accumulation in their biomass (Das et al., 2005).

Bioremediation of Arsenic in Soil-Plant Systems

- A study including selected arsenic (As)-contaminated soils from West Bengal assessed the capacity of indigenous soil bacteria, isolated from these soils, to convert hazardous inorganic arsenic into less toxic volatile arsenicals. Approximately 37 percent of As (III) (under anaerobic conditions) and 30 percent of As (V) (under aerobic conditions) were volatilized by these bacterial isolates within three days. Unlike the genetically engineered organism, the native soil bacteria successfully eliminated 16 percent of arsenic from the polluted soil over a 60-day incubation period in the presence of FYM (Majumdar et al., 2013a).
- Additional arsenic-oxidizing bacteria were recovered from arsenic-contaminated soils in West Bengal (Majumdar et al., 2013b), exhibiting tight phylogenetic relationships to various species of Bacillus and Geobacillus, as determined by their 16S rRNA gene sequences. They exhibited hyper-resistance to both As (V) (167–400 mM) and As (III) (16–47 mM). In these isolates, elevated rates of As (III) oxidation (278–1250 $\mu M \cdot h^{-1}$) and arsenite oxidase activity (2.1–12.5 $nM \cdot min^{-1} \cdot mg^{-1}$ protein) were recorded. Moreover, among the elite arsenic-oxidizers, the AMO-10 fully (100 percent) oxidized 30 mM of As (III) within 24 hours. The aoxB gene was confirmed in the examined isolates. The development of a phylogenetic tree based on the aoxB sequence indicated that the two bacteria, AGO-S_5 and AGH-0_2, clustered with Achromobacter and Variovorax, but the other two strains, AMO-10 and ADP-25, did not cluster. The elevated rate of hazardous As (III) oxidation to the less lethal As (V) forms by these indigenous strains may be utilized for the cleanup of arsenic in contaminated environments (Majumdar et al., 2013b).

Arsenic in Soil-Plant System and its Influence on Food-Chain

- As (III) was identified as the predominant arsenic species in grains of the transplanted autumn paddy, whereas As (V) was more prevalent in rice straw (Sanyal et al., 2015a). Organic soil amendment diminished arsenic buildup in both rice grain and straw of fall rice, evidenced by a decrease in inorganic arsenic (Sinha and Bhattacharyya, 2014a). Sinha and Bhattacharyya (2014b) examined the arsenic toxicity profile in rice cultivated in a contaminated region of rural Bengal, as well as the potential risk of food exposure. The distinctive nature of the anaerobic rice ecosystem leads to a substantial deposit of inorganic arsenic (i-As) in the soil and its subsequent concentration in rice. The recoveries of inorganic arsenic (i-As) were predominantly attributed to arsenic (III) in rice grains and arsenic (V) in rice straw, highlighting the elevated concentrations of the more dangerous arsenic (III) in the consumable part.
- The detection of organic arsenic species in rice grain and straw indicates the potential for methylation of inorganic arsenic within the plant system. The risk of dietary exposure to inorganic arsenic through rice, a staple meal in rural Bengal, presents a harm to human health comparable to that provided by contaminated drinking water. Organic amendments and enhanced phosphorus fertilization demonstrated significant potential in diminishing total and inorganic arsenic buildup in rice, hence mitigating dietary risk (Sanyal et al., 2015a).
- Limited reports exist that delineate daily arsenic (As) exposure via water and diet among populations residing in groundwater-contaminated areas and correlate this exposure with arsenic biomarkers. Demographic parameters and total daily arsenic exposure from water and diet were assessed in 167 people from an arsenic-endemic region (Group 1) and 69 participants from an arsenic-non-endemic zone (Group 2) in a study conducted in West Bengal. The results indicated a markedly elevated dietary arsenic consumption among residents of Nadia district in West Bengal, where contaminated groundwater was utilized for irrigation, in contrast to the much lower levels observed in Hoogly district, where groundwater remained uncontaminated. Despite reducing the arsenic concentration in drinking water to below 50 $\mu g\ As.L^{-1}$ (the allowed limit in India), substantial arsenic exposure persisted through water and dietary sources, as seen by the heightened levels of arsenic in urine, a biomarker, among individuals residing in the investigated arsenic-endemic region. Individuals with skin lesions exhibited elevated arsenic levels in urine and hair relative to those without skin lesions (Guha Mazumder et al., 2013, 2014).
- A recent study revealed a significant positive correlation between daily arsenic intake from both water and diet and urinary arsenic levels in an arsenic-endemic region of West Bengal, even among individuals utilizing

arsenic-safe water (< 50 μg As. L^{-1}) for drinking and cooking (Figure 17.1). Even when arsenic concentrations in drinking water were decreased to below 10 μg As. L^{-1} (the WHO acceptable limit), dietary intake remained substantially correlated with urine arsenic excretion. No substantial correlation was identified between arsenic dosage from drinking water and this group (Halder et al., 2012; Guha Mazumder et al., 2013; Sanyal et al., 2015a).

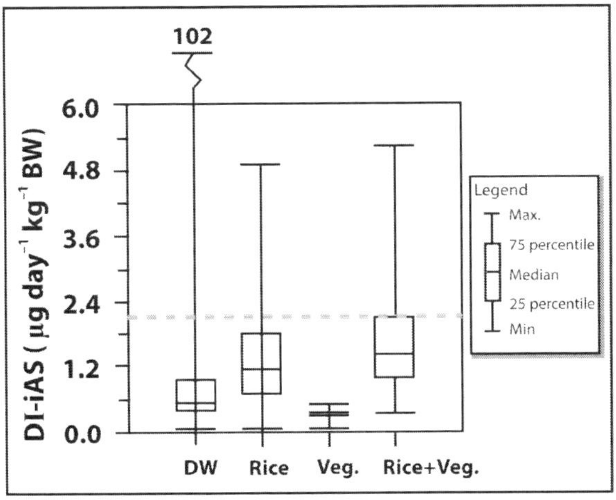

Figure 17.1: Comparison among Daily Intake of Inorganic As (DI-iAs) Due to Consumption of Drinking Water (DW), Rice and Vegetables (Veg.). The dotted line indicates WHO recommended provisional maximum tolerable daily intake (PMTDI) (PMTDIAs) value of 2.1μg As. kg body weight^{-1} day^{-1}.
***Source*: Halder *et al.* (2012).**

- Moreover, in people exposed to arsenic just by diet, urine arsenic concentration exhibited a positive association with dietary arsenic intake, accompanied by the presence of skin lesions; conversely, this link was insignificant in participants lacking skin lesions. It is therefore clearly demonstrated that providing arsenic-safe drinking water (< 10 μg As. L–1) to the rural Bengal population is insufficient to mitigate the risk of arsenic poisoning; the consumption of rice presents an additional potential route of inorganic arsenic (i-As) exposure that must also be addressed in effective remedial strategies. Any intervention to mitigate chronic arsenic toxicity in rural Bengal necessitates integrated strategies aimed at minimizing arsenic infiltration into the food chain while concurrently reducing arsenic levels in drinking water below safe thresholds.

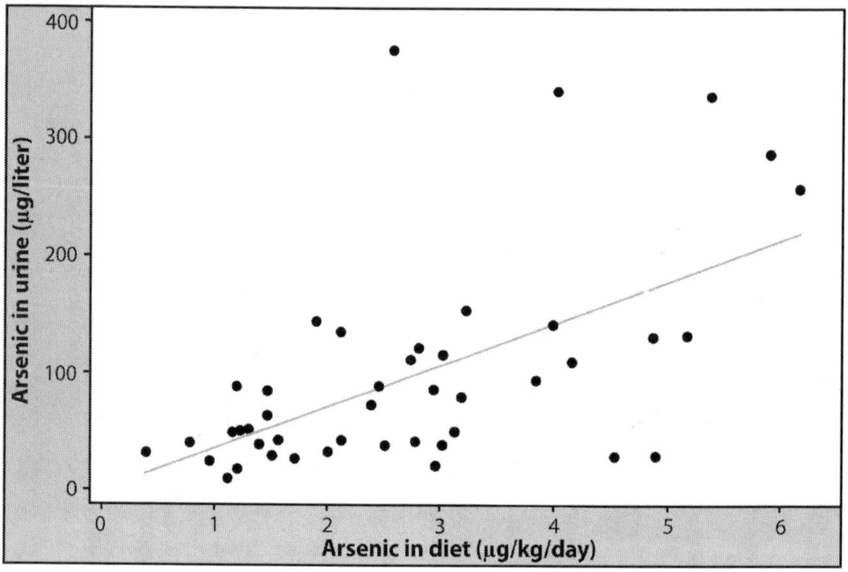

Figure 17.2: Correlation of Urinary Arsenic Concentration (µg As. L^{-1}) with Daily Dietary Arsenic Intake (µg As. kg^{-1}. day^{-1}) for Participants, Drinking Water with Arsenic Level < 0.3 µg As. L^{-1}, having Skin Lesion (n = 45), West Bengal, India (r = 0.573). ***Source*****: Guha Mazumdar *et al.* (2014).**

Remedial Measures

- ☆ Optimal conjunctive utilization of groundwater and surface water (e.g., collected rainfall) and the replenishment of groundwater resources with arsenic-free harvested rainwater.
- ☆ Identification and development of high-yielding, low arsenic-accumulating crop varieties, together with the promotion of low-water-requiring, farmer-attractive cropping sequences, particularly for the lean period from January to May, in arsenic-contaminated regions.
- ☆ Irrigation utilizing pond-stored groundwater, where partial arsenic decontamination is achieved through sedimentation and dilution with rainwater.
- ☆ Improving water use efficiency through optimal water management for groundwater irrigation, particularly for summer (boro) rice cultivation.
- ☆ Enhanced utilization of farmyard manure and other fertilizers, alongside green manure crops, in addition to the application of suitable inorganic additions such as zinc and iron salts when applicable.
- ☆ Identification and development of types of crops that accumulate minimal arsenic in the edible portions and have a low ratio of inorganic to organic arsenic forms.

- Formulating economical phyto- and bio-remediation alternatives. Establishment of widespread awareness: Mass campaigning, organizing farmers' day, conducting field demonstrations, and acknowledging socio-economic concerns (Sanyal et al., 2012, 2015a).

Selenium

Selenium Toxicity in Soil-Plant-Animal-Human System

- Selenium (Se), an element in the enzyme glutathione peroxidase, can be either advantageous or detrimental to plants, animals, and people, contingent upon its concentration. It is an inorganic carcinogen at elevated levels. The advised daily human intake of selenium is between 50 and 200 μg. (WHO, 1996). The "Kashin-Beck" illness (characterized by persistent arthritis and joint abnormalities in children and adolescents) and "Keshan" disease (a kind of cardiomyopathy) are severe conditions associated with selenium deficiency.
- The Food and Nutrition Board of the National Academy of Sciences, USA has established 5 mg Se. kg diet–1 (dry weight basis) as the threshold between hazardous and non-toxic animal feeds. Soils with over 0.5 mg Se per kg yield vegetation (e.g., fodder) with more than 5 mg Se per kg, and are classified as seleniferous soils (Sanyal and Dhillon, 2005). Selenium toxicity issues are observed in sporadically distributed selenium-toxic soils in the Great Plains and Rocky Mountains of the USA, the Prairie regions of Canada, Queensland in Australia, the Sangliao, Weihe, and Hua Bei plains of China, and the Punjab state in India. The Indian states of Haryana, sub-Himalayan West Bengal, Assam, and Meghalaya are also slightly impacted (Sanyal et al., 2015a). The productivity of animals and humans is intricately connected to the selenium content in plants and grains.

Location of Seleniferous Areas in Punjab

- Geogenic selenium toxicity has been documented in certain regions of Punjab, with soil concentrations reaching as high as 2.41 mg Se kg^{-1}. Over 1000 hectares of seleniferous soils are present in North Eastern Punjab, where toxic selenium sites are situated near the terminus of seasonal rivulets. The soils exhibit an alkaline reaction, are calcareous, have a texture ranging from silty loam to silty clay loam, and possess good drainage (Dhillon et al., 1992).
- The parent material of soils in the seleniferous region originates from the Upper Shiwalik rocks, predominantly consisting of polymictic conglomerates with diverse compositions, including several unstable minerals such as granite, basalt, and limestone. The rocks originate from the metamorphic region of the Himalayas (Sanyal and Dhillon, 2005).

- The toxicity of selenium producing endemic disease was documented in the Huabi province of China (Neal, 1995), where coal weathering released selenium into the soil. Additional instances of Se poisoning have been documented in Ireland and Turkestan. The use of groundwater for irrigation in seleniferous regions has resulted in the accumulation and toxicity of selenium in the soils of Punjab (Sanyal and Dhillon, 2005) and California (Mikkelsen et al., 1989). Selenium is distributed throughout the soil profile (up to 180 cm and beyond) in seleniferous regions, however its concentration across various strata lacks a definitive pattern. In the wheat season (upland conditions), 14 percent of the total selenium in surface soil was present as selenate selenium (SeVI form).
- Conversely, in submerged rice settings, merely 2.5 percent of the total selenium was identified as selenate-Se, while the majority existed largely as selenite-Se (SeIV form) (Sanyal et al., 2015a).

Selenium in Groundwater

- The concentration of Se in water is a critical factor in assessing its appropriateness for various applications. The USEPA water quality recommendations (Sanyal et al., 2012) provide that the maximum concentration level (MCL) of selenium (Se) in drinking water is 10 μg Se. L^{-1}, while the maximum permissible level (MPL) for irrigation water is 20 μg Se. L^{-1}. The existing ambient threshold water quality criterion for wildlife is 5 μg Se per liter.
- Considering the importance of bioaccumulation in relation to health risks for wildlife, Peterson and Nebekar (1992) established a toxicity threshold of 1 μg Se per liter of water for sensitive avian and mammalian species. The selenium content of groundwater and surface water is mostly influenced by the composition of specific geological formations, such as calcareous shale. Rainfall interacts with surface rocks, resulting in the transport of dissolved components via surface runoff into adjacent groundwater recharge zones.
- Water percolating through soil transports soluble elements into the surrounding groundwater (Sanyal and Dhillon, 2005). The solubility and mobility of selenium increase with pH. The inverse correlation between selenium and total dissolved salts indicates that selenium and other salts in groundwater do not derive from a common source (Sanyal et al., 2015a).

Chemistry of Selenium in Groundwater-Soil Environment

- In neutral or acidic soils, selenite ($SeI^{V}O_3^{2-}$) predominates, readily binding to Fe to create the less soluble Fe selenite. In well-aerated alkaline soils, selenate ($SeI^{V}O_3^{2-}$), which is more soluble and significantly more poisonous, predominates and is readily absorbed by plants (Sanyal and Dhillon, 2005).

- An elevation in phosphate concentration within the soil solution augmented selenium uptake by plants (Neal, 1995), whereas an increase in sulfate diminished selenium uptake due to competitive ligand-exchange preferences of the adsorbing soil colloidal surfaces (Mikkelsen et al., 1989). The incorporation of organic matter appears to reduce selenium availability, perhaps due to the established capacity of humic and fulvic colloids to chelate metallic cations and polyvalent oxyanions (Sinha and Bhattacharyya, 2011, 2014a; Ghosh et al., 2012).
- The selenium content differs among soils of various textural classes, indicating a higher seleniferous characteristic of heavier-textured soils (Sanyal and Dhillon, 2005). It is important to highlight that arsenites [As (III) forms] exhibit significantly greater toxicity and solubility compared to arsenates [As (V) forms]. Furthermore, selenates [Se (VI) forms] are more toxic and mobile than reduced selenites [Se (IV) forms] in soil and water (Sanyal et al., 2015a). Consequently, the phytomobilisation and phytotoxicity of arsenic (As) and selenium (Se) forms vary significantly under flooded and aerobic soil conditions. This is also evident in the accumulation of these toxicants in soils and crops, necessitating various cultural practices and corresponding management strategies (Sanyal and Dhillon, 2005).

Selenium in Fodders

- The selenium level in fodders varied from 0.16 to 37.3 mg Se. kg^{-1}, with a mean value of 3.21±4.66 mg Se. kg^{-1} (Sanyal and Dhillon, 2005). Fodders with ≤ 2 mg Se. kg^{-1} are deemed harmless; 2 to 5 mg Se. kg^{-1} are potentially hazardous, 5 to 10 mg Se. kg^{-1} are toxic, and >10 mg Se. kg^{-1} are classified as very toxic for animal consumption.
- Common indications of selenium toxicity in crops include snow-white or papery-white chlorosis accompanied by pink discoloration on the underside of leaves and sheaths in 45–60 day-old wheat shoots cultivated in selenium-affected areas (Sanyal and Dhillon, 2005; Sanyal et al., 2015a).

Selenium in Animal Hair

- The selenium level in animal hair varied from 0.016 to 37.10 mg Se. kg^{-1}, with a mean value of 0.905± 3.203 mg Se. kg^{-1} (Sanyal and Dhillon, 2005). Calves are deemed normal only if the maternal hair contains a minimum of 0.25 mg Se. kg^{-1}; however, if the hair contains more than 1 mg Se. kg^{-1}, the animals may experience Se poisoning. The concentration of selenium in animal hair appears to be contingent upon the selenium levels in the forages ingested by the animals.
- Animals afflicted by chronic selenosis had a 35 percent reduction in hemoglobin levels and experienced macrocytic and hypochromic anemia. The most prevalent clinical signs of selenium toxicity in animals are hepatic and renal dysfunction, alopecia, aberrant horn and hoof

enlargement resulting in fissures and detachment, tail necrosis, and disruption of the reproductive cycle.

Symptoms of Selenium Toxicity in Humans

- Individuals across all age demographics are impacted by selenium toxicity in the endemic area. Typical symptoms of selenium overdose in humans include hair loss from the body, particularly the scalp, deformity of fingernails and toenails, and progressive decline in overall health.
- In the endemic region, urine excretion of selenium was observed to be 17 to 22 times greater than that in the non-endemic region (Sanyal and Dhillon, 2005). Farmers in the impacted area noted that a temporary absence of about 3-4 months led to a significant recovery from selenium intoxication. Clearly, when the food source shifts from an endemic to a non-endemic zone, selenium intake diminishes.

Role of Groundwater to Selenium Enrichment of Soil- Plant System

The total quantity of Se extracted from the soil was greatest for mustard (732 g ha−1) and least for maize (92 g ha−1) (Sanyal and Dhillon, 2005). The addition of selenium through irrigation water was minimal for pearl millet (55 g ha−1), followed by mustard (74 g ha−1), contingent upon the irrigation water requirements. The incorporation of selenium via irrigation water was greatest in rice (406 g ha−1), far surpassing the selenium elimination by the crop. Conversely, mustard, Egyptian clover, pearl millet, and wheat exhibited significant selenium reductions from the examined soils. The incorporation and elimination of Se were determined to be equivalent in the instances of sugarcane and maize (Sanyal and Dhillon, 2005). This indicates that the cultivation of sugarcane and maize can contribute to the stability of the soil-plant system concerning selenium concentration in the current context. The balance of selenium was also found to be favorable for various rice-based cropping sequences. The findings indicate that rice farming in the seleniferous region exacerbates the selenium toxicity issue due to its significant irrigation requirements. Conversely, the hyper-accumulative characteristics of mustard indicate that a larger area should be cultivated with this crop to restore the soil of Se. Wild mustard is identified as the most appropriate crop for the phytoremediation of seleniferous soils.

Management of Seleniferous Soils

To improve selenium-contaminated soils, the following measures have been advised to farmers (Sanyal et al., 2015a):

- Gypsum amendment at 0.8 to 1 t/ha should be applied biennially, prior to seeding fodder crops, which decreases selenium absorption by crops by 60 to 70 percent.
- Utilizing oats, which exhibit minimal selenium absorption ability, as fodder on seleniferous soils.

- ☆ Avoid providing the initial cut of multi-cut fodders to animals, as it contains 2 to 3 times more selenium than the later cuts.
- ☆ Reduced utilization of selenium-contaminated groundwater for irrigation.

Consequently, farmers are recommended to implement maize-wheat rotation rather than rice-wheat. To limit selenium consumption, it is advised that fodder from seleniferous soils not be utilized for animal feed, nor should grains and vegetables from such soils be consumed by humans. Phytoremediation and bioremediation are frequently regarded as more effective methods for mitigating Se toxicity in water and soil, as they provide a more permanent extraction of Se from the soil matrix. Consequently, *Brassica* sp. serves as an exemplary phytoremediation crop. Nevertheless, during grain production, it loses nearly all its leaves at harvest in the field, which may include up to 50 percent of the total selenium collected from the soil (Dhillon & Dhillon, 2003). The disposal of Se-enriched vegetation presents a significant challenge. In this context, additional detailed knowledge on the microbial conversion of Se into its less hazardous or non-toxic forms is essential.

Fluoride

Effect of Excess Fluoride and its Spread on Health

- ☆ The presence of fluoride (F^-) in water, at a minimum concentration of 800 $\mu g\ F^-.L^{-1}$, is crucial for the prevention of dental caries and the fortification of bones; nevertheless, excessive fluoride can result in dental or skeletal fluorosis. The latter is a debilitating disease that impacts several regions, including the Rift Valley of East Africa, portions of Mexico, the former Soviet Union, and India (Sanyal et al., 2012).
- ☆ Toxic levels of fluoride disrupt calcium metabolism (due to the precipitation of CaF_2), leading to concurrent osteosclerosis of the spine and osteoporosis of the limb bones. Individuals' vulnerability to fluorosis may also be determined by renal impairment (Sanyal et al., 2015a).
- ☆ In India, this issue has reached serious levels in at least 17 states, primarily due to geogenic factors rather than industrial pollution. Approximately 50-100 percent of districts in former Andhra Pradesh, Tamil Nadu, Uttar Pradesh, Gujarat, and Rajasthan are impacted by fluoride toxicity; 30-50 percent of districts in Bihar, Haryana, Karnataka, Maharashtra, Madhya Pradesh, Punjab, Orissa, and West Bengal exhibit fluoride toxicity; less than 30 percent of districts in Jammu and Kashmir, Delhi, and Kerala are affected by fluoride toxicity (Rattan et al., 2009).
- ☆ Fluorotic ores are prevalent in extensive regions of Eastern and South-Eastern Rajasthan, situated within narrow synclinal zones in the core area of the Aravali synchronium. Secondly, the groundwater surrounding the mica mines is abundant in fluorides, and Rajasthan serves as a significant supplier of mica. of addition to India, regions of Pakistan, Bangladesh, Argentina, the United States, Morocco, Middle Eastern countries, Japan,

South Africa, New Zealand, Thailand, and others are also impacted by fluoride toxicity (Rattan et al., 2009).

Major Sources of Fluoride

- The three primary geochemical sources of fluoride in groundwater in India are fluorspars, phosphate rock, and phosphorites. The fluoride concentration in Indian aquifers ranges from 1000 to 25000 µg F^{-} L^{-1}, contingent upon the duration of contact between groundwater and fluoride-bearing rocks (Rao and Mamatha, 2004). Additionally, Rajasthan, being a significant supply of fluoride-rich mica, contributes to the fluoride enrichment of groundwater in the vicinity of the mica mines. The excessive extraction of groundwater for agricultural irrigation exacerbates the issue. The permissible fluoride concentration in water is influenced by temperature, as higher water consumption in warmer regions increases associated risks.PThe primary sources of fluoride in the human diet are marine seafood and potable water. Fluoride in potable water originates from the soil and the underlying lithology. Proximity to aluminium smelters has resulted in significant pasture pollution with fluoride from industrial emissions in numerous countries (Sanyal et al., 2012).

Occurrence in Soil

- Fluoride exists solely as the fluoride ion (F^{-}) in soils, where it forms strong complexes with metal ions such as Al^{3+} and Fe^{3+}. It may exist as a structural component of hydrous minerals, isomorphically replacing for structural OH. The chemisorption of fluoride onto clays and oxides by ligand exchange of surface hydroxyls, especially at low pH, results in decreased availability in soils.
- In acidic soils, the solubility and mobility of fluoride may be increased due to the formation of soluble Al-F (e.g., AlF_6^{3-}) cationic and anionic complexes; conversely, in calcareous soils, its solubility and mobility are restricted by its incorporation into insoluble Ca-minerals. In sodic soils, its mobility is enhanced by the production of highly soluble NaF (Rattan et al., 2009).

Remedial Measures

- Remedial measures for addressing fluoride toxicity in contaminated drinking groundwater include blending fluoride-rich water with low-fluoride water, fluoride removal through precipitation using appropriate coagulants (typically an alum-lime mixture as in the Nalgonda technique), adsorption on activated carbon or activated alumina, reverse osmosis, or ion exchange.
- The Nalgonda methodology, originating in this country, is an economical method that functions at the water-supply level, encompassing both piped water systems and hand pump units (Nawlakhe et al., 1975). Nevertheless,

the environmentally sustainable disposal of sludge from the Nalgonda process, as well as from processes utilizing activated alumina, frequently presents a significant issue.

- A recent technology developed at the Indian Institute of Science, Bangalore, effectively addresses this issue by utilizing magnesium oxide for the removal of fluoride from water (Rao and Mamatha, 2004).

Soil Contamination due to Agricultural Practices

A diverse array of contaminants can infiltrate the soil-crop system via chemical fertilizer residues, insecticides, herbicides, pesticides, farmyard waste, sewage sludge, composts, and wastewater, all of which pose potential hazards (Sanyal et al., 2015a).

Fertilizers

- Currently, the escalation of agricultural practices has subjected over 10 percent of the population in certain countries to nitrate concentrations in groundwater-sourced drinking water exceeding the guideline value of 10 mg NO_3^- -N L^{-1} (or 45 mg N as nitrate L^{-1}), thereby posing a risk of methaemoglobinaemia, particularly affecting infants (WHO, 1993). Intensive agricultural practices, reliant on substantial applications of nitrogen fertilizers (primarily urea, which converts to soluble nitrates upon nitrification), combined with extensive irrigation, promote nitrate leaching into groundwater, particularly in coarse-textured, highly permeable soils, thereby elevating the risk of nitrate contamination. Twenty percent of the examined wells in three areas of Punjab exhibited nitrate concentrations over the permissible threshold of 45 mg of N as nitrate per liter. Nitrate contamination in drinking water can significantly affect human health, particularly in infants. The primary health risks associated with consuming nitrate-contaminated water are blue-baby syndrome (methaemoglobinaemia) and cancer (Sanyal et al., 2015a).
- Excessive nutrients, such as NO_3^- from manure and fertilizers, can lead to eutrophication in aquatic systems, resulting in rampant algal proliferation and cyanobacterial blooms, as well as a decline in biodiversity. Eutrophication results in an unwelcome increase in the biological oxygen demand (BOD) and chemical oxygen demand (COD) of aquatic environments, adversely impacting various aquatic organisms and inducing detrimental alterations in ecosystems, frequently with significant economic repercussions (Sanyal et al., 2015b).
- The primary sources of phosphate in aquatic environments include residential sewage, which contains detergents and cleaning agents, agricultural runoff including fertilizers, and industrial effluents from the fertilizer, detergent, and soap industries. The utilization of synthetic detergents is escalating due to heightened urbanization, with many of

these products containing phosphate as a 'builder', hence augmenting phosphate loading rates in surface water bodies. The projected yearly usage of phosphate-based laundry detergents for India's existing population is around 2.88 million tonnes, with the total phosphorus discharge estimated at 146 thousand tonnes annually. Consequently, a primary concern regarding the assessment of eutrophication in aquatic systems, especially in vulnerable regions, is the mitigation of phosphorus inputs to surface waters (Kundu et al., 2015).

- Phosphorus is also depleted from agricultural soils through erosion or runoff. Consequently, non-point sources now represent a greater proportion of the nation's water quality issues than previously. The primary factors affecting phosphorus transportation can be categorized into transport factors and phosphorus source factors. Transport considerations encompass the ways via which phosphorus is distributed within a landscape. These are erosion and runoff caused by rainfall and irrigation. The determinants of the source and quantity of phosphorus accessible for transport include soil phosphorus concentration and the rate and mode of phosphorus application, whether through mineral fertilizers or organic forms (Sanyal et al., 2015b).
- As previously stated, phosphorus losses from cultivated areas can be mitigated through the implementation of prudent phosphorus management measures. Ongoing phosphorus fertilizer at rates beyond crop removal leads to phosphorus accumulation, frequently exceeding the levels necessary for crop productivity. The accumulation of soil test phosphorus at the surface, resulting from prior phosphorus applications, affects the concentration and loss of phosphorus in runoff. Substantial linear correlations are often observed between soil test phosphorus in the topsoil and dissolved phosphorus content in surface runoff. The implementation of soil test-based phosphorus fertilizer would be economically feasible and would prevent excessive accumulation in the soil (Sanyal et al., 2015b). As runoff flows into a stream channel and subsequently into a water body, there is often a gradual reduction of phosphorus load due to water dilution and sediment deposition. Particulate phosphorus in streams originates from eroding surface soil, plant debris, stream banks, and similar sources. The finer-sized fractions of the source material are preferentially eroded, resulting in the eroded particulate material having a higher phosphorus concentration and reactivity than the source soil.

Agrochemicals

- Pesticide spray drift may contaminate water bodies if treated crops and orchards are situated in proximity to rivers and streams. The primary hazard arises from inadequate storage and unintentional spills, allowing pesticides to infiltrate and pollute the groundwater. The non-biodegradable and bioaccumulative insecticides persist inside the food chain. Organochlorides induce alterations in the sexual and reproductive traits of wildlife. Top aquatic predators, specifically fish, are most impacted. The identification and assessment of pesticides in water is challenging and costly due to their presence in extremely low quantities (Sanyal et al., 2015a).

- In Maharashtra, where agricultural activities are heavy, the quantities of organochlorine (OClP) and organophosphate (OPP) pesticides in surface water were found to be greater than those in groundwater (Lari et al., 2014). During the monitoring study, α-HCH (0.39 μg L^{-1} in the Amravati region), α-endosulfan (0.78 μg L^{-1} in the Yavatmal region), chlorpyrifos (0.25 μg L^{-1} in the Bhandara region), and parathion-methyl (0.09 μg L^{-1} in the Amravati region) were commonly detected pesticides in groundwater. In contrast, α, β, γ-HCH (0.39 μg L^{-1} in the Amravati region), α, β-endosulfan (0.42 μg L^{-1} in the Amravati region), dichlorvos (0.25 μg L^{-1} in the Yavatmal region), parathion-methyl (0.42 μg L^{-1} in the Bhandara region), and phorate (0.33 μg L^{-1} in the Yavatmal region) were identified in surface water (Lari et al., 2014).
- Surface water was observed to be more contaminated than groundwater, exhibiting a greater quantity and concentration of pesticides. Water samples exhibited elevated amounts of organophosphate contamination compared to organochlorine chemicals (Sanyal et al., 2015a). Pesticide concentrations in surface water samples from the Bhandara and Yavatmal regions surpassed the EU allowed limit of 1.0 μg L^{-1} (total pesticide levels), although remained compliant with WHO criteria for individual pesticides (Lari et al., 2014).

Soil Pollution from Industries and Heavy Metals

- Industrial effluents and water runoff from waste and debris piles either flow straight into adjacent fields or reach local waterways, rivers, and ultimately the soil. The former results in toxicity to all life forms (Sanyal et al., 2015a). Cytotoxic, mutagenic, and carcinogenic heavy metals are introduced into soil by mining and smelting, the disposal of urban and industrial waste, and the application of fertilizers, pesticides, and cars, in addition to geogenic sources. The growth media (soil, air, nutrient solutions) of vegetable crops serve as primary sources of heavy metals, facilitating their absorption by the roots or leaves of the crops (Lokeshwari and Chandrappa, 2006).
- Research on crops irrigated with Bellandur Lake water in Bangalore revealed the presence of some heavy metals in rice and vegetables that exceed Indian regulatory limits. The transfer factors of metals from soil to vegetation are considerable for Zn, Cu, Pb, and Cd (Sanyal et al., 2015a). Upon comparing the concentrations of heavy metals in water, soil, and

vegetation with their normal levels, it was noted that the influence of lake water on vegetation exceeded that on soil. The findings indicated that the concentrations of Cd in spinach (4 $\mu g\ g^{-1}$) and radish (2.5 $\mu g\ g^{-1}$) exceeded the permissible limits. The accumulation occurs because cadmium is readily absorbed by food crops, particularly leafy vegetables. The average overall content of metals in plant foodstuffs, namely Fe, Zn, Cu, Cr, Pb, and Cd, was found to be 3.5, 2, 2.5, 15, 2.5, and 21 times greater than the natural concentration (Lokeshwari and Chandrappa, 2006).

- Chromium (Cr), a heavy metal, is predominantly introduced into the environment through the operations of tanneries, such as those located in Kanpur, near the Palar River in Tamil Nadu, among other sites. Consequently, 30-40 liters of tannery effluent are released for the processing of a single kilogram of skin or hide, with the effluent containing a chromium concentration ranging from 97 to 5125 mg Cr L^{-1}, resulting in contamination of soil, surface water, and groundwater. Typically, elevated deposition of Cd, Zn, and Pb in soils is attributable to mining activities and Zn- and Pb-smelting operations; additionally, Pb is contributed by vehicular activity. The soils irrigated with effluent from the zinc smelter in Debari, Udaipur (Rajasthan) had exceptionally elevated concentrations of metals (mg kg^{-1}), including Cd (0.32-214), Zn (40.7-21700), and Pb (3.50-3312). The average cadmium intake among individuals in the Jintsu Valley, Japan, was estimated to be approximately 600 µg per day, which is associated with the severe condition known as itai-itai disease (Sanyal et al., 2012). The combustion of fossil fuels and the disposal of waste materials (such as plastics, cadmium-containing batteries, and sewage sludge) further exacerbate the environmental load of heavy metals.

Soil Pollution from the Use of Solid Wastes in Agriculture

- The primary types of organic waste include domestic food waste, agricultural waste, and human and animal excreta. Organic recycling is essential for enhancing plant nutrients and preserving soil fertility, which is crucial for crop production (Jeevan Rao, 2005). Agenda 21, ratified in Rio in 1992, asserts that ecologically responsible waste management must encompass safer waste disposal or recovery and the transition to a more sustainable approach using integrated life cycle management principles (Sanyal et al., 2015a).
- Table 17.3 presents the nutritional potential and economic worth of biological and industrial wastes in India.

Table 17.3: Nutrient Potential and Economic Value of Biological and Industrial Wastes in India

Type of Wastes	*Total Quantity Available (mt)*	*Total Nutrients (000 t.yr^{-1})*				*Economic Value (million Rs**)*
		N	*P*	*K*	*Total*	
Cattle manure	280	2813	2000	2069	6882	30970
Crop residue	273	1283	1966	3904	7153	32188
Forest litter	19	100	37	100	237	1066
Rural compost	285	1431	862	1423	3715	16719
City refuse	14	98	84	112	294	1323
Sewage sludge	1	5	3	3	11	49
Press mud	3	33	79	55	168	756
Domestic waste water	6351*	318	140	191	648	2915
Industrial waste	66*	3	1	1	5	22

*: Million cubic meter per annum (Mm^{-3}. Yr^{-1}); **: Economic value compounded 0 Rs. 4500ft NPK.

Source: Juwarkar et al. (1992).

The accessible quantities of essential plant nutrients from organic sources have been periodically assessed, along with future forecasts, as illustrated in Table 17.4.

Table 17.4: Trappable Availability of Major Plant Nutrients (N + P_2O_2 + K_2O) from Organic Sources in India (mt. yr^{-1})

Source	*2000*	*2010*	*2025*
Human excreta	1.60	1.80	2.10
Livestock dung	2.00	2.10	2.26
Crop residues	2.05	2.34	3.39
Total	5.65	6.24	7.75

Source: Tandon (1997).

Soil Pollution from the Use of Wastewater in Agriculture

- ☆ The utilization of municipal wastewater in agriculture is a longstanding practice that is garnering additional focus due to the escalating scarcity of freshwater resources in numerous dry and semi-arid areas. Due to swift urbanization and increasing wastewater production, wastewater is extensively utilized as an economical substitute for traditional irrigation water; it sustains livelihoods and creates significant value in urban and peri-urban agriculture, notwithstanding the health and environmental hazards linked to this practice.

- In India, wastewater irrigation is progressively employed for crops such as vegetables, fruits, cereals, flowers, and fodder. Kolkata, previously known as Calcutta, has a longstanding tradition of employing wastewater stabilization tanks for aquaculture. over 2.4 tons per hectare (Sanyal et al., 2012) of fish is generated each year in Kolkata from over 3200 hectares of ponds, with an inflow rate of about 3 cubic meters per second. Industries in India recycle wastewater to diminish the demand for freshwater. This trend is driven by sectors in Saurashtra and Chennai. This technique, while widespread, is mostly uncontrolled in low-income nations, and its costs and benefits are inadequately comprehended.
- The properties of water are significant as the water supply might serve as the primary source of dissolved elements present in raw wastewater collected inside the sewage system. Per capita water consumption is a primary determinant for estimating the volume of wastewater to be released into the sewage system. Industrial flows may induce atypical fluctuations in flow rate and significantly contribute to specific elements.
- The metallic composition of urban wastewater is garnering significant attention and is a crucial consideration in the design of a terrestrial waste management system (Sanyal et al., 2012). Table 17.5 presents the specific features of wastewater from home and industrial sources in Ludhiana, an industrial city in Punjab.

Table 17.5: Selected Characteristics of Wastewater from Domestic and Industrial Locations in Ludhiana, Punjab

Location	*pH*	*Biological Oxygen Demand (mg. L^{-1})*	*Chromium Hexavalent (mg. L^{-1})*	*Nickel (mg. L^{-1})*	*Cyanides (mg. L^{-1})*
Electroplating industry	6.2-7.2	60-380	0.2-2.5	1.0-3.0	0.42-0.97
Sugar industry	7.1-7.9	1058-1540	-	-	-
Paper Industry	7.0-10.1	560-1113	-	-	-
Household	6.7-7.8	80-460	0.1-0.2	0.2-2.0	0.05-0.07
Maximum limits for disposal on agricultural lands	5.5-9.0	100	0.1	0.005	0.2
Maximum limits for disposal on agricultural lands	5.5-9.0	100	0.1	0.005	0.2

Source: Tiwana *et al.* (1987).

- A study examining the effects of wastewater irrigation since 1979 on metal concentration dynamics in the vadose zone of agricultural lands in the peri-urban region of New Delhi, specifically the western part of the National Capital Territory under the Keshopur Effluent Irrigation Scheme (KEIS), revealed a notable reduction in soil pH and electrical conductivity in wastewater-irrigated soils compared to their tube well-irrigated control counterparts (Deshmukh et al., 2015a).

- Sewage irrigation resulted in a significant decrease in soil sodicity. The application of sewage irrigation shown efficacy in enhancing soil organic carbon, yielding clear advantages. The sequence of extractable heavy metal accumulation in the vadose zone was Fe > Cu > Zn > Ni > Mn. Overall, the accumulation of none of the metals in extractable pools surpassed the phytotoxicity thresholds in soil. Periodic monitoring is necessary to evaluate the accumulation of metals in sewage-irrigated soils to safeguard human and animal health (Deshmukh et al., 2015a).
- Comprehension and quantification of geochemical processes in the vadose zone of sewage-effluent-irrigated soils are essential for forecasting the transport of metals and other ions to the food chain and groundwater. Consequently, an effort was undertaken to replicate the diverse geochemical processes occurring in the flow path of infiltrating sewage water within the vadose zone, utilizing Net Geochemical Reaction Along the Flow Path (NETPATH) in the same peri-urban region of New Delhi, specifically from the Keshopur Effluent Irrigation Scheme (KEIS). The findings revealed that groundwater from 20- and 10-year-old sewage-irrigated areas was marginally oversaturated concerning calcite and dolomite, while it was undersaturated regarding gypsum. Calcite dissolution was a prevalent characteristic in all sewage-irrigated soils. However, the dissolution of dolomite diminishes over time. The shallow groundwater in a 5-year sewage-irrigated field was determined to be under-saturated with respect to calcite, dolomite, and gypsum. Groundwater samples from a 20-year-old sewage-irrigated area exhibited decreased concentrations of Fe and Mn relative to the sewage effluent, likely attributable to the development of goethite and manganite in the vadose zone, as indicated by NETPATH simulations. Likewise, the elevation in Fe and Mn contents in groundwater from 10- and 5-year-old sewage-irrigated fields resulted from the dissolution of siderite and pyrolusite, respectively.
- The influence of wastewater irrigation on certain biological features of soil was examined in a region where treated sewage water has been provided to farmers since 1979 under the Keshopur Effluent Irrigation Scheme (KEIS). Three farms were chosen that had been irrigated with wastewater for the past 20, 10, and 5 years (Deshmukh et al., 2011). Sewage irrigation was observed to positively influence the fungal and bacterial populations in vadose zone soils.
- The prolonged application of sewage water augmented the bacterial and fungal populations in the soil at larger depths, facilitating the migration of microorganisms such as *Escherichia coli, C. freundii,* and *S. typhi.* Seasonal variations in groundwater pollution were seen to result

from fluctuations in the water table, which correlated with the depth of the vadose zone available for filtering, utilizing soil as a medium for the applied sewage until it reaches the groundwater. The organic content in the surface layer promotes microbial growth; however, the process of microbe adsorption onto the soil surface must also be considered during the infiltration of sewage water into the vadose zone. Swift identification of fecal contamination indicated that *C. freundii* and *Salmonella* predominated in the shallow groundwater, but *E. coli* was predominant in the deep groundwater sourced from the sewage-irrigated field. This indicated that the migration of *E. coli* in sewage-irrigated fields across the vadose zone was more pronounced than that of other species. The groundwater beneath sewage-irrigated fields in the research area was deemed unsuitable for human and animal use due to contamination by harmful microorganisms (Deshmukh et al., 2011).

Environmental Impacts from Solid Wastes

- ☆ Currently, the majority of solid waste is discarded in low-lying regions, resulting in the proliferation of rats, flies, and other domestic pests, as well as the contamination of surface and groundwater (Jeevan Rao and Shantaram, 1995).
- ☆ The persistent application of fresh trash to agricultural land for crop cultivation may lead to soil degradation; nevertheless, if managed appropriately, untreated solid waste (USW) can be regarded as a useful resource for use as manure in agriculture (Jeevan Rao and Shantaram, 1995).

Sewage-Sludge Utilization in Agriculture

- ☆ The incorporation of sewage sludge into soil offers a valuable source of plant nutrients, particularly nitrogen and phosphorus, as well as organic matter, which enhances the physical qualities of the soil. Sewage sludge has elevated levels of heavy metals, including Zn, Mn, Cd, Pb, and Cr, which can be hazardous to both animals and people when treated beyond specific thresholds.
- ☆ The nutritional potential of sewage in India is projected to exceed 350,000 tons of nitrogen, 150,000 tons of phosphorus, and 200,000 tons of potassium annually (Venkata Sridhar et al., 2006). The risk assessment, informed by numerous studies on metals in sludge, evaluated all potential impacts of these metals on human, animal, plant, and environmental health (air, water, and soil). The possible pathway of transfer is: Sludge → Soil → Plant → Human/Animal→Human (Sanyal *et al.,* 2015a).

Risk Assessment of Metal-Contaminated Soil

The risk evaluation of metal-contaminated soil relies on the accuracy of predicting metal solubility in soils. The risk of metal-contaminated soils can be evaluated by routinely estimating metal uptake by crops cultivated in these soils. To determine this, straightforward methods such as integrated solubility and free ion activity models can effectively forecast metal uptake in the edible parts of crops cultivated in polluted soils (Datta and Young, 2005). The fundamental principle is based on the premise that the reaction of plants and soil organisms to metal toxicity is predominantly influenced by fluctuations in free metal ion activity in soil pore water. Consequently, Datta and Young (2005) established a strategy for prescribing the hazardous limits of metals, predicated on extractable metals and soil properties, utilizing solubility and free-ion activity models. The solubility of metals in soil was forecasted utilizing the pH-dependent Freundlich equation (Jopony and Young, 1994) in accordance with the free ion activity of metals and metalloids (FIAM):

$$(M^{2+}) = M_C/[k_M\,(H^+)^{-n_M}]$$

In this context, (M^{2+}) denotes the activity of free metal ions in the soil solution within soil pores, MC represents the labile pool of soil metal, presumed to be solely adsorbed onto humus (mol. kg carbon–1), while kM and nM are empirical constants that illustrate the pH dependence of the metal distribution coefficient. Detailed theoretical analyses indicate (Datta and Young, 2005; Meena et al., 2016):

$$p(^{M2+}) = [p(Mc) + k_1 + k_2\,pH)]/N_f$$

In this context, k1 and k2 are empirical constants specific to each metal, indicating the pH dependency of the metal distribution coefficient, whereas Nf represents the power term derived from the Freundlich equation. The metal transfer factor from soil solution to plant biomass is given as:

Transfer factor = log [Mplant/(M2+)]

where, M_{plant} is the metal loading of plant biomass

Eqs. 8.5 and 8.6 can be combined to lead to Eq. 17.7 as follows:

$$p(M_{plant}) = C + \beta_1\,pH + \beta_2\,p(Mc)$$

C, β1, and β2 represent empirical constants specific to metals and plants. This model (FIAM) forecasts the free ion activity of trace metals and metalloids in soil solution based on labile soil extractable metal and pH, under the simplifying assumption that the total metal content (MC) is adsorbed onto soil humus.

Rang Zan et al. (2013) endeavored to forecast the free ion activity of Zn, Cu, Ni, Cd, and Pb in metal-contaminated soil, considering pH, soil organic carbon, and extractable metal content. Consequently, 0.005 M DTPA-, 0.05 M EDTA-, and 0.01 M $CaCl_2$-extractable Zn, Cu, Ni, Cd, and Pb served as indicators of the labile metal pool, whereas free metal ion activity, assessed using the Baker soil test, was utilized to estimate the free metal ion activity in the soil solution. Both Walkley–Black organic carbon (WBC) and $KMnO_4$-oxidizable organic carbon were utilized in the assessment of soil organic carbon. The solubility model was suitably parameterized

(Rang Zan et al., 2013). The authors (2013) observed that soil pH and organic carbon are significant variables that regulate the solubility of metals in polluted soils. The elements extractable by EDTA were determined to be more effective as inputs for the solubility model than those extractable by DTPA and $CaCl_2$. The Walkley–Black carbon served as a superior predictor variable for the solubility model compared to the labile pool of organic carbon. The solubility model, based on pH, organic carbon, and EDTA-extractable metals, was found to be reasonably effective in forecasting metal ion activity in contaminated soils (Rang Zan et al., 2013).

The risk to human health from metal intake by the consumption of green leafy vegetables has been quantified using the Hazard Quotient, HQgv. The former is provided as:

$$HQgv = (ADD/Rf\,D)$$

where, ADD = average total daily dose of metal intake from diet and drinking water (mg metal. kg body weight^{-1}. day^{-1}) and Rf D = the associated reference dose, established as the maximum permissible daily consumption of the specific metal that does not result in adverse health effects. Clearly, HQgv > 1.0 indicates a risk to human health. The Rf D values for Zn, Mn, Ni, Cd, Pb, and Cr (Meena et al., 2016) were 0.3, 0.14, 0.02, 0.001, 0.0035, and 0.003 mg kg body weight^{-1}. day^{-1}, respectively (IRIS, 2015). The provisional maximum tolerated daily intake (PMTDI) for copper (Cu) was established as 0.5 mg per kg of body weight per day, substituting the reference dose (Rf D) (Meena et al., 2016). The comparable PMTDI for arsenic is 2.1 µg per kilogram of body weight per day. As previously mentioned the joint FAO/WHO Expert Committee on Food Additives (JECFA) established a provisional maximum tolerable daily intake (PMTDI) of inorganic arsenic at 2.1 µg As. kg body weight^{-1}.day^{-1} for humans in 1983 and reaffirmed a provisional tolerable weekly intake (PTWI) of 15 µg As. kg body weight^{-1}.day^{-1} in 1988 (FAO/WHO, 1989). The daily consumption of wheat and rice by an individual weighing 70 kg was estimated to be 0.2 kilogram per day.

It follows that the Hazard Quotient on gravimetric basis (HQgv) is given by

$$HQgv = (Mplant.W.\,F)/(Rf\,D.70)$$

for a human of 70 kg body weight.

Here, Mplant = Metal content in plant (mg. kg^{-1}), W = Daily intake of green vegetables (mg metal. day^{-1}); F is the factor used to convert fresh to dry weight of plants.

Rattan et al. (2009) evaluated the Hazard Quotient (HQgv) for the ingestion of Zn, Cu, and Ni via the consumption of spinach (*Brassica oleracea*), gobhi sarson (*Brassica napus*), and Indian rape (*Brassica campestris*), cultivated on metal-contaminated soils subjected to sewage irrigation for 20 years. Despite Ni demonstrating comparatively elevated HQgv for all crops relative to the other two metals (Table 17.6), the majority of values remained significantly below 1, indicating that these green vegetables are unlikely to pose any health risks to humans regarding their metal content (Sanyal et al., 2015a). This risk assessment is incomplete, as green vegetables represent only a minor component of the human diet, and other pathways of metal exposure, including direct soil ingestion, inhalation, metal consumption through rice, wheat, other cereals, and drinking water, were not integrated.

Table 17.6: Risk Assessment to Human beings from Consuming Metal Contaminated Green Leafy Vegetables grown on Sewage-Irrigated Soil in Terms of Corresponding Metal Hazard Quotients (HQgv)

Crop	*HO*		
	Zn	*Cu*	*Ni*
Brassica napus	0.04.0.068	0.004-0.021	0.027.0.442
Spinach	0.035-0.152	0.008-0.015	0.046-0.502
Indian rape	0.027-0.053	0.004-0.014	0.016-0.429

Source: Rattan *et al.* (2009).

It is important to note that ADD pertains to the daily consumption of a certain metal from all dietary sources and drinking water. Consequently, it is clear that for every one food item (e.g., a vegetable or rice), the maximum value of HQgv will be below 1.0. Meena et al. (2016) contended that the acceptable thresholds for metals and metalloids in soil were determined by (i) their solubility in soil, (ii) their concentrations in rice and wheat grains, and (iii) the health risks posed by the ingestion of crops cultivated in metal-contaminated soils. To establish the hazardous threshold of extractable metals and metalloids in soils at a specific pH and organic carbon content, the critical value of HQgv employed by these authors was 0.5 for every one food item. The authors (2016) created a reference tool to calculate the allowable concentration of extractable metals and metalloids in soils, contingent upon pH and organic carbon levels. The allowable limits were established based on the anticipated HQgv as determined by the aforementioned solubility-FIAM. Meena et al. (2016) demonstrated that the permissible limit of DTPA-extractable cadmium in soil, when cultivating rice in cadmium-contaminated soil, is influenced by soil pH and organic carbon content, with the associated HQgv values (related to human metal intake via rice grain) remaining below the critical threshold of 0.5, as previously stated. At a pH of 6.0 and an organic carbon content of 0.25 percent in contaminated soil, the permissible limit of DTPA-extractable Cd can reach 0.15 mg kg^{-1}, with the corresponding HQgv remaining below the critical threshold of 0.5. Conversely, this limit may increase to 6.90 mg kg^{-1} when the soil pH is 8.0 and the organic carbon content is 0.5 percent. Consequently, suitable interventions may be developed to ensure that metal-contaminated soil remains within a safe range for cultivating foods intended for human consumption, thereby eliminating health risks.

Persistent Organic Pollutants

Persistent organic pollutants (POPs) are a category of chemicals that are either deliberately or unintentionally generated and released into the environment.

Owing to their stability and transport characteristics, they are now extensively disseminated globally, even in the most unexpected locations, such as the Arctic regions (Sanyal et al., 2012).

Symptoms of Contamination in Humans and Livestock, Fisheries and Birds

- Metal poisoning may result in biochemical (genetic) or physiological (environmental) problems. The former arises from (i) increased food consumption, (ii) heightened absorption, (iii) reduced excretion from the body, and (iv) diminished metabolism of the nutrients due to antagonism or metabolic obstruction (Sanyal et al., 2015a).
- Excessive chromium consumption is linked to renal failure, lung cancer, and eczematous dermatitis (Deckers and Steinnes, 2004). Hexavalent chromium (Cr^{6+}) is more hazardous than trivalent chromium (Cr^{3+}) due to its higher absorption rate by biological surfaces. Lead poisoning results in encephalopathy, reproductive failure, metabolic disorders, neuro-physical deficits in children, and impacts the hematologic and renal systems, leading to anemia and hypertension.
- Children are far more vulnerable to lead (Pb) poisoning than adults. Cereals and water constitute 35 percent of the daily lead consumption. Lead was the inaugural metal associated with reproductive failure (Sanyal et al., 2012). Elemental mercury (Hg) has low absorption following oral consumption, resulting in minimal toxicity. Methylmercury is nearly entirely absorbed via the gastrointestinal system and promptly enters the bloodstream. The resultant condition is Minamata disease, attributed to the ingestion of fish and other food items contaminated with methyl mercury, leading to fatalities (Sanyal et al., 2015a).
- Excessive cadmium (Cd) induces renal tubular failure, proteinuria, glucosuria, aminoaciduria, and the severe itai-itai illness. Cereals account for around 50 percent of cadmium ingestion. Aerosolized cadmium from cigarette smoke, industrialization, and population expansion is recognized as a rare contributor to hypertension, cancer, and immunological diseases (Sanyal et al., 2012). Excessive zinc disrupts reproduction, hinders embryonic development, and induces various forms of anemia.

Amelioration/Management Options in Agriculture

Nanotechnology for Pollution Control

- The purifying process of water and groundwater may employ Fe nanoparticles, nanofibers, nanoenzymes, and nanofiltration processes. Nanotechnology can be utilized to purify the air from hazardous carbon monoxide and volatile organic chemicals by the application of carbon nanotubes and nanoparticles as adsorbents (Sanyal et al., 2015a).

- Nanoparticles and nanotubes function as sensors for the detection of minuscule sizes and concentrations of dangerous compounds, which are challenging to identify using traditional methods (Pandya, 2015).

Phytoremediation

- Mandal et al. (2014) recently examined the present condition of heavy metal pollution in soil and water bodies, as well as the contribution of plant species to mitigating contamination within the Indian setting as a remediation method. Phytoremediation effectively mitigates the detrimental effects of toxic heavy metals; yet, it has not yet achieved commercial viability in India. The gradual process of biotechnological and classical hybridization, supported by interdisciplinary researchers, can yield more efficient hyper-accumulator species with desired features, thereby uncovering new possible mitigation options (Sanyal et al., 2015a).
- In situ remediation of mixed organic and inorganic contaminants by degradation by root-associated rhizosphere bacteria, known as rhizo-degradation, is likely the most significant process in the phytoremediation of petroleum hydrocarbons (Siciliano & Germida, 1998). Zayed et al. (1998) observed that root accumulation was 100 times more than that in shoots, irrespective of the chromium species provided, across many vegetable plants, including several Brassicaceae such as cauliflower, cabbage, and broccoli.
- Numerous studies have evaluated the capacity of maize to accumulate contaminants, specifically cadmium and arsenic, in polluted environments (Sanyal et al., 2012). Adhikari et al. (2010) noted that *Typha angustifolia* L. and *Ipomoea carnea* L. exhibit potential for the extraction of Pb from polluted wastewater, as they can sequester significant levels of Pb in their roots (1200 and 1500 mg Pb kg^{-1}, respectively) and shoots (275 and 425 mg Pb kg^{-1}, respectively) (Sanyal et al., 2015a).

Bio-remediation

- Microbes can diminish the activity of various metals or transform harmful metal forms into comparatively less toxic variants through diverse methods. The selection of microorganisms for bioremediation is contingent upon the availability of energy and carbon sources, environmental conditions, and the presence of harmful pollutants (Sanyal et al., 2015a). The acknowledged aerobic bacteria comprise the *Pseudomonas, Alcaligenes, Sphingomonas, Rhodococcus*, and *Mycobacterium* species. The primary enzyme in the aerobic degradation pathway utilized by Methylotrophs, specifically methane monooxygenase, exhibits a wide substrate specificity and is effective against many chemicals, including the chlorinated aliphatic trichloroethylene and 1,2-dichloroethane (Sanyal et al., 2012).

- As-oxidizing bacteria were extracted from the designated As-contaminated soils of West Bengal (Majumdar et al., 2013b), exhibiting tight phylogenetic relationships with several species of Bacillus and Geobacillus. The enhanced oxidation rate of highly lethal As (III) to the less toxic As (V) form by these native strains offers potential for successful bioremediation of arsenic-contaminated environments.

Other Management Practices

- One of the remediation technologies for metal-contaminated soil includes excavation of soil, followed by washing and subsequent disposal of the treated soils. However, this option is prohibitively expensive. A common method for immobilization of metals in soil is to apply amendments like lime, phosphates or organic manure (Ghosh *et al.,* 2012). Addition of other natural or synthetic chemicals like zeolites, beringite, palygorskite (Paulose *et al.,* 2007), hydrous oxides of Al, Fe and Mn are also known to facilitate the metal immobility in soils. Sadana and Bijay Singh (1989) reported that applying increasing levels of Zn to Cd-contaminated soils can effectively reduce Cd toxicity and ameliorate Zn deficiency in maize, and thus increase the crop yields. Chemicals like EDTA have been used for rendering the toxic elements in soluble form for their catalyzed removal by the hyper-accumulating plants (Sanyal *et al.,* 2015a).
- It is imperative that all stakeholders collaborate to develop solutions for remediating the contaminated soils and preventing further pollution (Sanyal et al., 2015a). To mitigate soil pollution or manage contaminated soils, it is imperative to establish standards for safe, marginal, and lethal concentrations of toxic elements in target human organs. Additionally, interdisciplinary research is necessary to create simulation models or transfer functions that quantify or predict the contributions of various toxic sources to the human body. Improvement of field and laboratory techniques for large-scale detection of geogenic pollutants, including their speciation in groundwater and soil, is necessary to determine the net toxicity within the soil-crop-animal-human continuum.
- The interactions of organic wastes or their microbial degradation products with metals require additional examination. Clearly, all such issues can be most effectively addressed through a multilevel stakeholder approach, engaging researchers, technologists, and planners with an emphasis on the actual beneficiaries, who should be empowered through education and training, thus enabling their active understanding and participation in soil pollution mitigation programs (Sanyal et al., 2015a). These efforts may necessitate a significant transition from a simply technical perspective to a holistic one, essential for guaranteeing a technically feasible, socially acceptable, commercially viable, and environmentally sustainable action plan.

Chapter 18

Remote Sensing, GIS, GPS and Soil Science

Comments

- The author is advised to maintain the content's relevance to the fundamentals of remote sensing, GIS, and GPS, as well as their applications exclusively in soil science and agriculture. The section detailing its applicability in other sectors such as finance and military should be eliminated.
- The contents lack coherence. The content appears to be a direct copy and paste, as it references figure and table numbers that are absent.
- Formulate findings and future prospects about the application of remote sensing, GIS, and GPS in soil research in the concluding section.
- Kindly create your own tables whenever feasible, rather than merely copying and pasting from source materials.
- The figure illustrating the spectral signature of soil should be included in the optical or hyperspectral remote sensing section.

Introduction

- Soil is an essential natural resource; it underpins human agriculture. From a practical standpoint, pedology and soil geography underpin soil surveys, which serve as the principal method for gathering, presenting, and archiving data on the spatial characteristics of soil in the USA and globally. The United States Department of Agriculture (USDA), in collaboration with various federal and state organizations, commenced soil survey activities in 1896. The joint initiative of state and federal governments was designated as the National Cooperative Soil Survey.

- Soil maps have been utilized to supply chemical and physical data for ecological and hydrological process models. The soil survey program was not intended to provide data for such uses. The advanced utilization of soil data has resulted in a heightened need for information regarding soil qualities that exceeds the capacity of traditional soil maps. Conventional soil survey methodologies rely on the qualitative identification of soil characteristics in connection to landscape and environmental factors.
- Soil systems, akin to other natural systems, exist in a state of dynamic equilibrium. Many changes occur gradually and are often undetectable, especially when considered within the context of a human lifespan. Catastrophic events, such as intense storms, can expedite erosion processes, leading to quantifiable alterations. The alterations mostly pertain to the structure and composition of the material, and these alterations are termed 'structural modifications.' Changes can be measured directly or indirectly, or may be inferred from the system's behavior. Numerous alterations pertain to soil utilization. These 'performance-related alterations' hold greater significance as they can be quantified, especially in terms of economic worth. A comprehensive understanding of soil types and their spatial distribution is essential for formulating a rational land use plan for agriculture, forestry, irrigation, and drainage.
- A soil resource inventory offers insight into the capabilities and constraints of soil for its efficient utilization. A soil survey offers a precise and scientific catalog of various soils, including their types, characteristics, and distribution, enabling predictions regarding their properties and potential. It also offers sufficient information regarding landforms, terraces, vegetation, and soil properties (e.g., texture, depth, structure, stoniness, drainage, acidity, salinity, etc.) that can be employed for planning and development. Over ninety percent of global food production relies on soil. A soil study based on climatic zones is recommended to establish a correlation between soil types and crop yields.
- In 1954, Dr. F.F. Raickens, a soil expert from the USA, was called by the Government of India to provide counsel on soil survey initiatives in India. Following Dr. Raickens' recommendations, the All India Soil Survey scheme commenced in 1956 at the Indian Agricultural Research Institute (IARI), establishing four regional centers in Delhi, Calcutta, Nagpur, and Bangalore to conduct reconnaissance soil surveys, correlate and classify soils, and produce small-scale soil maps. A soil survey manual outlining the methodologies for conducting soil surveys tailored to local conditions was produced and published in 1960. A new soil map of

India was published in 1962, based on the soil survey and classification conducted by the State and Central Soil Survey Organization. In 1969, the All India Soil and Land Use Survey Organization was divided based on developmental and research activities. A new entity, the "National Bureau of Soil Survey and Land Use Planning," was founded in Nagpur. The founding of the Indian Photo Interpretation Institute (IPI, now the Indian Institute of Remote Sensing) in 1966 offered training assistance to several soil surveyors in the utilization of aerial pictures. The preliminary soil surveys relied on either terrestrial methods or a systematic aerial photographic interpretation technique.

- The implementation of digital image processing for soil survey and mapping commenced with the founding of the National Remote Sensing Agency and Regional Remote Sensing Service Centres. The preliminary studies conducted by Venkatratnam (1980), Kudrat et al. (1990), and Karale (1992) illustrated the efficacy of digital image processing methods for soil survey.
- Data on soils about their characteristics, distribution, and spatial extent, as well as their potential and constraints, is essential for several applications, including agricultural development, engineering, sanitation, recreation, and aesthetics. Furthermore, this information is essential for modeling and environmental impact studies, namely to assess runoff and erosion through erosion simulation models or to evaluate agricultural chemical migration into groundwater using water quality simulation models. Soils can function as sources or sinks of primary greenhouse gasses. Systematic soil surveys yield this information. Despite being a non-renewable resource, the diverse applications of soil to satisfy the escalating demands for food, fuel, and fodder from a burgeoning population lead to soil health degradation, hence necessitating the regular updating of soil maps.
- Soil surveys serve as the primary information source for evaluating land suitability. Soil survey mapping units are delineated based on soil characteristics that influence management methods, including drainage, erosion control, tillage, and nutrition, encompassing the entire soil profile. Soil is a significant non-renewable resource that exists globally in a wide variety. Various soil kinds demonstrate distinct behaviors and physical characteristics. It offers vital assistance to ecosystems and to human existence and civilization. Consequently, it is essential to preserve soil functions and properties to support the environment and humanity. This prompted authorities to strategize and evaluate appropriate criteria for land utilization. The quality of land suitability evaluation and the dependability of land use decisions are significantly influenced by the quality of the soil information utilized in their derivation.
- In recent decades, the Earth's surface has experienced significant alterations in land usage. These alterations are expected to persist,

propelled by demographic pressures or climate change. In this environment, monitoring methods are essential for sustaining ecological integrity, enhancing soil conservation, and managing water resources effectively.

- ☆ Flooding, excessive runoff, soil erosion, and associated contamination and imbalance of the water and carbon cycles are, among other factors, critical challenges governed by soil surface features. The execution of sustainable agricultural, hydrological, and environmental management necessitates a deeper comprehension of soil at progressively finer scales. Traditional soil sampling and laboratory tests are inefficient in delivering this information due to their slow pace, high costs, and inability to capture all temporal and spatial variabilities.
- ☆ In this context, remote sensing has demonstrated significant potential in extracting soil characteristics during the past three decades. Various strategies have been suggested for estimating soil properties, utilizing diverse remote sensing sensors and techniques (both passive and active).
- ☆ In passive remote sensing, four primary categories of sensors can be identified:
  - ❖ Optical remote sensing with limited number of bands (*e.g.*, SPOT, ASTER, LANDSAT., *etc.*) particularly adapted for vegetation cover description, land use analysis,

Soil qualities in nature exhibit spatial variability; thus, they should be assessed as continuous variables rather than discrete point values to enhance accuracy and broaden applicability. Moreover, the conventional methods of soil investigation and interpretation are arduous and time-intensive, rendering them costly; hence, kriging and its variants have gained prominence as a significant spatial interpolation approach in land resource surveys. In this context, advancements in geographical information systems (GIS) and remote sensing technologies have facilitated the development of predictive soil mapping approaches. In situ point measurements of soil quality can facilitate regression analysis using comprehensive satellite-derived indices, and their correlation is employed to upscale these point data geographically over larger areas. The geographic maps serve as optimal input for spatially distributed models. Vegetation cover, slope, and erosion status obtained using remote sensing data to classify four primary categories of land degradation: undegraded, moderately degraded, deteriorated, and severely degraded. Likewise, remote sensing and GIS were effectively employed for natural resource mapping and soil taxonomic analysis by numerous studies. Remote sensing can be utilized to assess crop growth indicators that may be incorporated into agricultural models. It serves as a possible instrument for examining alterations in land cover, forest density, coastline morphology, reef state, and island biodiversity, even in remote locations.

- Hyperspectral sensor-based optical remote sensing, specifically tailored for soil texture characterization,
- Optical remote sensing utilizing the thermal infrared band, tailored for the assessment of soil temperature,
- Passive microwave remote sensing tailored for soil moisture and vegetation assessment.

Application of Remote Sensing in Agriculture

- The utilization of remote sensing is essential for the monitoring of agricultural fields, crop and soil health, water management and quality, and atmospheric conditions, with a focus on yield. Over the past two decades, remote sensing techniques have been employed to investigate agricultural applications, including crop discrimination, acreage estimation, condition assessment, soil moisture estimation, yield estimation, precision agriculture, soil surveys, agricultural water management, and agro-meteorological advisories.
- The application of remote sensing in agriculture, specifically regarding crops and soils, is exceedingly intricate due to the highly dynamic and intrinsic complexity of biological materials and soils. Nonetheless, remote-sensing technology offers numerous advantages compared to conventional approaches in the survey of agricultural resources. The primary advantages encompass (a) the ability to provide a synoptic picture, (b) the potential for rapid surveying, (c) the capacity for repetitive coverage to identify changes, (d) little cost participation, (e) enhanced accuracy, and (f) the utilization of hyperspectral data for augmented information.
- The primary applications of remote sensing in agriculture are outlined as follows:
    - ***In Crop Production Forecasting:*** Remote sensing is employed to predict anticipated crop production and yield in a specified area and to ascertain the quantity of the crop that will be harvested under particular conditions. Researchers can anticipate the yield of crops on a specific acreage during a designated timeframe.
    - ***In Assessment of Crop Damage and Crop Progress:*** In cases of agricultural damage or advancement, remote sensing technology can be utilized to assess the extent of damage to certain crops and evaluate the progress of the remaining crops on the farm.
    - ***In Crop Identification:*** Remote sensing has significantly contributed to crop identification, particularly in instances where the observed crop exhibits unusual traits. The obtained agricultural data will be analyzed in laboratories, where many characteristics of the crop, including its culture, are examined.

- ***In Crop Acreage Estimation:*** Remote sensing has significantly contributed to the assessment of acreage utilized for crop cultivation. This method is typically tedious when conducted manually due to the extensive sizes of the areas being assessed.
- ***In Crop Yield Modelling and Estimation:*** Remote sensing enables farmers and specialists to forecast anticipated crop yields from specific fields by assessing crop quality and land area. This is subsequently utilized to ascertain the overall anticipated yield of the crop.
- ***In Identification of Pests and Diseases Infestation:*** Remote sensing technology is crucial for identifying pests and diseases in agricultural fields and provides data on effective pest control methods to eliminate these threats.
- ***Soil Moisture Estimation:*** Measuring soil moisture can be challenging without the assistance of remote sensing devices. Remote sensing provides soil moisture data, facilitating the assessment of moisture levels in the soil and consequently, the identification of suitable crop types for cultivation.
- ***Soil Mapping:*** Soil mapping is a prevalent and crucial use of remote sensing technology. Soil mapping enables farmers and extension workers to identify the optimal soils for specific crops, as well as determine which soils necessitate irrigation and which do not. This data facilitates precision agriculture.
- ***Monitoring of Droughts:*** Remote sensing technology is employed to observe the meteorological patterns of a certain region. This system additionally tracks drought patterns in the region. Data obtained from remote sensing can predict rainfall patterns in a region and monitor soil moisture levels, vegetation health, and water supplies, hence facilitating the tracking of drought occurrence and progression.
- ***Water Resources Mapping:*** Remote sensing is essential for mapping agricultural water supplies across specific fields. Remote sensing enables farmers and extension workers to identify the availability and sufficiency of water resources across a specified area.

Optical Remote Sensing in Soil Survey and Mapping

- Satellite remote sensing data has become an essential instrument in soil resource assessment and information generation, facilitating the establishment of optimal land use plans for sustainable development across scales from regional to micro levels. One of the initial investigations in this domain was conducted by Mirajkar and Srinivasan (1975) utilizing Landsat image interpretation at a 1:1 million scale. The collection of soil information has been significantly improved by the increased speed and reliability of spatial and temporal data provided using remote sensing techniques, in a cost-effective manner.

- The surface characteristics depicted in satellite images offer sufficient data to precisely outline boundaries, achieved by methodical interpretation of satellite imagery. The soil variables that influence spectral reflectance include color, texture, mineral composition, organic matter, free carbonates, moisture content, and the presence of iron and manganese oxides and hydroxides. The data on these characteristic characteristics obtained through multi-spectral observation aids in reducing the need for extensive travel to strategically selected ground observation sites for the validation of soil borders. Remote sensing may provide opportunities to enhance current soil survey datasets. The data it supplies can be utilized in multiple manners. Initially, it may assist in delineating the terrain into internally homogeneous soil-landscape units, allowing for the evaluation of soil composition by conventional or sophisticated sample techniques.
- Secondly, remotely sensed data can be analyzed by physically based or empirical approaches to ascertain soil parameters. Additionally, remotely sensed photography can serve as a data source for digital soil mapping. Ultimately, remote sensing techniques enable the mapping of inaccessible regions by minimizing the necessity for significant, time-consuming, and expensive field surveys.
- Prior to the deployment of Landsat-1 in 1972, aerial pictures served as a remote sensing instrument for soil mapping, demonstrating their efficacy in analyzing physiography, land utilization, and erosion conditions. From 1972 onwards, both digital and analog satellite data have been employed to create small-scale soil resource maps depicting soil sub-groups and their associations. The high-resolution Landsat TM and Indian Remote Sensing Satellite (IRS) LISS II data, available from the mid-1980s, allowed soil scientists to map soils at a 1:50,000 scale for district-level planning. At this scale, soils could be classified at the association of soil series or family level. The SPOT and IRS-PAN data provided stereo capacity, enhancing soil mapping initiatives. The Indian Remote Sensing satellites (IRS-1A, 1B, 1C, and 1D) offer a sophisticated database for natural resource inventories. A multitude of studies has been undertaken to investigate the possibilities of LISS-I and LISS-II data for soil resource mapping at both 1:250,000 and 1:50,000 scales. Numerous research have commenced regarding the potential application of IRS-1C, LISS III, and PAN data for soil mapping, with the expectation that information at a scale of 1:25,000 to 1:12,500 might be produced through the integration of these datasets (Kudrat et al., 2000).
- The utilization of digital imaging and related geospatial information for the characterization and mapping of soils is rapidly increasing due to advancements in sensors, aircraft and satellite platforms, ortho-

rectification methods, mathematical models for integrating diverse spatial data sources, and the visualization of soil properties through traditional and web-based technologies.

- Remote sensing is the method of deducing surface characteristics from remote observations of the emitted electromagnetic radiation from the terrestrial surface (Schmugge et al., 2002). The radiation reflected or emitted by soil fluctuates based on various chemical and physical properties of the soil matrix (Anderson and Croft, 2009; Barnes et al., 2003; Mulder et al., 2011; Schmugge et al., 2002; Tang et al., 2009). This enables the differentiation of various soil surfaces by analyzing certain wavelengths or combinations of spectral bands. One can deduce the characteristics of a recognized soil surface from fluctuations in the recorded radiation. The primary components of soil that influence its reflectance properties for reflected solar radiation are (Jones and Vaughan, 2010; Mulder et al., 2011; Stoner and Baumgardner, 1981):
    - ***Moisture:*** Elevated soil moisture content reduces reflectance in water absorption bands as well as in other bands due to internal reflections within the water layer enveloping soil particles; hence, wet soils look darker (less reflecting) than dry soils.
    - ***Organic matter:*** Elevating organic matter content results in darker (less reflecting) soils.
    - ***Texture:*** sandy soils are more reflective than clay soils;
    - ***Surface roughness:*** Reductions in surface roughness Marginally enhance reflection: one instance is the formation of soil crust.
    - ***Iron content:*** Elevating the iron oxide content in numerous soils results in a transition to their distinctive brick-red hue, indicating an augmented reflection in the red spectrum and a reduction in the green spectrum (Ben-Dor, 2002).

Thermal Remote Sensing in Soil Mapping

- Thermal infrared remote sensing quantifies the thermal emission of the Earth within the electromagnetic wavelength range of 3.5 to 14 μm (Curran, 1985).

Microwave Remote Sensing

- Data on geographical and temporal variations in soil quality (soil characteristics) is essential for applications such as soil and crop management to enhance crop yield, sustainable land use planning, modeling soil erosion and runoff in watershed management, and precision agriculture.
- Radar is recognized for its sensitivity to several natural surface factors, including vegetation, surface roughness, and dielectric constant (e)

(Bell et al., 2001). The dielectric constant consists of the permittivity, representing the real component, and the loss factor, denoting the imaginary component. Research investigations suggest that soil salinity does not affect the real component of the dielectric constant, however the imaginary component is influenced and increases with rising salt across all three soil textures. Ulaby et al. (1986); Sreenivas et al. (1995). Bell et al. (2001) employed airborne polarimetric synthetic aperture radar for the mapping of soil salinity.

- Numerous research studies demonstrated that the spatial variability of soil texture and hydraulic parameters can be evaluated through temporal changes in brightness temperature and soil moisture content derived from microwave remote sensing (Camillo et al., 1984; Mattikalli et al., 1995; Mattikali et al., 1997; Mattikalli et al., 1998; Chang and Islam, 2000; Santanello et al., 2007). The findings of these research demonstrated a significant correlation between the space-time variations of remotely detected passive microwave brightness temperature and soil moisture, as well as soil types in relation to soil textural class.
- The utilization of Synthetic Aperture Radar (SAR) data, together with the integration of SAR and multispectral data, has been thoroughly examined for enhanced landform identification at a local level (Madhavan et al., 1997; Singhroy and Molch, 2004). Varied wavelengths of the SAR signal provide structural analysis of components within designated size categories, whereas polarization angles exhibit heightened sensitivity to directional structures. Additionally, SAR penetrates clouds, and the intensity of backscatter is influenced by the dielectric characteristics of surface materials (e.g., soil moisture content) and the co-polarization sensitivity to surface sediments, both of which enhance the classification of exposed surface sediments (Singhroy and Molch, 2004).

Hyperspectral Remote Sensing of Soils

- Hyperspectral sensors, or imaging spectrometers, are devices that capture images across multiple small, contiguous spectral bands within the visible, near-infrared, mid-infrared, and thermal infrared areas of the electromagnetic spectrum. Hyperspectral sensors can be classified as either along-track or across-track. A standard hyperspectral scanner captures over 100 bands, facilitating the creation of a continuous reflectance spectrum for each pixel.
- Laboratory and field investigations have demonstrated that specific soil qualities display distinct absorption bands, which can be utilized for their identification. Soil water has diminished absorption bands at 0.97, 1.2, and 1.77 μm, alongside significant absorption bands at 1.4 and 1.9 μm (Bowers and Hanks, 1965). Gypsum and montmorillonite display absorption bands at 1.8 and 2.3 μm, and between 0.52 and 1.0 μm, respectively

(Hunt et al., 1971). The majority of presently functioning optical sensors exhibit a coarse spectral resolution, typically ranging from 50 to 200 nm. Consequently, these measures are insufficient to detect minor variations in soil units resulting from differences in chemical composition. This has resulted in the creation of high spectral resolution spectrometers that can deliver many spectral bands with bandwidths typically ranging from 1 to 5 nanometers.

- The acquisition of soil parameters at a spatial scale by image spectroscopy will enhance digital soil mapping and the comprehension of soil processes on a spatio-temporal scale. The extrapolation of relationships formed at the laboratory level to larger scales presents numerous challenges related to potential confounding factors, including (i) variations in soil roughness, moisture, illumination, and visibility conditions; (ii) sensor attributes such as spectral and spatial resolution, and radiometric calibration, which may alter the relationships between measured reflectance and actual soil properties; and (iii) potential atmospheric influences. A number of research have examined the application of HRS imaging data for soil categorization and parameter estimation. The majority of these research have been conducted as part of a national-level initiative on hyperspectral remote sensing applications by the Space Applications Centre (SAC), ISRO, in Ahmedabad. With the increasing accessibility and decreasing cost of hyperspectral imagery, the utilization of partial least squares (PLS) regression on soil spectral reflectance data offers an efficient approach for generating high-resolution raster maps of critical soil properties such as texture, pH, and carbon and nutrient content. This information can subsequently assist farmers in decision-making, facilitate precise environmental management of agricultural fields, enhance sustainable crop production, and mitigate nutrient, sediment, and carbon losses from agricultural systems.

Airborne Topographic Lidar

- Light Detection and Ranging (LIDAR) is an innovative geospatial technique enhancing our understanding of terrestrial terrain. Advantages compared to other types of remotely sensed data encompass the collection of spatial data in three dimensions, geo-referencing during collecting, and the capability to classify three-dimensional elements within point clouds into user-defined surface and above-surface features (Renslow, 2012). Enhanced representations of the Earth's surface, structural features, and reflectance intensity facilitate the extensive application of LIDAR technology for mapping terrain derivatives and landscape conditions essential for soil research. Elevated horizontal and vertical precision facilitate the mapping of terrain characteristics that enhance our understanding of soil properties and dynamic processes across many scales.

- LIDAR, at an appropriate resolution, facilitates the identification of nuanced topographic influences on soil variability that are typically overlooked at coarser scales. Topography governs water redistribution in the landscape, which subsequently influences pedogenesis throughout geological time and the distribution of soil across the terrain. These scientific principles are familiar to soil resource inventories. Nevertheless, data such as LIDAR and the previously stated techniques offer spatially explicit representations of soils and soil processes in a measurable manner. Digital soil mapping procedures quantify and document soil patterns influenced by topography, parent materials, and other soil-forming elements (Jenny, 1941; McBratney et al., 2003), presenting this information in a digital format for computer-based applications. The amalgamation of remotely sensed data, delivery technology, and conceptual scientific comprehension enhances soil resource management.

Several issues may arise when mapping soil using thermally remotely sensed data.

- Identifying, categorizing, and mapping soils can be a difficult process often reliant on soil qualities that are not visible to the naked eye and necessitate field or laboratory investigations (e.g., pH).
- Soil constitutes a multifaceted three-dimensional entity. Most remote sensing technologies exclusively assess the surface or, under optimal conditions, the shallow layers of soils. Often, the surface properties may not accurately reflect the characteristics of the underlying soil profile (e.g., soil organic carbon concentration diminishes with depth).
- Soil characteristics can fluctuate significantly in both space and time within a limited area.
- The upper surface may undergo frequent modifications due to tillage, precipitation, erosion, crusting, and other surface processes. Vegetation cover conceals the majority of soils for the majority, if not all, of the time. Soil during agricultural cultivation will be revealed following tillage. Soil beneath natural plants may remain perpetually unexposed.
- The signal captured by the sensor is the outcome of a confluence of various interrelated soil parameters. Such combinations frequently obscure the signal from the feature being examined.

The spectral resolution of sensors is inadequate for mapping soil parameters, as it does not encompass diagnostic spectral areas and is primarily oriented towards vegetation observation.

Geoinformation

An Information System, comprising a compilation of data and tools for data manipulation, encompasses data in either analog or digital format pertaining to real-world occurrences. Our perception of the world, through selection, generalization, and synthesis, provides us with information, and the representation of this information, namely the data, forms a model of those phenomena. The database is a physical repository of many perspectives on the real world, encapsulating our knowledge at a specific moment in time.

Information is extracted from individual data pieces within a database; it is evident that information is generated from data through our cognitive processes, institutions, or other sources based on our expertise. Consequently, within a database environment, the concepts of data, information, and knowledge are distinguished. The data is crucial and enhances value as we transition from data to information, and subsequently to knowledge. The data, originating from various sources and formats, may include any of the following:

- Real, for example, the terrain conditions, *etc.*
- Captured, *i.e.*, recorded digital data from remote sensing satellites or aerial photographs of any area.
- Interpreted, *i.e.*, land use from remote sensing data.
- Encoded *i.e.*, recordings of rain-gauge data, depth of well data, *etc.*
- Structured or organized such as tables about conditions of particular watershed.

Concepts of Space and time in Spatial Information Systems

- Spatial information pertains exclusively to geographic space, specifically large-scale environments. This refers to the area external to the human body, encompassing the surrounding geographic environment. Within this place, we continuously traverse, maneuver, and conceptualize it in various manners. Geographic space encompasses the topography, land use/land cover, climatic, cadastral, and various other characteristics of the geographic realm. Geographic Information System technology is utilized to manipulate entities within geographic space and to get insights from spatial data.
- Geographic space differs from small-scale space, or tabletop space. In other terms, items that are diminutive compared to us and can be manipulated on a tabletop pertain to small-scale space and do not capture our interest.
- The comprehension of space by humans, shaped by language and cultural context, significantly impacts the design and utilization of instruments for spatial data processing. Spatial information is intrinsically linked to geographic space and, similarly, to geographic time, which manifests in the observable changes within the geographic world. Our focus is not on abstract philosophical or physical theories regarding time or space-time, but rather on the observable spatio-temporal effects that can be characterized, quantified, and archived in information systems.

Spatial Information Systems

- The management of spatial data typically encompasses data collecting, storage and maintenance, analysis, and output. For numerous years, this has been accomplished with analogue data sources, human processing, and the generation of paper maps. The advent of current technologies has resulted in a heightened utilization of computers and information technology across all facets of spatial data management. The technology employed in this field is geographic information systems (GIS).
- An overarching rationale for employing GIS can be exemplified by the following instance. Planning tasks typically require many maps and additional data sources. In accordance with a traditional analogue methodology, it is important to gather the requisite maps and documentation prior to commencing the research. The initial challenge we face is the necessity to gather maps and data from various sources at disparate locations (e.g., mapping agency, geological survey, soil survey, forest survey, census bureau, etc.), which are presented in differing scales and projections. To amalgamate data from maps, they must be transformed into functional documents of identical scale and projection. This task must be executed manually, necessitating significant time and financial resources.
- Utilizing a GIS, maps can be digitally recorded in a database using global coordinates (meters or feet). This renders scale transformations superfluous, and the conversion between map projections can be executed effortlessly using the program. The spatial analytic capabilities of the GIS are subsequently utilized to execute the planning activities. This can expedite the process and provide straightforward revisions to the analytical method.

GIS, Spatial Information Theory and the Geoinformatics Context

- The management of spatial data encompasses numerous fields. We can identify disciplines that formulate geographic concepts, offer methods for the acquisition and processing of spatial data, establish a formal and theoretical framework, are application-focused, and facilitate the administration of spatial data in legal and administrative contexts.
- Table 18.1 presents a taxonomy of several of these disciplines. They are categorized based on their handling of spatial information. The list is not intended to be comprehensive. The field that encompasses all facets of geographic data management is referred to as geoinformatics.

Table 18.1: Classification of Disciplines

Characteristics of Disciplines	*Sample Disciplines*
Development of spatial concepts	Geography
	Cognitive Science
	Linguistics
	Psychology
Means for capturing and processing spatial data	Remote Sensing
	Surveying Engineering Cartography
	Photogrammetry
Formal and theoretical foundation	Computer Science
	Expert Systems
	Mathematics
	Statistics
Applications	Archaeology
	Architecture Forestry
	Geo-Sciences
	Regional and Urban Planning Surveying
Support	Legal Sciences
	Economy

Geoinformatics: Meaning and Scope

- Geoinformatics, also referred to as Geoinformation, is defined as "the science and technology concerned with the structure and nature of spatial information, encompassing its capture, classification, qualification, storage, processing, representation, and dissemination, along with the infrastructure required for its optimal utilization" (Groot, 1989) or "the art, science, or technology focused on the acquisition, storage, processing, production, presentation, and dissemination of geoinformation" (Ehlers and Amer, 1991).
- The term geoinformatics is utilized worldwide with diverse meanings, although it originated and is predominantly used in Europe. The International Institute of Geo-Information Science and Earth Observations (ITC, Netherlands) asserts that technologies facilitating the acquisition, analysis, and visualization of spatial data are the foundation of geoinformatics (ITC, 2001). Other scholars regard geoinformatics as an emerging subject that synthesizes components from other fields related to geographic or geospatial data, namely data associated with particular locations on the Earth's surface. The disciplines encompass surveying, remote sensing, geographic information systems (GIS), cartography,

geosciences, informatics, computer science, statistics, and management (Virrantaus and Haggrén, 2000).

- The phrases Geo-information Science and Spatial Information Science are largely synonymous. A concept often used interchangeably with geoinformatics is geomatics. Geomatics includes geoinformatics, however it also emphasizes surveying. It was initially launched in Canada and gained significant popularity in Francophone countries. Laurini and Thompson (1992) characterize it as "the amalgamation of concepts from geo-sciences and informatics." The term geomatics has not been widely embraced in the United States, where geographical information science is favored. Goodchild (1992) characterizes GIS research as the investigation of overarching issues related to the utilization of GIS technology, obstacles to its effective application, or insights derived from comprehending its potential possibilities.
- Geoinformatics fundamentally encompasses the technologies that facilitate the acquisition, analysis, and visualization of spatial data. Geomatics and geoinformatics both encompass and depend significantly on the principles and practical applications of geodesy. Geoinformatics integrates geographic analysis and modeling, the creation of geospatial databases, information systems architecture, human-computer interaction, and both wired and wireless networking technologies. Geoinformatics employs geo-computation and geo-visualization to analyze geoinformation.
- Surveying and mapping have recently evolved from discipline-specific technologies, such as geodesy, surveying, photogrammetry, and cartography, into a methodology-oriented integrated discipline of geoinformation, utilizing GPS positioning, remote sensing, and digital photography for data acquisition, and GIS for data manipulation and output. The fundamental principles of remote sensing involve the properties of electromagnetic radiation and its interaction with matter. These radiations are captured by sensors on platforms in either analogue or digital format, resulting in images that undergo image processing. In photogrammetry, the stereo-concept is employed for the spatial localization of information in three dimensions. The emergence of high-resolution stereo satellite systems has heightened the significance of analytical photogrammetry, which transforms two-dimensional picture data into three dimensions, thereby integrating remote sensing with photogrammetry. The outcome of the restoration serves as a direct input into geographic information systems, either in vector or raster format. The implementation of these is feasible at the global, regional, and local levels. Geocoding facilitates data integration, with the GPS satellite positioning system assuming an increasingly significant role.

- The disciplines of geoinformatics encompass Photogrammetry, Cartography, Geodesy, Remote Sensing, Geographical Information Systems (GIS), Global Positioning System (GPS), and Web Mapping. The intrinsic characteristics of Earth system science and geoinformatics create a synergistic relationship between them. Geoinformatics facilitates the integration of geographic data, models, and knowledge, offering valuable decision-making tools for the planning, monitoring, management, and optimal exploitation of resources at national, regional, and global scales.

Surveying

- The American Congress on Surveying and Mapping (ACSM) defines surveying as the science and art of conducting essential measurements to ascertain the relative positions of points and/or physical and cultural features above, on, or below the Earth's surface, and to represent them in a functional format, or to determine the positions of points and/or features. Geodesy primarily aimed to ascertain the Earth's dimensions and form, as well as to establish a framework for the orientation of following terrestrial surveys, whereas survey technology sought to furnish instruments for the precise location of specific items.
- Currently, it is feasible to amalgamate GPS receivers and PC graphic functionalities into a singular device, termed the 'Electronic Field Book,' which can be utilized efficiently for the enhancement of vector graphics or for navigation in the field through digitized and displayed images. In contemporary standards, basic surveying instruments like the barometer are deemed insufficiently accurate for height measurements. Spirit level devices have been progressively supplanted by instruments equipped with automated compensators that guarantee a horizontal line of sight by gravitational force. The operator's optically observed level rods have been automated with digitally coded reading devices to facilitate expedited levelling operations.

Photogrammetry

Photogrammetry is the technology used to extract geometric information from things using image measurements. The advancement of photogrammetry is evidently contingent upon the overall progress of science and technology. The four principal phases of photogrammetry are intrinsically linked to the technological advancements in photography, aviation, computing, and electronics.

Application of Geoinformatics

- Numerous domains derive advantages from geoinformatics, encompassing urban planning and land use management, vehicular navigation systems, virtual globes, public health, local and national gazetteer administration, environmental modeling and analysis, military applications, transportation network planning and management, agriculture, meteorology and climate change, oceanography and integrated ocean-atmosphere modeling, business site selection, architecture and archaeological reconstruction, telecommunications, criminology and crime simulation, as well as aviation and maritime transport.

- The significance of the spatial dimension in evaluating, monitoring, and modeling diverse challenges pertaining to the sustainable management of natural resources is acknowledged globally.
- Geoinformatics has emerged as a crucial technology for decision-makers across various sectors, including disciplines, industries, the commercial sector, environmental agencies, local and national governments, research institutions, academia, national surveying and mapping organizations, international entities, the United Nations, emergency services, public health and epidemiology, crime mapping, transportation and infrastructure, information technology industries, GIS consulting firms, environmental management agencies, the tourism industry, utility companies, market analysis and e-commerce, and mineral exploration.
- Numerous governmental and non-governmental organizations have been utilizing geographical data for the management of their daily operations.

Geographical information Systems (GIS)

Introduction

- A Geographic Information System (GIS) is a computer-based system utilized to digitally represent and analyze the geographic features on the Earth's surface and the events (non-spatial qualities associated with the geography under examination) occurring therein. A geographical information system (GIS) is, in a strict sense, a computer system designed for the input, processing, storage, and output of digital spatial data. It is a digital system for the collecting, management, analysis, and visualization of spatial data aimed at planning, managing, and monitoring the natural and social environment. It signifies a comprehensive digital representation of geography.
- A GIS comprises a system for data entry in vector, raster, and alphanumeric formats, a CPU housing the applications for data processing, storage, and analysis, as well as capabilities for data display and hard copy output. A GIS encompasses data handled by an administration or project unit for inventory, analysis, and presentation to facilitate administrative or decision-making support.
- The term "GIS" appears challenging to describe as it holds varying meanings for individuals across many fields. Definitions of GIS vary from highly technical to simplistic, encompassing both functional and abstract descriptions of its components and aspects. Below are example definitions of GIS:

- A geographic information system (GIS) is a computer-based instrument for mapping and analyzing entities and occurrences on Earth. GIS technology amalgamates standard database functions, like querying and statistical analysis, with the distinctive visualization and geographic analytical advantages provided by maps. These capabilities differentiate GIS from conventional information systems, rendering it invaluable to various public and private organizations for elucidating events, forecasting consequences, and formulating plans. – ESRI, 1997.
- GIS is an organizational body that combines technology with a database, experience, and sustained financial support throughout time. Carter, 1989.
- An advanced suite of instruments for the collection, storage, retrieval, transformation, and visualization of spatial data from the physical world for specific objectives. – Burrough, 1986.
- Geographic Information Systems, commonly referred to as GIS, denotes computer systems utilized for managing spatially referenced data, enabling the interrelation of datasets, performing analytical functions, and presenting the results. — Chorley Report, 1987.
- Geographic Information Systems (GIS) encompass the systematic endeavor through which individuals: quantify elements of geographic phenomena and processes; depict these quantifications, typically within a computer database, to highlight spatial layers, entities, and interrelations; manipulate these depictions to yield additional measurements and uncover novel relationships by synthesizing diverse sources; and adapt these representations to align with alternative frameworks of entities and relationships – Chrisman, 1997.

- Certain definitions highlight the differentiation between Geographic Information Systems (GIS) and Geographic Information Science (GISc), suggesting that the former refers to an amalgamation of software, hardware, and user engagement that facilitates the management of spatial data and analyses, whereas the latter denotes a scientific discipline focused on the methodologies concerning the development, representation, and analysis of spatial data (Goodchild, 1992). It is widely acknowledged that GIS includes both domains, and therefore words such as spatial data analysis, spatial statistics, and spatial modeling are either encompassed by this definition or are closely associated, regardless of whether these methods are incorporated as tools in GIS software packages.
- Alternative definitions emphasize the objectives of utilizing GIS and characterize it based on its capabilities. Supplementary details concerning the actions facilitated by GIS elucidate the utility of the tools. It is beneficial to examine each letter of the acronym separately:

- *G: (geographic or geographical) This terminology is employed because GIS predominantly addresses 'geographic' or 'spatial' attributes. These are entities that can be associated with a particular spatial place. Those who are more pedantic might insist that this data be recorded in established coordinate systems, whether a latitude-longitude system, an easting and northing pair, or x- and y-coordinates, to be deemed 'geographical.' The artifacts may possess physical, cultural, or economic characteristics. Map features are graphical representations of spatial entities in the physical world. Symbols, colors, and line styles are employed to depict the many spatial characteristics on the two-dimensional map.*

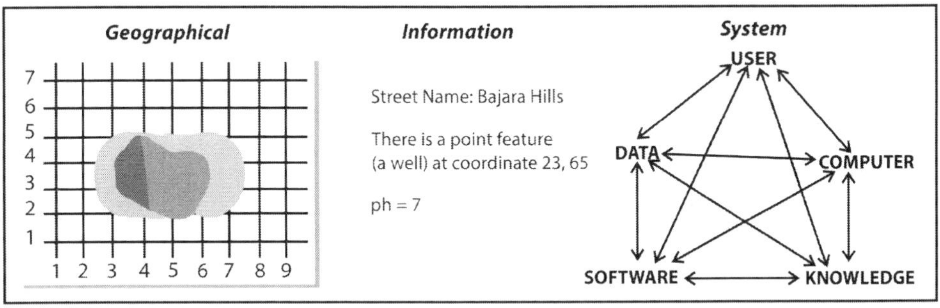

Figure 18.1: Geographical Information Systems.

- I: Information indicates that we possess certain data measures inside the framework of a meaningful system. Knowledge emerges from information. At least one component of the information must be associated with the G; specifically, there must be some geographical data (for instance, obtained from map coordinates). Additional information may pertain to qualities (such as those obtained from a name or label) or topology (offering insights into the immediate environment). Traditionally preserved as paper documents, computer technology has facilitated significantly more effective information processing and management through automated database management systems.
- S: (system) indicates that we are connecting distinct items through links. The entities may include computer hardware, software, data, and the user. The amalgamation or connection of these elements results in a system of interactions and interdependencies. This word denotes the systems approach employed by GIS, in which complicated settings are deconstructed into their constituent elements for enhanced comprehension and management, while still being regarded as an integrated whole. Computer technology has facilitated and rendered essential this methodology, resulting in the predominance of computer-based information systems.

- Computer systems are essential for the storing and processing of growing data volumes, the management of intricate spatial algorithms, and the integration of data across many scales, projections, and formats. All of these are fundamental to GIS.
- The entirety of GIS surpasses the aggregate of its constituent elements. Geographic Information Systems have more components than merely the software and technology that facilitate their operation. The fundamental components constituting the entirety of GIS.

Characteristics of GIS Technology

To gain insight into the possibilities of GIS, it is essential to comprehend the following properties of the technology:

- The data input or collection component of a GIS can transform data from existing formats into a form compatible with the system. Geographic information, sometimes referred to as georeferenced data or geodata, was formerly available solely through paper or electronic maps and attribute tables. Currently, pictures predominantly sourced from aerial photography and satellites are integral to numerous GIS processes.
- Geographic characteristics comprise four primary components: geographic location, qualities, spatial relationships, and temporal context (i.e., the location, nature, relationships with other geographical features, and the temporal existence of the condition or feature). A GIS can include non-spatial qualities, which are information not represented by locational data, in addition to georeferenced data.
- The conversion of data into a format compatible with a GIS is costly, labor-intensive, and frequently challenging. This process is especially burdensome when data originate from diverse sources, exhibiting varying geographic sizes and degrees of dependability. Despite the emergence of new survey technologies that provide digital data, a substantial backlog of surveys and maps still need conversion for utilization in a GIS. Conversion often accounts for 75 percent or more of the overall expense associated with GIS implementation.
- For a GIS to operate effectively, relationships must be articulated in a format compatible with computer processing. A unified coordinate system must be employed for all data. Practically, it is infeasible to retain comprehensive information regarding all potential spatial relationships. In a GIS, only certain geographic relationships are clearly established, while others are either computed as necessary or are unavailable.
- The output or reporting function of GIS programs has greater variability in quality, accuracy, and usability than in capacity. Reports can be presented as maps, tables, or text in either physical or digital formats.

- The analytical functions of a GIS can be utilized to address inquiries regarding current conditions or historical occurrences. Furthermore, they can be utilized to forecast future events.
- The data management component of a GIS include the capacity to store and retrieve information from the database. The techniques employed to execute these functions influence the operational efficiency of the system. The configuration of data and the arrangement of the database impose limitations on the manner and velocity of data retrieval.

Benefits of using GIS

- Implementing a specific application on a GIS offers several advantages compared to manual development or deployment on alternative computer systems.
- The primary advantage and the justification provided by numerous large organizations for investing in GIS technology.
- Nevertheless, the additional benefits of GIS are often intangible and hence typically excluded from conventional cost-benefit evaluations; however, they remain significant and are likely more crucial in the long term.
    - ***Existing Tasks More Efficiently:*** A GIS offers the significant benefit of substituting paper maps and documentation that have historically been replicated across an organization and necessitate regular updates and replacements. Substantial economic advantages can be realized by eliminating the necessity for paper documents and transitioning to GIS technology. GIS-based maps may be rapidly updated, modified, printed, or copied, while traditional maps need extensive manual labor over several days to do the same tasks.
    - ***New Tasks Not Previously Possible:*** A GIS offers technology to execute operations that were previously infeasible due to excessive time requirements or impracticality prior to automation. Examples of such advanced analysis include multiple map overlays, the construction of buffer corridors, surface interpolation, and the visualization of terrain models.
    - ***Data Management:*** A GIS offers numerous benefits of regulated information management, including data sharing among multiple users, minimizing data duplication, and enhancing security, correctness, integrity, and validity of data. Centralized access to a database is significantly more efficient than distributing many copies of identical data for unrestricted viewing and modification. Inconsistencies in the data emerge, leading to errors that diminish the reliability and utility of the information.
    - ***Scenario Modelling:*** A GIS offers functionalities to conduct modeling scenarios and evaluate 'what if?' inquiries. This is a highly effective

instrument for planners that allows for the rapid and efficient testing of many potential outcomes arising from alterations to the input parameters. The capacity for more informed decision-making is thereby enhanced.

- ***Value-Added Processing:*** A GIS enables the generation of new information from existing data via selection and combination analytic techniques. Individual data layers can be amalgamated in various manners to generate additional information layers. The integration of slope, drainage, soil, and vegetation data layers could yield a graded map indicating erosion potential or agricultural suitability, contingent upon the method of their combining.
- ***Information Provision:*** A GIS is predicated on the understanding that information is a valuable resource to be maximized. GIS offers very efficient tools for data administration, retrieval, and analysis; nevertheless, its most significant advantage lies in its capacity to visualize geographical features and relationships. Understanding the location, attributes, and interrelations of items is essential for efficient management, planning, and investment decisions. It is increasingly becoming a legal need for organizations to furnish precise and current information for various purposes; hence, the demand for GIS and other automated data management systems is expected to rise.
- ***Who can benefit from GIS?***
  - Planning and development agencies and regulatory bodies: local, regional, state, and federal.
  - Real estate developers.
  - Special purpose agencies: transportation authorities, utility companies.
  - Waste management companies.
  - Conservation and environmental protection agencies and groups.
  - International lending agencies.
  - Donor agencies.

Components and Functional Elements of GIS

- A geographic information system broadly comprises hardware, software, personnel, data, procedures, and the organizational context in which it operates. In a narrow context, we define a GIS as a software system, focusing on its architecture and functional components.
- A GIS is defined as comprising modules for the input, storage, analysis, display, and output of spatial data. For a specific GIS, each module may offer numerous or only a limited number of functionalities. Nonetheless, if any of these features were entirely absent, the system should not be designated as a geographic information system.

- A delineation of the many purposes of the components can furnish a functional characterization of a GIS. In addition to data intake, storage, maintenance, analysis, and output, geoinformation processes also encompass dissemination, transfer, exchange, and organizational considerations. The latter delineate the context and regulations governing the acquisition and processing of geoinformation.
- Geographic Information Systems (GIS) is an interdisciplinary field. From a systems perspective, the primary components of GIS are hardware, software, personnel, data, and procedures.

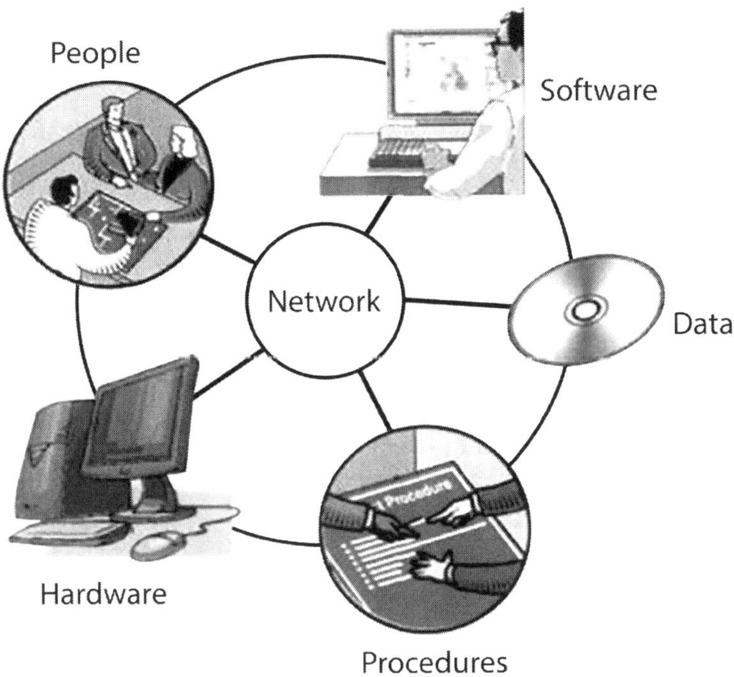

Figure 18.2: Components of GIS.

GIS software comprises directives for the computer that will be translated into actions. An example instruction could be 'Launch the software package,' which will cause the GIS software's initial interface to appear on the computer screen, displaying geographic information or data. Effective GIS software necessitates user-friendliness, functionality, compatibility, updatability, comprehensive documentation, and cost-effectiveness. GIS software can be integrated with external databases or statistical software programs to enhance its functionality.

- These are the most dynamic components of GIS. The individuals may signify a person, a department, or an organization. This component of the system governs all decisions and activities. To obtain meaningful insights from GIS, the user should possess expertise in spatial data, geographical ideas, GIS software, and the relevant domain of interest. A GIS analyst possessing expertise in planning and demographics would be an exemplary GIS user for an exercise in delineating urban expansion zones.

Table 18.2: Elements of GIS and Their Specifications

Component	*Details*
Hardware	*Type of Computer Platforms* ✰ Modest Personnel Computers ✰ High performance workstations ✰ Minicomputers ✰ Mainframe computers
Software	*Input Devices* ✰ 4 Scanners ✰ Digitizers ✰ Tape drivers -.4 CD ✰ Keyboard ✰ Graphic Monitor *Output Devices* ✰ Plotters ✰ Printers
Data	Input Modules
	Editing
	MRP Manipulation/ Analysis Modules
	Modelling Capability
	Attribute Data
	Spatial Data
	Remote Sensing Data
	Global Database
People	People responsible 'or digitizing, implementing using GIS. Trained resources

✰ The efficacy of GIS technology is constrained without the individuals who oversee and formulate strategies for its application to real-world issues. The breadth of GIS users includes technical specialists who create and manage the system, as well as those who utilize it to facilitate their daily tasks. The distinction between GIS specialists and end users is frequently essential for the effective adoption of GIS technology. This is referred to as 'brainware,' which is as crucial as hardware and software. Brainware pertains to the aims and objectives, offering the rationale and justification for employing GIS.

GIS Applications

- Geographic Information Systems (GIS) provide multiple purposes, much to duct tape. Contemporary professional GIS software is available for all categories of users. Certain applications, such as land record management or natural resource inventories, are conventional and offer useful, timely information. Other uses, such as forecasting flood risk or assessing possible hurricane damage, are more experimental and even test the limits of GIS capabilities.
- A key distinction among GIS applications is whether the spatial phenomena analyzed are anthropogenic or natural. Implementing a cadastral information system or utilizing GIS for urban planning requires an analysis of anthropogenic factors, specifically: parcels, roads, sidewalks, and, more broadly, suburbs and transit lines, all of which are human-created. Such entities often have, or are believed to have, clear boundaries; for instance, we can delineate the termination of one parcel and the initiation of another.
- Conversely, geomorphologists, ecologists, and soil scientists frequently focus on natural processes as their subjects of research. They may be examining geological formations, tectonic plates, the distribution of natural flora, or soil classifications. Frequently, these entities lack distinct boundaries, and there are transition zones where one form of vegetation is progressively supplanted by another.
- It is not unusual to see GIS applications that encompass both natural and anthropogenic phenomena. Instances are prevalent in domains where we examine the impact of human activity on the environment. Railroad construction encompasses various factors, including government land

It is essential to evaluate whether GIS technology is the most suitable and pertinent for any specific planned application. Not all application types benefit from GIS treatment. Projects necessitating geographical analysis of many data sets, requiring frequent access and query capabilities, involving regular updates, and catering to several workers are suitable for GIS implementation due to the significant advantages of automation. Nonetheless, when an application pertains to a restricted audience, encompasses minimal data layers, and serves a singular purpose, the relevance of GIS technology becomes questionable. The processes of data collecting and conversion often constitute a significant amount of the overall time required for complete GIS deployment, and the associated expenses may not justify the investment for certain types of projects.

reclamation, environmental impact assessments, and other limits, such as avoiding seasonally flooded areas and adhering to slope limitations in steep regions.

- A second divergence in GIS applications arises from the overarching objectives for utilizing the system. A typical application of GIS is inside a research project that has a well stated purpose. Such undertakings often possess a predetermined duration. Feasibility studies, such as site suitability assessments and simulation studies like erosion modeling, provide as examples. We refer to all of them as project-based GIS applications.
- Unlike the previously stated GIS applications, there exist institutional GIS applications. They can be delineated in multiple manners. The lifespan of these apps is either indefinite or not predetermined a priori. Their objective is typically to furnish foundational data to others, rather than to resolve a specific research problem. Exemplary instances of this category include monitoring systems such as early warning systems for food and water scarcity, as well as systems that monitor meteorological patterns. Our El Niño case is most appropriately categorized under this area, as the SST and WS readings persist. A different category of examples is located in governmental entities such as national topographic surveys, cadastral organizations, and national census bureaus. They perceive their responsibility as managing geographic alterations, with their primary objective being to remain current and supply data to others, historically through printed materials like maps and more recently through digital data.

Major Applications

The applications of GIS are limitless, applicable wherever geographic features require modeling and analysis. Below are many prevalent examples.

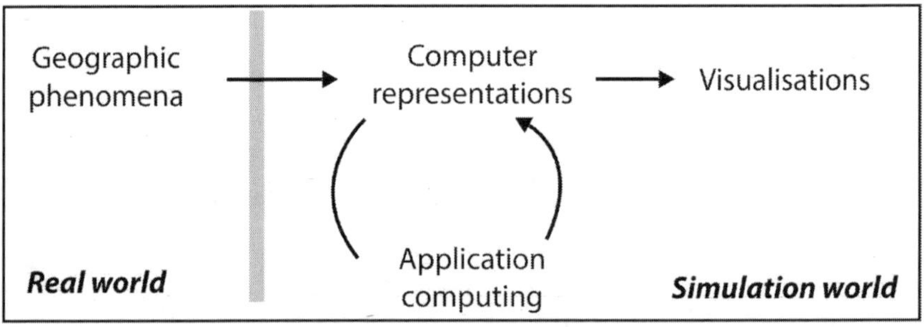

Figure 18.3: The Three Ways in which we can Look at the Objects of Study in a GIS Application.

Managing Business Activities

Geographic Information Systems are relevant to commerce. A rising market for business geographics is developing, as GIS can impact profitability, improve product quality, and generate new opportunities. Numerous examples illustrate how diverse sectors within the business domain are employing GIS.

- ***Banking:*** GIS applications in banking and financial organizations include regulatory compliance (such as Home Mortgage Disclosure Act regulations), client prospecting, and the identification of new branch and ATM sites.
- ***Business Locations and Customer Behavior:*** Comprehending the geographical distribution of your clientele facilitates more effective business placement judgments. Your closeness to persons whose buying habits correspond with your products greatly increases your chances of success. Companies can assess their market share relative to adjacent firms and adjust to changing demographic circumstances.
- ***Insurance Claims Adjustment:*** Geographic Information Systems (GIS) can perform all conventional business and marketing operations, including identifying potential clients and evaluating risk considerations. As an insurance broker, imagine the capacity to correlate storm and flood insurance applications with a map of the 100-year floodplain.
- ***Journalism:*** The media, particularly television news, use flythroughs, zoom-ins, and visual map overlays on images, creating a visually engaging and rich product.
- ***Real Estate:*** Real estate agents, encompassing both residential and commercial sectors, utilize a Geographic Information System (GIS) via the multiple listing service to identify and select properties that align with client requirements. Certain appraisers utilize GIS to do simultaneous bulk valuations across entire regions.
- ***Trucking and Delivery:*** The cost of transporting products and materials is ever increasing. Reducing route lengths decreases expenses, accelerates delivery times, and enhances customer satisfaction. Geographic Information Systems possess capabilities specifically engineered for the analysis of road and rail networks.

Planning City Operations and Expansion

Urban, county, and regional planning has historically employed GIS to monitor development, designate land use, evaluate available resources, and strategize for future expansion. GIS enables planners to assess master plans, track expansion and traffic patterns, forecast changes, monitor demographics, and determine the optimal location for the new government planning office. Numerous GIS activities globally are referred to be planning departments rather than GIS departments. However, they frequently utilize GIS as their principal instrument.

Emergency Services, Planning and Routing

The identification of optimal sites for emergency care centers can be enhanced through GIS analysis of criteria including road accessibility, population density, and health indicators. Law enforcement, social services, and emergency response agencies, such as the Enhanced-911 call-routing service, are increasingly adopting Geographic Information Systems (GIS). Crime analysts can pinpoint criminal hot spots and deploy officers accordingly, corrections personnel can monitor parolees, hospitals can be strategically located to address demand, and dispatchers can efficiently direct emergency services (such as ambulances and fire units) to their respective locations, all utilizing the capabilities of GIS. Network analysis can be employed to determine optimal routes, such as the shortest or fastest, for the navigation of emergency response vehicles.

Highway Maintenance and Accident Monitoring

Roads and motorways require maintenance and surveillance for accident-prone areas. Geographic Information Systems (GIS) are optimal for illustrating the spatial relationships among segments and for keeping the pertinent information associated with each road segment. Maintenance records can be integrated into the GIS, thereby offering current representations of the road network's condition and identifying parts necessitating rapid maintenance.

Provision of Health, Educational or Retail Services

Prior to the effective placement of new facilities or the allocation of new services, it is essential to consider the spatial distribution of various population sectors, their health and socio-economic attributes, transport route accessibility, and the location of existing facilities.

Facility Management for the Utilities

The utility sectors typically possess numerous facilities to oversee in order to deliver efficient and reliable service to extensive customer regions. Utilities such as gas, water, electricity, and sewage possess extensive land, buildings, cables, pipelines, and other physical infrastructure that require monitoring, maintenance, and management to deliver good service.

Market Analysis

The spatial distribution of the population, especially for various age demographics and socio-economic sectors, is crucial information for market analysts seeking optimal locations for product launches or brand sales. The efficacy of any certain marketing plan can also be modeled and assessed.

Population Analysis and Prediction

The spatial distribution and projected population levels are critical data for planners and developers in determining the necessary facilities to establish for future population needs. Census data thus serve as a crucial input for GIS.

Land Management and Conservation

The initial significant use of GIS oversaw the vast array of Canada's natural and mineral resources. This program can monitor fires and deploy firefighters, oversee disease and pest outbreaks, regulate land use and inventory, designate set asides and easements, track and manage animals, prepare for tourist, and more. The integration of today's growing array of Earth-sensing satellites will undoubtedly enhance the importance of GIS in land management.

Military and Defense-Related Tasks

The military and intelligence sectors are utilizing the GIS toolkit, which comprises tools specifically designed for their users. By integrating classified satellite data and visual assessments from unmanned aerial vehicles (UAVs) with the capabilities of GIS and pre-existing datasets, defense agencies may analyze troop movements, direct artillery fire, simulate scenarios, conduct supply and logistics operations, and oversee border security. The military and intelligence sectors frequently utilize the same geospatial technologies accessible to the public; yet, they possess exclusive access to specific data and sources, along with advanced technological solutions.

Treasure Chest of Possibilities

The possible applications of GIS are limitless. Jack Dangermond, president of the Environmental Systems Research Institute (ESRI), stated, "The application of GIS is solely constrained by the user's imagination." I have observed GIS employed to reconstruct battles of the American Revolutionary War, trace the Salem witch trials, monitor debris from descending space apparatus, aid hotline psychics in profiling their callers, track honey bee movement, manage agricultural machinery, advise ranchers on livestock relocation, and strategize engaging entertainment venues.

Although conventional applications of GIS technology dominate the market, the utilization of GIS continues to broaden. The capabilities of GIS are continually expanding as humanity's understanding of geography advances and individuals encode this knowledge into databases and GIS algorithms. If you are now entering the field of GIS, you will likely devote considerable time to understanding the fundamental concepts of GIS and navigating the continuously emerging software, often as you get proficient with your current tools. Entering the word "geographic information systems" into an Internet search engine, such as Google (www.google.com), yields nearly 11,000,000 results. The abundance of information may seem daunting; nevertheless, you may stay abreast of the numerous changes by consulting GIS faculty at academic institutions, GIS corporate websites, and online GIS communities. An Internet search for the phrase GIS yields approximately 50,000,000 results; however, many pertain to U.S. military GIs rather than geographic information systems.

GIS in Urban Planning

- Geographic Information Systems (GIS) were established in the late 1960s; yet, during their initial phase, few planning departments adopted them due to the exorbitant hardware costs and the software's restricted functionalities. Early software systems primarily concentrated on computer mapping, with limited analytical capabilities. The most potent software at that time was grid-based (e.g., IMGRID). The following decline in the prices of hardware, computer storage, and peripherals, coupled with enhancements in hardware and software performance, as well as advancements in data structures and associated algorithms of vector-based GIS, has rendered GIS more economical, efficient, and practical. Since the early 1980s, there has been a significant rise in the implementation of GIS across various levels and departments of municipal and regional governments in industrialized countries, particularly in Europe and North America. As the cost of computer hardware and software continued to decline, the application of GIS in urban management and planning became prevalent in developing countries during the 1990s. Geographic Information Systems (GIS) are becoming more accessible to planners and have emerged as a crucial instrument for urban planning in both wealthy and developing nations.
- Geographic Information Systems (GIS) represent a formalized computer-based information system that integrates data from diverse sources to furnish essential information for effective decision-making in urban planning. Additional information systems for urban planning encompass database management systems (DBMS), decision support systems (DSS), and expert systems.
- GIS functions as both a database and a toolkit for urban planning. A database-oriented GIS utilizes the georelational paradigm to store and link geographical and textual data. Contemporary GIS facilitates efficient data retrieval, querying, and mapping. Planners can pull data from their databases and enter it into other modeling and geographical analysis systems. Geographical information, whether integrated with data from other tabular databases or specifically performed surveys, can facilitate good planning decisions. GIS serves as a toolkit for planners to do spatial analysis through geoprocessing capabilities, including map overlay, connectivity assessment, and buffering. Among all geoprocessing functions, map overlay is arguably the most valuable tool. This is due to the longstanding tradition of planners use map overlays in land suitability research, a crucial element of urban planning.
- The primary applications of GIS in urban planning and management include database management, visualization, spatial analysis, and spatial modeling. Geographic Information Systems (GIS) facilitate the storing of land use maps and plans, socio-economic data, environmental data,

and planning applications. Planners can retrieve valuable information from the database via spatial queries. Mapping offers the most potent visualization capabilities in Geographic Information Systems (GIS). It can be utilized to examine the distribution of socio-economic and environmental data, as well as to present the outcomes of geographical research and modeling endeavors. Spatial analysis and modeling are employed for spatial statistical analysis, site selection, identification of planning action areas, land suitability analysis, land use transport modeling, and effect assessment. Interpolation, map overlay, buffering, and connectivity assessment are the most commonly utilized GIS functions in spatial analysis and modeling. The use of the aforementioned functions differs based on various tasks and phases of urban planning.

- The many benefits in using GIS in urban planning and management include (Royal Town Planning Institute 1992):
  - Improved mapping-better access to maps, improved map or accuracy, more effective thematic;
  - Mapping, and reduced storage cost;
  - Greater efficiency in retrieval of information;
  - Faster and more extensive access to the types of geographical information important to planning and the ability to explore a wider range of 'what if' scenarios;
  - Improved analysis;
  - Better communication to the public and staff;
  - Improved quality of services, for example speedier access to information for planning application processing.

GIS in Environmental Monitoring and Disaster Management

- The influence of environmental pollution and natural calamities on contemporary life is escalating swiftly. In the last forty years, the incidence of natural disasters has increased threefold, but inflation-adjusted economic damages have risen ninefold. The primary causes include heightened urbanization, the establishment and industrialization of particularly susceptible areas, the susceptibility of contemporary technology, and anthropogenic alterations to the environment.
- Effective disaster mitigation necessitates comprehensive understanding of the anticipated frequency, kind, and intensity of hazardous events in a region, alongside the susceptibility of individuals, structures, infrastructure, and economic activity within a potentially perilous area. Geographic Information Systems (GIS) play a crucial role in environmental monitoring and catastrophe management. This is mostly due to the necessity of employing GIS capabilities, and given that GIS is increasingly user-friendly and more affordable, envisioning a future for environmental monitoring systems devoid of it is challenging.

- Recent advancements in spatial database technology have enabled the development of dynamic GIS applications capable of executing operations on distant datasets. Internet-based applications utilizing technologies like 'data digging' are now feasible. Such applications will increasingly disregard the original intent of the searched database.

GIS in Environmental Monitoring

Geographic Information Systems (GIS) play a crucial role in environmental monitoring. Geographic Information Systems (GIS) are optimally utilized for the visualization of data obtained from dispersed measurement stations, such as field-based water quality sensors. Regrettably, most GIS exhibit significant deficiencies in managing the common data derived from such measurements, namely time series data. Time series consist of extended sequences of data, such as temperature recorded every thirty minutes at a specific location. Most typical GIS software programs lack sufficient features for managing temporal data, necessitating extensions or the integration of supplementary applications. The time series data regularly gathered for environmental monitoring frequently presents two major analytical and display challenges. Initially, they usually pertain to points, but data based on areal units would frequently be more advantageous (a challenge of spatial object classification). Secondly, the frequency of temporal sampling is typically excessive, leading to substantial data quantities. Geographic Information Systems provide comparatively straightforward solutions for addressing each of these issues. Initially, concerning spatial resolution, a measurement point might be permanently allocated to a representative area using a straightforward point-to-area transformation, such as those utilizing Thiessen tessellations. An alternative approach is to employ basic GIS gridding and contouring techniques to interpolate between point measurements throughout the whole study area. Secondly, concerning temporal resolution, standard database selection functionalities can be employed to retrieve data associated with a time value relevant to other comparative data. Alternatively, conventional descriptive statistical functions can be employed to diminish data quantities. Any of these methods may be employed individually or in conjunction. Figure 4.141 illustrates the data flow and hardware architecture for an integrated monitoring and GIS system. The depicted system is a complicated example executed as a multi-user, client-server setup. In addition to illustrating the fundamental system architecture, the figure also depicts the several functionalities that may be incorporated in a GIS-based monitoring system. It is essential to recognize the extent to which the system is devoted to data collection, processing, and storage, as opposed to direct spatial analysis and visualization tasks. Fundamentally, GIS is utilized exclusively at user workstations, while the majority of the system is allocated to other functions. This distribution is typically seen in the associated implementation and operational expenses. Typically, at least 90 percent of the expenses associated with establishing an environmental monitoring system go to the measurement program, quality control, and data processing; a negligible portion is allocated to the programming of GIS-based analytical and display features. The basic data analysis functions offered by conventional GIS software are frequently inadequate for addressing the data monitoring needs associated with environmental issues.

GIS in Disaster Management

Numerous natural calamities have transpired throughout human history and will persist in occurring. Approximately 95 percent of fatalities associated with natural disasters transpire in developing nations, which are home to over 4.2 billion individuals. Economic losses due to natural hazards in underdeveloped nations may account for up to 80 percent of their gross national product. No technology or understanding can avert the occurrence of natural dangers. Nonetheless, technology advances daily, enhancing our understanding of the factors contributing to these hazards. In disaster management, specialists strive to monitor situations, properly simulate complex natural phenomena to develop improved prediction models, propose suitable contingency plans, and establish spatial databases. A comprehensive strategy for disaster management is necessary to effectively mitigate the impact of natural disasters, sometimes referred to as the disaster management cycle. Geographic Information Systems (GIS) integrated with remote sensing have demonstrated significant value in several phases of disaster management. Due to its intrinsic attributes—spatial continuity, consistent accuracy and precision, multi-temporal coverage, and comprehensive site coverage—remotely sensed data can be utilized efficiently in disaster management. Remote sensing and GIS offer a database for interpreting evidence from past disasters, which can be integrated with additional information to produce hazard maps that identify potentially risky areas.

Global Positioning System (GPS)

- The Global location System (GPS) is an emerging technology that offers unparalleled precision and versatility in location for navigation, surveying, and GIS data acquisition. The NAVSTAR-GPS (Navigation System with Time and Ranging-Global Positioning System) is a satellite-based navigation, timing, and positioning system that has been in development by the US military since 1973. Since 1993, the system has comprised twenty-one operational satellites and three reserves. The satellites traverse an orbit around the Earth at a height of 20,200 kilometers. Their orbits are configured so that at least four are in direct line of sight from any location on the Earth's surface, 24 hours a day, unless obscured by structures, vegetation, or rugged terrain. GPS technology has been advanced for real-time navigation and geodetic positioning.
- The GPS offers constant three-dimensional positioning around the clock globally. The technique appears advantageous for the GPS user group, providing accurate data up to around 100 meters for navigation, meter-level precision for mapping, and millimeter-level accuracy for geodetic positioning. GPS technology has a vast array of uses in GIS data collection,

surveying, and mapping. The capacity to geocode GIS data in both vector and raster formats currently relies on contemporary geopositioning technology.

- The Global Positioning System (GPS) was created by the U.S. Department of Defense (DOD), Ivan Getting, and the Massachusetts Institute of Technology (MIT). The GPS, officially designated as NAVSTAR (Navigation System with Timing And Ranging), initially comprised 11 orbiting satellites and was launched by the Department of Defense in 1978 only for military purposes. The concept for the system originated somewhat earlier. The fundamental concept of GPS navigation may be traced back decades to the initial traveler who inquired, "Where am I?" The origins of GPS technology, however, date back to the 1950s.
- In 1957, the Soviet Union launched a satellite, Sputnik, into orbit. Following the launch, MIT researchers observed that Sputnik's radio signal fluctuated in intensity as it orbited the Earth. As it neared their location, the signal power intensified. Upon the satellite's departure from the MIT researchers' location, the signal strength diminished. The fluctuations in the radio transmission enabled the MIT researchers to ascertain Sputnik's precise orbit. The acknowledgment that radio waves from a satellite, or "artificial star," could ascertain specific locations on the Earth's surface was indeed the catalyst for the GPS system.
- Prior to then, the foundations for GPS technology were being established. In 1951, Dr. Ivan Getting, an MIT alumnus and Rhodes Scholar from Oxford, applied his Ph.D. in astrophysics at Raytheon. The Air Force solicited a guiding system for a planned intercontinental ballistic missile transported by railroad. In answer to this request, Dr. Getting created the inaugural three-dimensional positioning system utilizing time difference of arrival. This system served as the foundation for the subsequent GPS technology.
- The deployment phase of GPS history commenced in 1973 with the resolution to create a satellite navigation system utilizing the existing technologies of the U.S. Air Force and U.S. Navy. The system underwent rigorous testing over the subsequent three years. In 1977, the initial transmitters were erected on Earth's surface and tested, despite the absence of any deployed satellites. The transmitters were designated as Pseudolites (pseudo satellites).
- From 1978 to 1985, eleven satellites were deployed into space and positioned accordingly. In 1979, a decision was taken to augment the number of satellites to 18. The inaugural Block I satellite was launched in 1980. This "bird" was equipped with sensors specifically engineered to detect atomic detonations and was positioned in orbit to monitor the Soviet Union's adherence to the 1963 treaty with the United States prohibiting nuclear testing.

- In 1980, the onboard atomic clocks, the most precise timekeeping devices globally, were enabled. Atomic clocks, designed by scientists, quantify time based on the fluctuations in energy levels of electrons. These clocks are reliable, continuous, and precise to a nanosecond, or one billionth of a second.
- In 1983, the GPS transitioned from a purely military system to one accessible for public use. President Ronald Reagan declassified the GPS system following a catastrophe involving a civilian aircraft. Korean Air flight 007 was lost over Soviet territory and subsequently shot down by Soviet forces. The Challenger space shuttle disaster in 1986 suspended shuttle launches, consequently postponing the deployment of the GPS system, as the shuttles were designated to carry the new Block II satellites. Ultimately, Delta rockets, the initial conveyance, were reinstated for satellite launches, and in 1988, the decision was made to augment the satellite count to 24, as functionality with 18 satellites remained constrained.
- Although the GPS system was accessible to the public, its use was temporarily suspended during the first Gulf War in 1990 due to the military's requirement for additional receivers. Public access was reinstated in 1993, accompanied by the determination that the GPS system will be provided at no cost to the global community. Full Operational Capacity (FOC) was attained in July 1995 with the deployment and activation of the final satellite in the constellation of 24.
- Since 1995, the history of GPS has experienced significant technological progress. An increased number of satellites have been deployed into orbit, enhancing both accessibility and precision.
- Historically, GPS could identify a subject's location within a 100-meter radius. The precision has enhanced to locating persons within 10-15 meters and can currently be measured within centimeters in certain instances. Availability has expanded from military and position-specific sectors such as aviation, nautical navigation, and land surveying to encompass several personal uses. Employers utilize a cost-effective receiver to monitor their car fleet, parents employ the system to oversee young children and unruly teenagers, while pet enthusiasts use it to watch their canines.

Basic Concept of GPS

- A GPS receiver determines its location by accurately measuring the time of signals transmitted by satellites orbiting the Earth. Every satellite consistently broadcasts messages that encompass:
  - The time the message was transmitted.
  - Precise orbital information (the ephemeris).

- ❖ The general system health and rough orbits of all GPS satellites (the almanac).

- ☆ The receiver analyzes the received signals to ascertain the transit time of each message and calculates the distance to each satellite. The distances and the satellites' locations are utilized, potentially with trilateration, depending on the technique employed, to determine the receiver's position. The position is subsequently presented, potentially utilizing a dynamic map display or coordinates of latitude and longitude; elevation data may also be incorporated. Numerous GPS devices display derived data, including direction and speed, computed from positional alterations.
- ☆ Three satellites may appear sufficient to determine position, given that space comprises three dimensions and a location near the Earth's surface can be presumed. Nonetheless, even a little clock inaccuracy, when amplified by the immense speed of light—the velocity at which satellite signals travel—yields a significant location error. Consequently, receivers utilize four or more satellites to determine the receiver's position and time. The precisely calculated time is largely obscured by most GPS applications, which utilize merely the position data. Several specialized GPS applications utilize time, including time transfer, traffic signal synchronization, and the coordination of cell phone base stations.
- ☆ While four satellites are necessary for standard functionality, fewer may suffice under exceptional circumstances. If one variable is known, a receiver can ascertain its position using merely three satellites. A vessel or aircraft may possess a known altitude. Certain GPS receivers may employ supplementary cues or assumptions (such as utilizing the last known altitude, dead reckoning, inertial navigation, or integrating data from the vehicle's computer) to provide a less precise (degraded) position when fewer than four satellites are observable.

Components of GPS

- ☆ The GPS use satellites and computers to determine locations globally. The GPS operates on satellite triangulation. The Earth's position is ascertained by calculating the distance from a constellation of satellites in orbit. The fundamental premise of GPS is straightforward, despite the system utilizing some of the most advanced technology ever created. To comprehend the fundamentals of GPS, the system can be delineated into five logical steps:
  - ❖ Triangulation from the satellite is the basis of the system.
  - ❖ To triangulate, the GPS measures the distance using the travel time of the radio message.
  - ❖ To measure travel time, the GPS need a very accurate clock.
  - ❖ Once the distance to a satellite is known, then we need to know where the satellite is in space.

- ❖ As the GPS signal travels through the ionosphere and the earth's atmosphere, the signal is delayed.

☆ To determine a position in three-dimensional space, four satellite observations are required. The GPS employs a trigonometric method to determine positions. The GPS satellites are positioned at significant altitudes, resulting in highly predictable orbits, and each satellite is outfitted with a precise atomic clock.

GPS Structure

The present GPS comprises three principal components. The components include the control segment (CS), the space segment (SS), and the user segment (US). The U.S. Air Force designs, sustains, and manages the space and control components. GPS satellites emit signals from orbit, which each GPS receiver utilizes to determine its three-dimensional position (latitude, longitude, and altitude) as well as the current time. The space segment consists of 24 to 32 satellites in medium Earth orbit, together with the payload adapters necessary for their launch launchers. The control segment consists of a primary control station, a secondary control station, and numerous dedicated and shared ground antennas and monitoring stations. The user demographic consists of hundreds of thousands of U.S. and allied military personnel utilizing the secure GPS Precise Positioning Service, alongside tens of millions of civil, commercial, and scientific users of the Standard Positioning Service.

GPS Positioning Types

Absolute Positioning

The placement method depends on a solitary receiver station. It is sometimes known as 'stand-alone' GPS, as, in contrast to differential positioning, ranging occurs alone between the satellite and the receiving station, without the involvement of a ground-based reference station for error correction computations. The positions obtained in absolute mode are susceptible to the inherent inaccuracies associated with satellite positioning. The absolute placement precision is deemed to be no better than 50 meters according to Ackroyd and Lorimer, and ±100 meters as per the U.S. Army Corps of Engineers.

Differential Positioning

Relative or Differential GPS advances triangulation techniques by incorporating a second receiver at a predetermined reference location. This configuration requires the gathering of an error-correcting message from the reference receiver to enhance the determination of a point's position relative to a known point on the Earth's surface. Differential-mode positioning depends on a designated control point. The reference station is situated at the control point, a triangulated location, according to the control point's coordinates. This enables the computation and application of a correction factor to other roaming GPS units utilized in the same region and temporal series. Errors in the control point's coordinates are directly additive to inaccuracies in the satellite positioning procedure. The error corrections generated by the reference station fluctuate fast, as the factors influencing positional inaccuracies are not constant over time. This mistake correction enables the mitigation of a substantial proportion of errors, potentially up to 90 percent.

Satellite Frequencies

All GPS satellites transmit on two identical frequencies: 1.57542 GHz (L1 signal) and 1.2276 GHz (L2 signal) (Table 18.3). The satellite network employs a CDMA spread spectrum method in which low-bitrate message data is encoded using a unique high-rate pseudo-random (PRN) sequence for each satellite. The C/A code, designated for civilian use, transmits data at a rate of 1.023 million chips per second, whereas the P code, intended for U.S. military applications, transmits at a rate of 10.23 million chips per second.

The L1 carrier is modulated by both the C/A and P codes, but the L2 carrier is exclusively modulated by the P code. The P code can be encrypted into a P(Y) code, accessible solely to military apparatus possessing the appropriate decryption key. Both the C/A and P(Y) codes provide the exact time of day to the user.

Table 18.3: Overview of GPS Frequencies

Band	*Frequency*	*Description*
L1	1575.42 MHz	Coarse-acquisition (C/A) and encrypted precision (P(Y)) codes, plus the L1 civilian (L1C) and military (M) codes on future Block III satellites.
L2	1227.60 MHz	P(Y) code, plus the L2C and military codes on the Block IIR-M and newer satellites.
L3	1381.05 MHz	Used for nuclear detonation (NUDET) detection.
L4	1379.913 MHz	Being studied for additional ionospheric correction.
L5	1176.45 MHz	Proposed for use as a civilian safety-of-life (SoL) signal

The L3 signal, operating at a frequency of 1.38105 GHz, is employed by the United States Nuclear Detonation Detection System (USNDS) to detect, localize, and report nuclear detonations (NUDETs) within the Earth's atmosphere and near space. One application is the implementation of nuclear test ban treaties. The L4 band at 1.379913 GHz is under investigation for enhanced ionospheric correction. The L5 frequency band at 1.17645 GHz was incorporated during the GPS upgrade process. This frequency resides inside an internationally safeguarded spectrum for aircraft navigation, ensuring minimal to no interference under all conditions. The inaugural Block IIF satellite, which would deliver this signal, was launched in 2009. The L5 comprises two carrier components that are in phase quadrature. Each carrier component is modulated using bi-phase shift keying (BPSK) with an own bit sequence. L5, the third civil GPS signal, will ultimately facilitate safety-of-life applications in aircraft while enhancing availability and precision.

GPS Accuracy

The real-time GPS mining system can yield four fundamental levels of accuracy or sorts of solutions, as delineated in Table 18.4.

GPS satellites emit signals over five distinct carrier frequencies, with each frequency containing specific information or codes. It might be conceptualized as three distinct radio stations transmitting various programs. Specific codes are modulated onto carrier waves, enabling the measurement of distances from a designated satellite to a receiving antenna on a GPS receiver. The receiver coordinates can be determined within a shared reference frame using the distances calculated from three known satellite positions.

Table 18.4: GPS Accuracy Levels

GPS System	*Accuracy*
Common Accuracy	15-100 meters
Differential GPS (DGPS)	0.5-5 meters
Real-Time Kinematic Float (RTK Float)	20 cm-1 meter
Real-Time Kinematic Fixed (RTK Fixed)	1 cm-5 cm

Application of GPS

- Initially a military initiative, GPS is regarded as a dual-use technology, signifying its substantial military and civilian applications. A notable and distinctive characteristic of Global Positioning Systems is the availability of positioning signals to users at any location globally and at all times. A properly working GPS system can serve a wide and expanding community due to its diverse applications, including surveying, mapping, navigation, and GIS data collecting. The GPS will soon integrate with the comprehensive functionality of technology.
- Numerous GPS applications exist; among these, several are noteworthy:
  - **Surveying and Mapping:** The exceptional accuracy of GPS carrier phase measurements, coupled with suitable adjustment algorithms, offers a sufficient instrument for diverse surveying and mapping activities. Employing DGPs methodologies enables precise and prompt mapping of nearly any subject. The GPS is utilized to delineate cut blocks, road alignments, and environmental threats including landslides, forest fires, and oil spills. Applications requiring a high level of precision, such as cadastral mapping, can also be executed using high-quality GPS receivers. Continuous kinematic methods are applicable for topographic surveys and precise linear mapping.
  - **Navigation:** Utilizing GPS for navigation can conserve numerous hours in the field. Any feature, regardless of its underwater status, can be pinpointed up to one hundred meters by scaling coordinates

from a map, inputting waypoints, and proceeding directly to the location. Examples encompass road crossroads, corner posts, plot centers, accident sites, geological formations, and similar entities. GPS navigation in helicopters, automobiles, or ships offers a convenient navigation solution with significant cost reductions.

- **Remote Sensing and GIS:** GPS positioning can also be integrated with remote sensing techniques, including photogrammetry, aerial scanning, magnetometry, and video technology. Utilizing DGPS or kinematic methods, contingent upon the requisite accuracy, real-time or post-processing will yield locations for the sensor that can be projected onto the ground, rather than projecting ground control into an image. GPS is increasingly a useful instrument for GIS data acquisition. The GIS user community gains advantages from employing GPS for the acquisition of locational data in many GIS applications. The GPS may be effortlessly connected to a laptop in the field, and with suitable software, users can consolidate all their data on a common platform with little distortion. Consequently, GPS can assist in several facets of constructing precise and prompt GIS databases.
- **Geodesy:** Geodetic mapping and other control surveys can be conducted efficiently utilizing high-quality GPS technology. Particularly in scenarios involving helicopters or when line of sight is obstructed, GPS can establish new benchmarks for precision and efficiency.
- **Military:** The GPS was originally designed for real-time military positioning. Military uses encompass:
  - ***Navigation:*** GPS enables soldiers to locate objectives, even in darkness or unfamiliar terrain, and to coordinate troop and supply movements.
  - ***Target tracking:*** Numerous military weapon systems utilize GPS to identify possible ground and aerial targets prior to designating them as hostile. These weapon systems provide target coordinates to precision-guided munitions, enabling accurate target engagement. Military aircraft, especially in air-to-ground operations, utilize GPS to identify targets.
  - ***Missile and projectile guidance:*** GPS facilitates precise targeting of many military armaments, including intercontinental ballistic missiles, cruise missiles, and precision-guided munitions. Artillery munitions. Embedded GPS receivers capable of enduring accelerations of 12,000 g, equivalent to approximately 118 km/s Two have been designed for utilization in 155 millimeter (6.1 inch) howitzers.

- ***Search and Rescue:*** Downed pilots can be located faster if their position is known.
- ***Reconnaissance:*** Patrol movement can be managed more closely.

Other Applications

- ***Geofencing:*** Vehicle tracking systems, personal tracking systems, and pet tracking systems utilize GPS technology to ascertain the whereabouts of a vehicle, individual, or pet. These gadgets are affixed to the car, individual, or pet collar. The application offers real-time tracking and mobile or Internet notifications if the target exits a specified zone.
- ***GPS tours:*** The location dictates the content to be shown, such as details regarding an imminent point of interest.
- ***Phasor measurements:*** GPS facilitates precise timestamping of power system measurements, allowing for the computation of phasors.
- ***Robotics:*** Autonomous robots equipped with GPS sensors that determine latitude, longitude, time, speed, and heading for self-navigation.
- ***Recreation:*** For example, geocaching, geodashing, GPS drawing and waymarking.
- ***Telematics:*** GPS technology integrated with computing and mobile communication systems in automobile navigation.

Future of GPS Technology

Unless substantial new obstacles arise from S/A (Selective Availability) imposed by the DOD, the GPS business is expected to persist in its advancement inside the civilian sector. A multitude of vendors produce GPS receivers, with a prevailing trend towards smaller, more affordable, and user-friendly systems. Although very accurate, portable (hand-held) receivers are currently accessible, prevailing conjecture anticipates affordable and comparably precise 'wristwatch locators' and navigational systems for vehicles. Nonetheless, a significant future trend pertinent to the GIS user community is the emergence of community base stations and regional receiving networks, alongside advancements in GPS management and technology that will enhance the ease and precision of GPS surveying.

Terminology

Absorption: The physical absorption of water and/or ions/energy by a material.

Acid soil: Soil exhibiting a pH value below 7.0.

Acidification: The process by which soil becomes acidic (pH < 7) occurs due to the presence of acid parent material or in areas with significant rainfall, leading to soil leaching. Human activities, such as the application of fertilizers and the deposition of industrial and vehicle pollution, can expedite acidification.

Acidifying: The process by which soil pH decreases, resulting in increased acidity of the soil. It assumes the characteristics of vinegar, which are detrimental to numerous crops.

Adsorb: To retain cations in the soil solution as a thin coating on the surface of soil particles. Aggregates are clusters of soil particles that are loosely bonded.

Adsorption: It is the process via which atoms, molecules, or ions are adhered to the surfaces of solids via chemical or physical bonding.

Aeration of soil: Volume of air-filled pores in the soil, calculated as the difference between total porosity and actual soil moisture content. Optimal soil aeration is 30 percent, but it is significantly influenced by the structure and compaction of soil particles; 15–20 percent is generally adequate for the cultivation of grasses and cereals; levels below 10 percent are detrimental to plant growth.

Aggregate: Soil aggregation composed of two or more soil particles interconnected by diverse forces. Soil structural units adhere more firmly to one another than to adjacent particles and are created through natural soil formation processes.

Aggregation: The mechanism by which primary soil particles (sand, silt, clay) get aggregated, typically through natural forces and chemicals originating from root exudates and microbial activity. Soil aggregates are organized into soil peds, which are units of soil structure categorized by size, shape (platy, prismatic, columnar, angular, subangular, blocky, granular), and grade (single-grain, huge, weak, mild crumb, strong). From an agronomic perspective, the most significant soil aggregates vary from 1 to 3 mm.

Amino acids: Nitrogen-containing organic acids combine to make proteins, which are the fundamental building blocks of all life.

Anaerobic: In the absence of oxygen in the environment. Anaerobic.This situation arises in soil as a result of excessive moisture.

Anion exchange capacity: Total of exchangeable anions that a soil may retain. Typically quantified in centimoles or millimoles of charge per kilogram of soil or other adsorptive materials, such as clay.

Anion: An ion with a negative charge. See also ion, cation.

Aniones: Negatively charged particles.

Antibiotic: A natural compound that suppresses or eradicates the proliferation of germs. A substance that inhibits or ceases the growth of bacteria, fungus, and other microbes.

Arable land: Cultivated agricultural ground that is ploughed to a depth of typically 20 to 30 centimeters. Exceeding 30 cm signifies deep ploughing.

Archaeologist: An archaeologist specializing in ancient human civilizations.

Arid: Dry, like a desert.

Bedrock: Mass of solid rock underlying the soil. It can serve as parent material if it is sufficiently near the surface to undergo weathering into soil. It is the solid rock located at the Earth's surface or beneath the soil.

Black Earth: Term synonymous with Chernozem employed (e.g. in Australia) to denote self-mulching black clays.

Blocky: A soil composition. Soil particles are organized into forms that resemble little cubes with either sharp or rounded edges.

Bog: A wetland characterized by the absence of substantial inflows or outflows, which sustains acidophilic mosses, especially Sphagnum, and where peat accumulation occurs. Comparable to: fen, marsh, pocosin, swamp, and wetland.

Boulder clay: Unstratified glacial deposits formed directly beneath the ice or released from the surface after melting; boulder clay and till are similar terms for this unsorted material, which varies from rock flour to large rocks and boulders, depending on the characteristics of the bedrock.

C:N ratio: The ratio of carbon and nitrogen present in the soil or organic materials. A reduced fraction facilitates the absorption of nitrogen by soil organisms and plants.

Calcification: It is the process of calcium carbonate accumulation in a certain horizon of the soil profile.

Capillary water: It is the water retained in the minute capillary pores at a tension exceeding 60 cm of water. Capillary water can ascend against gravitational forces. The water situated between field capacity and the hygroscopic coefficient is termed capillary water. It is tethered to the earth by capillary forces.

Carbon cycle: A series of processes in which carbon dioxide is converted into organic forms by photosynthesis or chemosynthesis, cycled through the biosphere (with partial integration into sediments), and finally reverted to its original condition through respiration or combustion.

Cation exchange: The exchange of a cation in solution with another cation at the interface between the solution and the surface of negatively charged substances, such as clay or organic matter.

Cation: A positively charged atom or atomic group.

Clay coating/film: Coatings of orientated clay on the surfaces of peds and mineral grains, as well as lining pores, are referred to as clay skins, clay flows, illuviation cutans, or argillans.

Clay loam: A loam with a predominance of clay is categorized as clay loam. Soil texture classification. Refer to soil texture as well.

Clay minerals: Secondary minerals, either crystalline or amorphous, that are less than 2 microns in diameter and possess considerable surface area. Clay-sized hydrous aluminum silicates with substantial interlayer spacing capable of retaining considerable quantities of water and other materials; they possess a wide surface area facilitating expansion and contraction; examples include montmorillonite, smectite, and kaolinite.

Clay: Clay is a surface-active fraction characterized by a high degree of chemical and physical activity.The finest soil particles (<0.002 mm). Frequently exhibit plate-like forms. Exhibits a sticky texture when moist. It also denotes a soil texture comprising a minimum of 40 percent clay particles.

CLORPT: The five factors that determine soil formation are climate, organisms, relief, parent material, and time.

Coating: A layer of a substance that entirely or partially covers a soil surface; coatings may consist of clay, calcite, gypsum, iron, organic material, salt, etc.

Colloid: The particle measures from 1 and 1000 nanometers in diameter. Clay and soil organic matter are frequently referred to as soil colloids due to their particle sizes being within or nearing colloidal proportions.

Colluvial: Concerning materials or processes related to transportation and/or deposition through mass movement (direct gravitational influence).

Columnar: A soil composition. Soil particles are organized into elongated vertical structures or columns, typically featuring rounded apexes.

Compost: It is a combination of decomposed organic matter from plants and animals, utilized to nourish soil and enhance its structure and water retention capacity.

Compounds: The amalgamation of two or more components. Hydrogen and oxygen combine to form water.

Crop rotation: Sequential cultivation of several crops on the same parcel of land. Successive cultivation of various crops in the same field. For instance, maize is cultivated in the first year, beans in the second year, and the land is kept fallow in the third year. The agricultural rotation consists of maize, beans, and fallow.

Crust formation: Development of a superficial crust on the soil due to the impact of raindrops. Raindrops disrupt soil aggregates and create a thin, compressed layer, known as a crust, on the soil surface.

Decalcification: Leaching of calcium carbonate or calcium ions from the soil.

Decomposition: The process involves the chemical decomposition of a molecule (mineral or organic) into simpler compounds facilitated by microbes.

Deficiency: Lacking of some necessary quality or element.

Denitrification: This is the process whereby nitrates and nitrites are transformed into gaseous compounds by soil microorganisms in low-oxygen conditions.

Developed soil: A soil that has had a long time to form, such as most tropical soils. A mature soil.

Dormant: A condition in which a plant or animal is not undergoing growth.

Drainage: Water movement from the soil profile via surface flow or subsurface flow.

Ecologist: A scientist who examines the interactions between organisms and their surroundings.

Electrical conductivity (EC): The electrical conductivity of a distilled water-saturated soil extract is typically stated in siemens or decisiemens per meter at 25°C.

Eluviated horizon (E horizon): An eluvial horizon is defined as a horizon that has lost soil due to the process of eluviation.

Enzyme: Proteins synthesized within the cells of living creatures that serve as catalysts in their biochemical processes.

Erosion: The erosion of the land surface by rain, irrigation water, wind, ice, or other natural or human-induced processes that abrade, separate, and transport geological parent material or soil from one location on the Earth's surface to another.

Estuary: An estuary is a region where a freshwater river or stream converges with the ocean.

Feldspar: Feldspars are non-ferromagnesian minerals that function as reservoirs for sodium, calcium, potassium, and various trace elements in soils.

Fen: Level and marshy terrain, typically at a low elevation and akin to a bog or wetland.

Fertility: The capacity of soil to provide vital nutrients to plants.

Fertilizer: Any organic or inorganic substance, whether natural or synthetic, introduced to soil to provide one or more critical nutrients for plant growth.

Field capacity: The volume of water held in the soil once the downward flow from a pre-saturated condition has significantly halted.

Fixation: The mechanism via which accessible plant nutrients become less available or entirely unavailable in the soil.

Fluvioglacial deposits: Material displaced from glacier peripheries and subsequently classified and deposited by streams emanating from the melting ice.

Fungus: A collection of spore-producing organisms that derive sustenance from organic matter, encompassing molds, yeasts, mushrooms, and toadstools. These are uncomplicated organisms devoid of chlorophyll, consisting of cellular filamentous structures referred to as hyphae.

Geomorphology: An examination of the Earth's surface characteristics and their correlation with geological formations.

Gibbsite: Aluminum hydroxide. It is the predominant aluminium compound in soil, particularly prevalent in the severely worn soils of tropical environments that support Oxisols.

Glacial drift: Glacial drift refers to sedimentary material that has been conveyed by glaciers. It comprises clay, silt, sand, gravel, and boulders.

Glacier: A glacier of restricted breadth advances gradually across the terrain. It perpetually transitions from elevated to diminished levels and is confined within defined valley boundaries.

Glaciofluvial deposits: Glaciofluvial deposits are sediments deposited mainly by fluids originating from ice sheets and glaciers.

Glaciolacustrine deposits: Sediments deposited into lakes that have come from glaciers are called glaciolacustrine deposits.

Gley soil: A Gley is a hydric soil that, if not drained, remains saturated with groundwater for extended durations, resulting in a distinctive gleyic color pattern. The soil evolved under conditions of inadequate drainage, resulting in a loss of iron and other elements.

Granular: All axes are uniformly developed with identical lengths, featuring curved and uneven surfaces.

Green fertilizer: Green fertilizers are natural substances which added to the soil to increase soil fertility.

Ground water: The segment of water under the Earth's surface that is under a pressure equivalent to or above atmospheric pressure. Refer to the water table as well.

Gully: A narrow, steep-sided valley that may arise naturally or result from accelerated erosion.

Horizon: A generally homogeneous layer that extends laterally (continuously or discontinuously) throughout the pedon; it runs approximately parallel to the ground surface and exhibits distinct differences in various soil qualities compared to the associated horizons.

Humification: Humification is the degradation of organic matter resulting in the production of humus.

Humus: The well-decomposed, relatively stable component of soil organic matter, resulting from the breakdown of raw organic material.

Hydromorphic soils: These soils are formed under conditions of excess water, waterlogging, or anaerobic environments.

Immobilization: It is the process of transforming an element from its inorganic to organic form within microbial or plant tissues.

Infiltration: It is the infiltration of water at the soil-air interface resulting from sorption and the vertical movement of water through the soil profile.

Ion: An atom or group of atoms that carries a positive or negative electric charge as a result of having lost or gained one or more electrons.

Karst: Karst is a landscape created by the breakdown of soluble rocks, including limestone, dolomite, and gypsum.

Landslide: The descent of a substantial amount of soil or rock from a mountain slope.

Leaching: The leaching of soil constituents in solution and/or in suspension as water percolates from the surface to the subsurface.

Loam: A soil texture characterized by about equal proportions of sand, silt, and clay. Such soils offer an ideal connection among soil, air, and water for plant growth.

Loess: A parent material, transported and deposited by wind, primarily composed of silt-sized particles originating from arid locations.

Macronutrients: Macronutrients required by organisms in substantial amounts.

Map scale: The ratio of ground distances to map distances, for instance, a country depicted on a map with a scale of 1:1,000,000.

Marsh: Occasionally inundated or persistently flooded lowland region with a surface that is not extensively submerged, predominantly populated by sedges or other hydrophytic vegetation.

Massive: A structure with no visible peds or aggregates.

Microbes: Microscopic creatures, including bacteria and fungi. Microbes constitute the predominant organisms in soil.

Microbiologist: A scientist specializing in the study of microscopic creatures, known as microbes.

Micro-organisms: Organisms that are imperceptible to the naked sight due to their diminutive size. They are only visible using a microscope.

Minerals: Minerals are solid entities made of atoms arranged in an organized and systematic manner.

Mottles (Mottling): Variations in color, manifested as spots or streaks, emerged within the predominant soil matrix color as a result of intermittent reduction (waterlogging) and oxidation conditions.

Mulch: Any organic or inorganic substance, including leaves, straw, plastic film, and sawdust, that is applied on the soil surface to safeguard the soil and plant roots from the impacts of raindrops, soil crusting, and evaporation.

Munsell Color System: The Munsell soil color chart comprises three fundamental variables: hue (color wavelength), value (lightness or darkness), and chroma (purity or intensity).

Nitrogen fixation: Nitrogen fixation is the process of capturing almost limitless atmospheric nitrogen gas (N_2) to synthesize ammonia (NH_3), which is easily assimilated by plants and microorganisms.

Nutrient balance: It is an indeterminate theoretical ratio of many plant nutrient concentrations for optimal growth rate and production.

Nutrients: Nutritionally essential components without which plants cannot finish their life cycle.

Organic matter: Carbonaceous biological matter originating from the decomposition of flora and fauna, typically located in natural ecosystems.

Organic soil material: It remains waterlogged for extended durations unless artificially drained, containing 18 percent or more organic carbon (by weight) if the mineral fraction exceeds 60 percent clay, more than 12 percent organic carbon if the mineral fraction lacks clay, or between 12 to 18 percent carbon if the clay content of the mineral fraction ranges from 0 to 60 percent. i) not saturated with water for more than a few days and containing above 20 percent organic carbon.

Organic soil: A soil where the cumulative thickness of layers consisting of organic components often exceeds that of mineral layers.

Organisms –Single living organism, including bacteria, fungus, plants, or animals.

Parent material: The unconsolidated material from which the soil layer emerges through soil-forming variables and processes. The C-horizon is commonly referred to as the parent material of soil.

Pasture: Grassland used for grazing of mainly domestic herbivores.

Peat: It is an organic soil material formed by partial decomposition of organic matter.

Peatland: A broad designation for any wetland characterized by the accumulation of partially decomposed plant material; mire, moor, bog, fen, and marsh are words employed for peatlands in Europe and Canada.

Ped: A unit of 3-dimensional soil structure formed by natural processes.

Pedogenesis: The process of soil formation and development is influenced by various soil-forming factors: climate (primarily temperature and precipitation), parent material, biotic organisms (flora and fauna), terrain, temporal considerations, and other unidentified environmental variables.

Pedologist: A scientist who investigates the origins, composition, and distribution of soils, as well as the materials from which soils are derived.

Pedon: The minimal three-dimensional volume that can be classified as soil.

Perennials: Perennial plants, which survive for over two years, in contrast to annuals that germinate from seeds each year and biennials that exist for only two years.

Periglacial: Physical processes associated with the freezing and thawing of terrestrial materials along the proximal edges of glaciers and ice sheets, primarily situated above permafrost or influenced by frost-action mechanisms.

Permafrost: A dense, permanently frozen subterranean layer of earth that persists at or below 0°C year-round, created during frigid glacial epochs.

Permanent grassland: Land is utilized permanently for grazing primarily grasses or other herbaceous forages for a continuous period of five years or more, either through cultivation or spontaneously.

Phosphorus (P): Macronutrient vital for the overall health and vitality of plants; it participates in energy conversion, root development, sugar and starch transformation, nutrient transport throughout the plant, and enhancement of crop uniformity and quality.

Photosynthesis: Photosynthesis is the process through which green plants, certain algae, and specific bacteria synthesize carbohydrates from carbon dioxide and water, utilizing sunlight and chlorophyll, while concurrently producing oxygen as a by-product.

Platy: A soil structure characterized by the arrangement of soil particles in flat plates or sheets stacked horizontally upon one another. It is typically located in forest soils or in underground soils that are prone to leaching or compaction.

Ploughing (tillage): Physical and mechanical alteration of agricultural soils to achieve an optimal depth (10-15 cm) of pliable arable soils for enhanced crop development.

Pocosin: A marshy terrain created in shallow indentations, characterized by inadequate drainage.

Pores: It pertains to the volume of soil spaces that can be occupied by water and/or air, facilitating the penetration of plant roots into the soil.

Potassium (K): It is a vital macronutrient necessary for optimal plant development and reproduction; it significantly contributes to the regulation of stomatal opening and shutting, enzyme activation, osmoregulation, drought resistance, and starch synthesis, among other functions.

Primary mineral: Primary minerals are those generated by the crystallization of molten magma.

Prismatic: A soil structure characterized by vertical pillars of soil particles divided by distinct vertical fissures. The pillars are angular, while the tops of the prisms are unclear and typically flat.They are typically located in the B-horizons or subsoils.

Productive: A designation for land capable of yielding a profusion of crops.

Protection of soil: A deliberate process essential for the preservation of soil and its properties, implemented at many levels (individual, local, national, continental) utilizing data acquired from soil research. Sustainability is the outcome of this process.

Recovery: The proportion of chemical fertilizers utilized that is assimilated by the crop.

Regolith: The unconsolidated mantle of weathered rock and soil on the Earth's surface, often seen as loose earth components situated above solid rock.

Relief: The vertical and horizontal measurements of the ground surface that indicate the disparity between its maximum and minimum elevations.

Respiration: The process by which plants decompose carbohydrates generated during photosynthesis, along with oxygen, to generate energy for growth.

Root exudates: Substances in the rhizosphere, emanating from the root systems of living plants and microbially altered products, consist of modest quantities of low-molecular-weight organic molecules.

Root zone: The region of soil encompassing the roots of a plant includes the topsoil.

Runoff: Water from precipitation or irrigation that does not infiltrate the soil profile but instead runs off the surface and enters streams and rivers.

Saline soil: A non-sodic soil with adequate soluble salts that negatively impacts the growth of the majority of crop plants. The standard lower threshold for the electrical conductivity of saturation extract in such soils is established as 4 dS m−1 (at 25°C).

Saline-sodic soil: Soils influenced by salinity, characterized by an exchangeable sodium percentage (ESP) over 15 percent, a sodium adsorption ratio (SAR) larger than 13, and a pH typically below 8.5, are generally unsuitable for agricultural use.

Salinization: Salinization refers to the accumulation of soluble salts in the soil.

Salt-affected soil: Soil negatively impacted by soluble salts, regardless of the existence of elevated exchangeable sodium levels.

Sand: Soil particles ranging from 0.05 mm to 2 mm (with some nations defining the lower limit as 0.06 mm), characterized by a low specific surface area, are classified as a textural class or soil type. Sandy soils possess a fragile structure characterized by inadequate water retention, elevated permeability, and heightened susceptibility to compaction, resulting in numerous detrimental effects.

Secondary mineral: The minerals that originate from the alteration of primary minerals or are produced by weathering processes.

Sediment: Any fragment of soil or rock that has been deposited by hydrological, aeolian, glacial, or gravitational processes.

Sewage: It is a category of wastewater discharged from residential residences, flowing through drains (such as those in bathrooms, kitchens, and laundry rooms) to a treatment facility or septic system. It is primarily defined by its physical properties, chemical and poisonous components, and its bacteriological profile, including the types and amounts of organisms present.

Silt: Soil particles ranging from 0.002 mm to 0.05 mm (with some nations designating 0.06 mm as the highest limit) possess a high or medium-high specific surface area that affects soil structure stability; this classification is also known as a texture class for medium and medium-fine soil components.

Single-grained: A structureless dirt. Soil particles are not interconnected, similar to beach sand.

Slope: A sloped or inclined terrain.

Sludge: Semi-solid, viscous, moist substance, often the byproduct of an industrial or refining procedure, remaining after sewage treatment in a facility. Can be utilized as a fertilizer in some circumstances

Sod: The ground's surface, with grass anchored by roots, is cohesive. It can be segmented into blocks and utilized as a construction material, also referred to as grass.

Sodic soil: Soil characterized by excessive sodium content, a pH typically between 8 and 10, an exchangeable sodium percentage (ESP) exceeding 15 percent, a sodium adsorption ratio (SAR) greater than 13, and a severely compromised soil structure. These soils require specialized care and are unsuitable for agricultural use.

Soil biology: A scientific discipline centered on the biotic components of soils, chiefly including bacteria, fungi, protozoa, nematodes, arthropods, earthworms, and mammals.

Soil chemistry: A scientific profession that examines the chemical properties of soils, primarily influenced by mineral composition, the presence of organic matter, and associated environmental conditions.

Soil classification: The systematic classification of soils into groups or categories based on their attributes.

Soil compaction: The method involves augmenting soil density by compacting soil particles, hence diminishing the pore space between them. Compacted soils possess minimal big holes, reduced overall pore volume, and thus, increased density, which diminishes the rates of both water infiltration and drainage. Furthermore, the gas exchange diminishes, necessitating that roots use increased force to penetrate the compacted layer. Anthropogenic compaction results from the movement of heavy machinery and extensive soil exploitation.

Soil degradation: Adverse impacts on soil health, leading to a reduced capacity of the soil ecosystem or overall soil degradation, such as compaction, erosion, waterlogging, salinity, acidity, alkalinity, and the influence of toxic chemicals and pollutants, are frequently exacerbated by human activities, including improper soil management and construction practices.

Soil depth: Soil profile depth from the surface to the parent material, bedrock, or root obstruction layer. It varies considerably among distinct soil types. It is a fundamental criterion employed in soil categorization. Soils can be categorized as very shallow (less than 25 cm), shallow (25 cm to 50 cm), moderately deep (50 cm to 90 cm), deep (90 cm to 150 cm), and extremely deep (greater than 150 cm).

Soil fertility: A metric reflecting soil's capacity to supply plants with adequate nutrients and water, resulting in sustained and consistent high-quality harvests. It is an appropriate medium for root development to provide optimal plant growth and maturation.

Soil geography: Scientific field focused on the distribution of soil types across landscapes, articulating this distribution in accordance with geographical principles.

Soil micro-organisms: These are diminutive life forms that can exist as unicellular entities or may aggregate into colonies, such as protozoa, viruses, bacteria, fungi, and algae. The majority of microorganisms inhabit topsoil, particularly in the vicinity of plant roots, known as the rhizosphere, where shed cells and root-released compounds offer abundant nutrients. Micro-organisms are classified into autotrophic and heterotrophic categories based on their nutritional and oxygen requirements, further categorized as aerobic or anaerobic. Microorganisms serve as effective indicators of soil condition and quality.

Soil monitoring: Systematic monitoring and quantification of certain soil qualities and functions, primarily to examine alterations in soil conditions.

Soil morphology: Form and arrangement of pedological features.

Soil physics: The scientific discipline that examines the dynamics of physical soil components, their phases, and qualities such as density, porosity, water retention, permeability, and hydraulic conductivity.

Soil porosity: The volume of water and air retained in soil; the ratio of void volume to total soil volume.

Soil profile: A vertical profile of soil horizons from the surface to the parent material, depicting the arrangement of soil horizons typical of particular soil types used for classification purposes.

Soil sorption: Selective process, which occurs on soil particles smaller than 0.002mm (<2μm); these small particles have colloidal properties, are able to hold and exchange ions, water or gases.

Soil structure: The configuration of main and secondary particles into a specific structural pattern

Soil texture: It is the relative composition of sand, silt, and clay.

Soil: Soil is the unconsolidated uppermost layer of the Earth's crust, composed of minerals, organic materials, water, and air. It is physiologically active and offers structural support for plant development.

Stoniness: The relative proportion (volume percentage) of coarse particles (beyond 2 mm in diameter) in the soil or on the soil surface; a concentration exceeding 15 percent of stones might impede cultivation and diminish the soil's water retention ability. Subsoil (B horizon): The layer of soil situated directly beneath the surface soil, characterized by a concentration of minerals that have been eluviated or leached from the overlying horizons. Absent in all soils.

Surface soil: The upper 15 to 20 cm of the soil layer at the surface, known as topsoil; the section of soil often cultivated. Swamp: A wetland frequently partially or intermittently inundated with water, particularly characterized by the prevalence of woody plants.

Swelling and shrinking: Swelling is the phenomenon that occurs when interacting clay platelets separate. Shrinking is the process by which soil material diminishes in volume due to water loss.

Tidal flats: Almost level regions, intermittently inundated and exposed by tidal fluctuations, unsuitable for agricultural purposes.

Till: Unstratified glacial sediment deposited directly by glacial ice.

Topsoil: The uppermost soil horizon, typically 5–10 inches deep, is where the majority of the earth's biological soil activity transpires, owing to its elevated organic matter content.

Transform: The transformation of one kind of energy into another or the change of one state into another, such as liquid to gas.

Tundra: A broad, level, treeless Arctic expanse spanning Europe, Asia, and North America, characterized by continuously frozen subsoil.

Uptake: The capacity of a plant to assimilate water and nutrients from the soil.

Vadose water: Water within the vadose zone.

Vadose zone: It is the portion of soil that extends from the ground surface to the upper layer of the permanent water table, where groundwater is under air pressure.

Water retention: The capacity of soil to retain water for a duration exceeding infiltration, typically 48 hours in well-draining soil. It is heavily contingent upon the bulk density of soil organic matter and soil texture.

Water table: The uppermost layer of groundwater or the depth at which water is maintained at atmospheric pressure.

Weathering: It refers to the mechanical disintegration and chemical alterations of rocks and minerals at or near the Earth's surface, caused by physical, chemical, and biological forces or their mixtures.

Wetland: A transitional zone between aquatic and terrestrial ecosystems that is flooded or saturated with water for extended durations, resulting in hydric soils and supporting hydrophytic plants. The depth at low tide does not surpass six meters.

Wilting point: It denotes the soil moisture content at which plant roots are unable to collect water at a pace adequate to satisfy transpirational demands. At this moisture level, plants lose cellular turgor and exhibit withering signs.

Index

I

L

M

R

S